Materials for the Hydrogen Economy

Materials for the Hydrogen Economy

Edited by
Russell H. Jones
George J. Thomas

CRC Press
Taylor & Francis Group
Boca Raton London New York

CRC Press is an imprint of the
Taylor & Francis Group, an **informa** business

CRC Press
Taylor & Francis Group
6000 Broken Sound Parkway NW, Suite 300
Boca Raton, FL 33487-2742

First issued in paperback 2019

ISBN-13: 978-0-8493-5024-5 (hbk)
ISBN-13: 978-0-367-38788-4 (pbk)

Library of Congress Cataloging-in-Publication Data

Materials for the hydrogen economy / editors, Russell H. Jones, George J. Thomas.
p. cm.
Includes bibliographical references and index.
ISBN-13: 978-0-8493-5024-5
1. Hydrogen--Industrial applications. 2. Materials--Research. 3. Hydrogen as fuel--Research. 4. Hydrogen industry. I. Jones, Russell H. II. Thomas, George J.

TP245.H9M387 2007
665.8'1--dc22 2007020789

Visit the Taylor & Francis Web site at
http://www.taylorandfrancis.com

and the CRC Press Web site at
http://www.crcpress.com

I thank my wife, Cathy, for her support during the time needed to assemble this book, the contributors who helped make it a reality, and the publisher for their patience with this project.

— *Russell H. Jones*

Contents

Preface

The purpose of this book is to provide the reader with a comprehensive overview of materials being developed or considered for a hydrogen-based energy economy, the state of their development, and issues associated with their successful deployment. It is our hope that this book will be useful to both the newcomer to this field and the experienced engineer and researcher desiring to know more about the broader topic. It is expected that students, professors, engineers, scientists, and managers will find this book useful.

Introduction

Will a hydrogen-based energy economy, with its promise of clean, sustainable energy, become a reality? This is clearly a complex issue involving economic and societal drivers (such as energy independence, energy costs, global warming, pollution), politico-economic decisions (such as infrastructure investment, R&D investment), and exogenous developments (such as advancements in the performance of other energy systems, military conflicts).

These economic and societal issues will play out on the national and global stage; however, the use of hydrogen as an energy carrier may ultimately hinge upon the performance achieved in hydrogen production, distribution, storage, and propulsion systems and components. The performance of these, in turn, is highly dependent on technological advancements, particularly on the properties of the materials used in their manufacture. In other words, *materials are a key enabling technology to a viable hydrogen economy.*

Even a cursory examination of the material properties needed in various hydrogen components suggests that a multidisciplinary approach will be needed to overcome some of their current limitations. However, before examining materials issues in detail, a brief look at just two of the factors that may lead society to consider alternative energy supply systems, such as hydrogen, will help us better understand the scope and impact of the problem.

Oil Supply

Technologically advanced countries are prodigious consumers of oil. World oil consumption (ca. 2004) is about 80 million barrels a day and close to the world's production capacity for this resource.[1] The U.S. alone consumes nearly 20 million barrels a day, much of which supports the transportation sector of the economy.[1] Globally, the use of oil and its derivatives continues to increase with no slowing in sight. Since the standard of living correlates closely with energy consumption in the modern world, the emerging economies of China and India, coupled with the reticence on the part of large oil-consuming nations (particularly the U.S.) to reduce energy consumption, can only point to an ever-increasing demand for oil.

What the future holds for oil supply and demand has been the subject of a number of recent books,[2–6] with conflicting viewpoints expressed on such issues as the timescale for reaching the Hubbard Peak[2] in oil production, current oil reserve estimates, the ability of economic forces to drive exploration and expansion of oil reserves, and the role of more difficult to obtain and expensive sources of oil (such as tar sands) to satisfy demand. A related issue to natural oil supply is the dependence of many nations, including the U.S., on imported oil for the bulk of their energy requirements and the impact this has on their national security. Historically, in terms of national policies, energy security has typically translated into short-term military security, rather than the long-term approach of concerted technology development of alternative energy sources.

Global Climate Change

Equally important, or, depending on your point of view, even more important, to diminishing oil reserves is the potential for global climate change. It is generally agreed that important indicators of man-made climatic changes have emerged recently.[7] This is not surprising. In little more than a hundred years, the geologic time span equivalent of the blink of an eye in human time, we have burned a significant fraction of the earth's oil reserves and have also burned large quantities of other hydrocarbons, such as coal and natural gas. The carbon released into the atmosphere during this interval, in the form of CO_2, had been locked up for millions of years within the earth's crust. The process is, of course, continuing, with over 24×10^9 metric tons of CO_2 (2002 values) released to the atmosphere every year,[1] potentially altering the energy balance in the earth's atmosphere. Furthermore, there is a complex, dynamic interplay among carbon levels in the atmosphere, the oceans, the ground, and the earth's flora and fauna. These interactions are not completely understood. Should we wait to observe ever larger changes in the world's ecosystems before acting? One must remember that a reduction in atmospheric CO_2 levels would take many years even if emissions were drastically reduced today.

THE CASE FOR HYDROGEN

There are differing opinions on the impact, if any, that the factors above will play in the world's geopolitical arena, or if other factors might emerge that will trigger an aggressive move away from hydrocarbon fuels on a large scale. It is clear, though, that alternative energy sources will eventually be needed to satisfy the world's ever-increasing energy requirements. Since such a transition would be revolutionary, rather than evolutionary, it will require a significant investment in research, development, and infrastructure over a relatively long period. In other words, it is not too soon to pursue the development of alternative fuels.

It is our belief that a transition of the world's energy infrastructure from the current one, based almost entirely on hydrocarbon fuels, to a diversified mix, including significant use of renewable sources (e.g., solar, wind, geothermal), increased use of nuclear energy, as well as petroleum, coal, and natural gas, would likely include hydrogen as an energy carrier, particularly in the transportation sector.

It is in the transportation sector, in fact, that hydrogen could have the greatest impact. For more than 100 years, gasoline- and diesel-fueled internal combustion engines have been used to supply motive power for a wide range of vehicle sizes, shapes, and applications. These vehicles are supplied with fuel by an efficient and pervasive petroleum-based infrastructure that produces a fuel with high energy density and consistent performance. The challenge, then, for alternative fuels is to supply equivalent, or nearly equivalent, vehicle performance, vehicle cost, and operating costs.

Furthermore, these requirements must be met on a scale sustainable at the levels expected for global automotive use. For example, the transportation share of U.S. petroleum consumption is about two-thirds of the total petroleum used, or about 13 million barrels per day. In energy equivalence, this corresponds to roughly 550 million kg of hydrogen per day. Taking into account the expected efficiency improvement with hydrogen vehicles, daily hydrogen consumption might be expected to be in the range of 200 to 250 million kg H_2, roughly 9×10^{10} scf H_2.

The major advantage of hydrogen as a transportation fuel, particularly with hydrogen fuel cell vehicles, is that it simultaneously addresses many issues associated with current petroleum-based vehicle technologies, including (1) reduced greenhouse gas emissions, (2) reduced pollutant emissions, (3) diversification of fuel feedstocks, (4) energy independence, and (5) on-board fuel efficiency. As stated earlier, however, there are significant technical challenges, many of them materials related, that must be overcome. Each of the stages in the hydrogen fuel chain—production, distribution, storage, utilization (e.g., fuel cell, internal combustion engines)—employs components and systems that require unique and sometimes extraordinary material properties. It is these properties and the current efforts in developing these materials that are described in this book. A brief introduction to the material issues for each of these areas of hydrogen use follows.

Hydrogen Production

Hydrogen is used primarily for petroleum refining and ammonia production with about 3.2×10^{12} scf produced in 2003. Most of this H_2 was produced by steam methane reforming. While this process can be expanded to produce more H_2 to meet the needs of a hydrogen-based economy, it does not eliminate greenhouse gas emissions or the U.S. dependence on foreign fossil fuels. There are a number of processes that can produce H_2 by the dissociation of water or steam. These include low- and high-temperature electrolysis, solar and photoelectrochemical processes, and thermochemical processes such as the sulfur–iodine process. The source of the energy to dissociate water is a key to whether these processes will reduce greenhouse gases and dependence on foreign fossil fuels. Nuclear energy as a source of electrical and thermal energy offers a significant opportunity to achieve both goals.

Steam methane reforming is performed in a high-temperature, high-pressure reaction chamber typically operating between 1,250 to 1,575°C at pressures of 20 to 100 atmospheres. Materials issues are the same as those of high-temperature, high-pressure vessels where creep of corrosion-resistant materials is important for the containment vessel and durability of alumina, chromia, or SiC refractory lining materials is critical to the performance of the system.

Electrolytes are a critical material in the performance of electrolyzers. Low-temperature electrolysis of water relies on proton exchange membrane (PEM) cells using sulfonated polymers for the electrolytes. Key issues for all electrolyzers are the kinetics of the system that is controlled by reaction and diffusion rates. Catalysts such as platinum, IrO_2, and RuO_2 are used to improve the reaction kinetics, but they also contribute to the cost of the system, which is also an issue. Steam electrolysis is also a possibility at a temperature of about 1,000°C using ceramic membranes.

Materials issues surround the kinetics of the electrode processes and durability of the interconnect materials in the high-temperature, oxygen-rich environments. Thermochemical water-splitting processes such as the S-I process offer high efficiency when coupled with an efficient source of heat, but have significant issues associated with corrosion of system materials. Materials being considered include Hastelloy B-2, C-276, Incoloy 800H, SiC, and Si_3N_4 with and without noble metal coatings.

Use of solar energy to produce H_2 is another route for reducing greenhouse gas emissions from fossil fuels while also reducing our dependence on foreign fossil fuels. Photoelectrochemical and photobiological processes are two examples that are solar energy driven. Photobiological hydrogen production is a process where microorganisms (algae or cyanobacteria) function as photocatalysts. Algae or cyanobacteria use photosynthesis to split water into O_2, protons, and electrons. Materials issues associated with this process are sketchy since this process has not developed beyond the exploratory stage. The low energy density of sunlight will dictate a system that covers a large area, so material costs will be a critical issue in the economics of this process. A concentrating reactor system will require light-transmitting elements from the dish-concentrating collector into the reactor. An overall list of material properties that will be critical to the operation of this type of H_2 production system includes transmittance, outdoor lifetime (i.e., durability to sunlight), biocompatibility, H_2 and O_2 permeation rates, and physical and mechanical properties.

Hydrogen Distribution

The distribution of hydrogen from a central production facility may be done with pipelines, trucks, or other carriers, but will very likely involve some off-board storage capability as well. Therefore, the primary materials issue associated with distribution deal with H_2 effects on pipeline and vessel materials. Transport of H_2 in a carrier such as ammonia, a hydrocarbon, or other form or local production of H_2 could alter some of the issues but is not likely to totally eliminate them. The safety of hydrogen distribution is a primary issue that affects material choice. The closer to population centers, the higher the risk and the more conservative the design.

Hydrogen storage and transport in steel pipelines have been done successfully in the industrial gas and petroleum industries. A key difference will be the gas pressures needed for commercial distribution of H_2 for the hydrogen economy. Materials are more susceptible to hydrogen effects with increasing pressure. Hence, there will be key issues related to safety and economy. Yet it is well known that steels can be susceptible to hydrogen-induced crack growth and embrittlement. Methods to reduce these effects include modifying the gas composition to reduce H_2 uptake and modifying the steel to reduce its susceptibility. The addition of impurity concentrations of O_2 is one option for reducing H_2 uptake, while manganese and silicon additions to the steel are possible routes for reducing the susceptibility of gas pipeline steels to H_2 effects. Considerable effort is needed to verify that these changes can be done effectively and that they provide the needed operational safety.

Hydrogen Storage

A key technical impediment to the deployment of hydrogen as a transportation fuel is the relatively low energy density for on-board hydrogen storage systems. Physical approaches, such as compressed gas and liquid hydrogen systems, are the only near-term options available, but these have limitations in terms of volumetric energy density or cryogenic requirements. In the long term, better storage alternatives will be needed, and current research efforts are focused on materials and chemical approaches, where the chemical bonding between hydrogen and other elements increases the volumetric density beyond the liquid state. With the recent launch by the U.S. DOE (Department of Energy) of a national "Grand Challenge" for hydrogen storage development, a number of exciting new research directions have appeared that have shown good progress over the last few years. Similar efforts have been launched in other countries as well, and formal international agreements, such as International Partnership for a Hydrogen Economy (IPHE) and IEA Task 22 (hydrogen storage), have led to good cooperation between researchers.

In contrast to the earlier development work in the 1970s, where intermetallic hydrides were intensively studied, recent work has focused on materials with high hydrogen capacity. The FreedomCAR and Fuel Partnership (an industry–government partnership) has established very challenging system-level performance targets for storage, for example, gravimetric energy density targets of 6 wt% H_2 for 2010 and 9 wt% H_2 for 2015.[7] Since these targets include the mass of system components, the storage materials must have even higher hydrogen capacities. System-level volumetric targets are equally as challenging.

Generally speaking, high-capacity materials often have thermodynamic properties (e.g., enthalpy of formation, operating temperature, stability, reversibility) or kinetic properties (e.g., absorption, desorption rates) that render them unsuitable for use in storage systems. Thus, research efforts are directed at (1) searching for new storage materials using rapid combinatorial screening methods and computational techniques; (2) improving the performance of storage materials through alloying, using catalysts and nano- or mesoscale structural modifications; and (3) examining alternate reaction pathways to overcome thermodynamic barriers.

Hydrogen Fuel Cells

Proton exchange membrane (PEM) fuel cells are the primary choice for transportation systems, but they can also be useful for stationary power production or local hydrogen production. Most of the challenges of PEM fuel cell commercialization center around cost and materials performance in an integrated system. Some specific issues are the cost of catalyst materials, electrolyte performance, i.e., transport rates, and water collection in the gas diffusion layer (GDL).

The anode and cathode electrodes currently consist of Pt or Pt alloys on a carbon support. Two low-cost, nonprecious metal alternative materials for anode catalysts are WC_x and WO_x. Pt alloyed with W, Sn, or Mo has also been evaluated for anode catalyst materials. Some non-Pt cathode catalysts that are being evaluated include $TaO_{0.92}$, $N_{1.05}ZrO_x$, pyrolyzed metal porphyrins such as Fe- or Co-N_x/C and

Co–polypyrrole–carbon. However, none of these have matched the catalytic performance of Pt.

The electrolyte membrane presents critical materials issues such as high protonic conductivity over a wide relative humidity (RH) range, low electrical conductivity, low gas permeability, particularly for H_2 and O_2, and good mechanical properties under wet-dry and temperature cycles; has stable chemical properties under fuel cell oxidation conditions and quick start-up capability even at subfreezing temperatures; and is low cost. Polyperfluorosulfonic acid (PFSA) and derivatives are the current first-choice materials. A key challenge is to produce this material in very thin form to reduce ohmic losses and material cost. PFSA ionomer has low dimensional stability and swells in the presence of water. These properties lead to poor mechanical properties and crack growth.

Solid-oxide fuel cells (SOFCs) are being developed for distributed power such as home power units and large power production units. They are not being considered for transportation, although that is conceivable with some difficulties. SOFC electrolytes are ceramic and operate at temperatures of up to 1,000°C, while PEM fuel cells operate at around 100°C or less. A key to the power production with SOFCs, as with PEM fuel cells, is the ability to produce thin electrolyte layers. Considerable development effort has resulted in cost-effective methods for producing thin and dense layers of ytrria stabilized zirconia (YSZ) that exhibit sufficient stability in the air/fuel environment. Doped CeO_2 is a leading candidate for operating temperatures below 600°C.

A primary limitation of YSZ is its low ionic conductivity. To overcome this, thinner electrolyte layers have been developed and yttria has been replaced with other acceptors. Even with these developments, the electrolytes must operate at temperatures exceeding 600°C. CeO_2 materials have a higher ionic conductivity than YSZ and can operate in the temperature range of 500 to 700°C but suffer from structural instability in the reducing atmosphere of the cell.

Interconnects are used to electrically connect adjacent cells and to function as gas separators in cell stacks. High-temperature corrosion of interconnects is a significant issue in the development of SOFCs. Ferritic stainless steels have many of the desired properties for interconnects but experience stability issues in both the anode and cathode environments. The dual environments cause an anomalous oxidation for which a mechanistic understanding has yet to be determined. Protective coatings from non-chromium-containing conductive oxides such as $(Mn,Co)_3O_4$ spinels look promising but need further development.

REFERENCES

1. Davis, Stacy C. and Diegel, Susan W. *Transportation Energy Data Book*, 25th ed., ORNL-6974. Oak Ridge National Laboratory, Oak Ridge, TN, 2006.
2. Deffeyes, Kenneth S. *Hubbert's Peak: The Impending World Oil Shortage*, rev. ed. Princeton University Press, Princeton, NJ, 2001.
3. Klare, Michael T. *Resource Wars: The New Landscape of Global Conflict.* Henry Holt and Company, New York, 2002.
4. Smil, Vaclav. *Energy at the Crossroads: Global Perspectives and Uncertainties.* MIT Press, Cambridge, MA, 2003.

5. Roberts, Paul. *The End of Oil: On the Edge of a Perilous New World*. Houghton Mifflin, Boston, 2004.
6. Goodstein, David. *Out of Gas: The End of the Age of Oil*. W.W. Norton and Co., New York, 2004.
7. Storage target values with footnotes can be accessed at http://www1.eere.energy.gov/hydrogenandfuelcells/storage/current_technology.html.

Contributors

Wade A. Amos
Gremlin Hunters, Inc.
Boulder, Colorado

U. (Balu) Balachandran
Energy Technology Division
Argonne National Laboratory
Argonne, Illinois

James P. Bennett
National Energy Technology Laboratory
U.S. Department of Energy
Albany, Oregon

Daniel M. Blake
National Renewable Energy Laboratory
Golden, Colorado

S. E. Dorris
Energy Technology Division
Argonne National Laboratory
Argonne, Illinois

Bin Du
Plug Power, Inc.
Latham, New York

S. Elangovan
Ceramatec, Inc.
Salt Lake City, Utah

John F. Elter
Plug Power, Inc.
Latham, New York

Maria L. Ghirardi
National Renewable Energy Laboratory
Golden, Colorado

Qunhui Guo
Plug Power, Inc.
Latham, New York

J. Hartvigsen
Ceramatec, Inc.
Salt Lake City, Utah

C.H. Henager, Jr.
Materials Science Division
Pacific Northwest National Laboratory
Richland, Washington

T. H. Lee
Energy Technology Division
Argonne National Laboratory
Argonne, Illinois

Paul A. Lessing
Materials Department
Idaho National Laboratory
INL Research Center
Idaho Falls, Idaho

Leng Mao
Plug Power, Inc.
Latham, New York

Richard Pollard
Plug Power, Inc.
Latham, New York

Zhigang Qi
Plug Power, Inc.
Latham, New York

Chris San Marchi
Sandia National Laboratories
Livermore, California

Michael Seibert
National Renewable Energy Laboratory
Golden, Colorado

Prabhakar Singh
Pacific Northwest National Laboratory
Richland, Washington

Brian P. Somerday
Sandia National Laboratories
Livermore, California

Jeffry W. Stevenson
Pacific Northwest National Laboratory
Richland, Washington

G. J. Thomas
U.S. Department of Energy
Washington, DC

Paul Trester
General Atomics
San Diego, California

Bunsen Wong
General Atomics
San Diego, California

Zhenguo Yang
Pacific Northwest National Laboratory
Richland, Washington

Xiao-Dong Zhou
Pacific Northwest National Laboratory
Richland, Washington

Abstract

Materials play a key role in whether a hydrogen-based energy economy, with its promise of clean, sustainable energy, will become a reality. The viability of a hydrogen-based economy is a complex issue involving technical, economic, and societal factors, but materials will be a key to its technical feasibility. Energy independence and global warming are two very important drivers for pursuing a hydrogen-based economy. Materials will be crucial in the production, distribution, storage, and utilization of hydrogen.

Today, hydrogen is produced primarily by steam methane reforming. It is used for petroleum refining and ammonia production. However, it is also possible to produce hydrogen from coal, petroleum, petroleum by-products, petroleum coke, refinery sludge, hydrocarbon fuels, refinery gas, natural gas, liquid waste, agricultural and municipal wastes, plastic, and tires. However, these are all carbon-based feedstocks. Noncarbon methods for producing hydrogen include electrolysis (methods exist for high- and low-temperature electrolysis), thermochemical production from water, and photobiological methods.

Hydrogen distribution and off-board storage have related materials issues, while on-board hydrogen storage has a unique set of criteria because of storage volume and weight restrictions. Distribution of hydrogen through steel pipelines is currently done for commercial purposes, but there is a need to ensure the safety of this distribution method in areas close to population centers. Distribution in trucks is also a possibility, but the economics of this are less favorable. There are also methods to distribute hydrogen in a carrier such as natural gas, ammonia, or petroleum.

Physical approaches, such as compressed gas and liquid hydrogen systems, are the only near-term options available for on-board storage, but these have limitations in terms of volumetric energy density or cryogenic requirements. In the long term, better storage alternatives will be needed, and current research efforts are focused on materials and chemical approaches, where the chemical bonding between hydrogen and other elements increases the volumetric density beyond the liquid state. With the recent launch by the U.S. DOE (Department of Energy) of a national "Grand Challenge" for hydrogen storage development, a number of exciting new research directions have appeared that have shown good progress over the last few years.

Proton exchange membrane (PEM) fuel cells are the primary choice for transportation systems, but they can also be useful for stationary power production or local hydrogen production. Most of the challenges of PEM fuel cell commercialization center around cost and materials performance in an integrated system. Some specific issues are the cost of catalyst materials, electrolyte performance, i.e., transport rates, and water collection in the gas diffusion layer (GDL).

Solid-oxide fuel cells (SOFCs) are being developed for distributed power such as home power units and large power production units. They are not being considered for transportation, although that is conceivable with some difficulties. SOFC electrolytes are ceramic and operate at temperatures of up to 1,000°C, while PEM fuel cells operate at around 100°C or lower. A key to power production with SOFCs, as with PEM fuel cells, is the ability to produce thin electrolyte layers. Considerable development effort has resulted in cost-effective methods for producing thin and dense layers of ytrria stabilized zirconia (YSZ) that exhibit sufficient stability in the air/fuel environment.

Editor

Dr. Russell H. Jones has 38 years of experience in materials development, evaluation, and characterization. Dr. Jones has extensive experience in the fields of stress corrosion cracking, radiation effects on materials, corrosion, and high-temperature composites. His work in stress corrosion cracking includes evaluation of the effects of hydrogen, aqueous, high-temperature, and nuclear environments on crack growth behavior of iron, nickel, aluminum, and magnesium alloys, and ceramics and ceramic composites. Dr. Jones's nuclear experience includes development of materials for advanced nuclear reactors and irradiation-assisted stress corrosion cracking for light water reactors. Specific corrosion experience includes evaluation of the effects of interface, grain boundary, and surface chemistry on corrosion of materials, including Yucca Mountain waste container materials. Dr. Jones has been instrumental in the development of SiC_f–SiC composites for advanced nuclear reactor applications, including high-temperature properties, corrosion, and radiation stability.

Dr. Jones has performed R&D in support of several energy technologies and the automotive industry. He has worked on materials for advanced nuclear reactors, light water reactors, nuclear waste containment, gas pipeline, and steam turbines and gas turbine blading. Dr. Jones has worked on lightweight materials, hydrogen materials compatibility issues, and hydrogen storage for advanced automotive applications.

Dr. Jones worked for the Westinghouse Research and Development Center from 1971 to 1973 doing materials development of steam and gas turbine materials; Pacific Northwest National Laboratory from 1973 to 2005 doing research on a wide range of materials, with an emphasis on corrosion and stress corrosion cracking; Exponent, Inc., from 2005 to 2006 as a senior managing engineer in the Bellevue office, working in mechanics and materials practice; and GT-Engineering from May 2006 to the present.

1 Issues in Hydrogen Production Using Gasification

James P. Bennett

CONTENTS

1.1 INTRODUCTION

Gasifiers are used commercially to react a carbon-containing material with water (or steam) and oxygen under reducing conditions (shortage of oxygen), producing chemicals used as feedstock for other processes, fuel for power plants, or steam for other processes. Gasifiers used in industry for chemical processing are high-temperature, high-pressure reaction chambers, typically operating between 1,250 and 1,575°C, and with pressures between 300 and 1,200 psi. Ash, originating from impurities in the carbon feedstock, is a by-product of gasification. It can exist as a powder or melt to form a slag, depending on the carbon source, the gasification temperature, and the ash melting point.[1] Gasifiers were first used in industry around 1800, but the high-temperature, high-pressure units currently used by the chemical, petrochemical, and power industries were first developed and put into commercial service in the 1950s and 1960s, and are greatly improved technologically over those of the past.[1] They are widely used in industries such as the petrochemical to process heavy oil by-products into H_2 used in refining or fertilizer manufacture, and are considered a potential source of H_2 for fuel cells or the emerging hydrogen economy. An example of an air-cooled slagging gasification system with the ability to produce a variety of products, from power generation to chemicals, is shown in figure 1.1. The gasification process, the types of gasifiers used, and how gasification is or can be used to generate H_2 are important in understanding its future role in the world.

In its simplest form, a gasifier is nothing more than a containment vessel used to react a carbon-containing material with oxygen and water (steam) under reducing conditions (shortage of oxygen) using fluidized bed, moving bed, or entrained flow technology. The gasification process produces CO and H_2 as the primary products, along with by-products of CO_2 and minority gases. Because the gasification process is intentionally conducted with a shortage of oxygen needed for complete combustion of the feedstock carbon, the partial oxidation reaction shown in equation 1.1 occurs:

$$C + H_2O\,(\text{gas}) + O_2 \rightarrow CO + H_2 + CO_2 + \text{minority gases} + \text{by-products} \qquad (1.1)$$

Note: By-products include mineral impurities in the carbon feedstock that become ash or slag.

A number of carbon feedstocks can be used as the carbon source, with the most common being coal, methane, or by-products/tails from the petrochemical industry. Since gasification occurs in an environment with a shortage of oxygen (reducing),

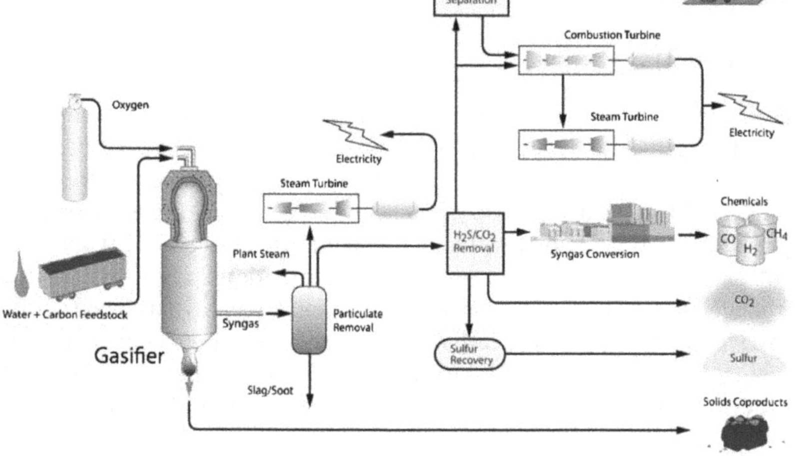

FIGURE 1.1 An air-cooled slagging gasification plant

the general balanced chemical reaction that occurs throughout the gasification process for a hydrocarbon can be written as in equation 1.2:

$$C_xH_y + x/2\ O_2 \rightarrow x\ CO + y/2\ H_2 + (\text{heat}, C^*) \tag{1.2}$$

Gasification is considered a noncatalytic process that involves a number of endothermic and exothermic reactions, with the overall process being exothermic.[1] In the reducing environment of gasification, excess carbon from the feedstock becomes a by-product of the process. Ideally, the amount of excess carbon should be small, about 1.0 wt%,[2] but is dependent upon variables such as the gasifier type, carbon feedstock, O_2/C ratios, and the level of carbon beneficiation. In the reducing environment of gasification, between 20 and 30% of the O_2 required for complete combustion of the C and H_2 in the carbon feedstock[3] is supplied. Because of the controlled oxygen shortage, gasification produces a primary product of CO and H_2, called synthesis gas (shortened to syngas), that is commercially valuable, along with a number of by-products that depend on the process and impurities in the carbon feedstock. The by-products can include excess C, sulfur, ash, soot, metal oxides, and gases (common gaseous impurities include CO_2 and H_2S; low-level impurity gases include CH_4, NH_3, HCN, N_2, and Ar). The type, quantity, and amount of any impurity are determined by the composition of the gasifier feedstock, the gasification temperature, and other factors, such as temperature or O_2 level. Depending on the application for the syngas, most impurities are removed at the gasification facility, which is accomplished using a variety of chemical processing techniques run downstream from the gasifier.

A number of different gasifier designs are used commercially, with the residence time for the carbon feedstock in the gasifier as short as seconds to hours, depending on the gasifier design. A gasifier thermally breaks down organic matter in the carbon feedstock, while inorganic materials (impurities) remain as discrete particles (ash). Some gasifier types operate at a low temperature to keep the ash as discrete particles (below their fusion temperature), while other designs operate at elevated temperatures that cause the ash to become molten and flow down the sidewalls of the gasification chamber as slag. Flux is often added to high melting point, high-viscosity ash to make a slag that melts at a lower temperature and has a lower viscosity, allowing the gasifier to operate at a lower temperature and slag to be fluid and flow down the gasifier sidewalls. As a process, gasification differs from incineration in that it creates valuable products (syngas or steam) used by a number of industries—it does not oxidize carbon into CO_2 as is done by incineration or a conventional power plant.[1] Gasification uses a variety of carbon feedstock, ranging from methane to carbon materials of low commercial value.

The earliest use for syngas generated from modern gasifiers was in the 1950s to 1960s for chemical synthesis.[4] By 1989, syngas was commonly used in other industries, causing chemical usage to be reduced to about half of all added production capacity, and with new gasification facilities added since 1990 favoring power generation. Coal and petroleum coke make up about 80% of the gasifier feedstock on units built between 1990 and 1999, with higher percentages after 2000 (94%).

* C originates from excess feedstock carbon in the reducing gasification environment.

Gasification can produce a range of H_2/CO ratios, depending on variables such as the feedstock, the amount of O_2 supplied during gasification, the operating temperature, and the gasifier design. Gasification is reported to be very energy efficient, with[3] up to 80% of the original energy in a coal feedstock available in the CO and H_2 syngas, about 15% recovered in the form of steam, and 5% lost as process energy during gasification. The new proposed integrated gasification combined cycle (IGCC) power generation facilities claim to have an overall efficiency of about 40%.[5]

1.2 CARBON FEEDSTOCK FOR GASIFICATION

A number of carbon feedstocks have been used or are being evaluated for use in gasification. These include coal (all ranks—anthracite to lignite, and the liquidification residues); petroleum (including heavy oil, high-sulfur fuel oil, and Orimulsion®—high-viscosity petroleum mixed with approximately 26 to 30% water and a dispersant[6]); petroleum by-products (known as oil distillates, residual oil, heavy oil, asphalt, visbreaker tar, refinery tar, heavy refinery feedstock, petroleum coke (delayed and fluid), refinery sludge, hydrocarbon fuels, refinery gas, bunker C-oil, vacuum residue, vacuum flashed cracked residue, miscellaneous liquid waste, and excess refinery products); natural gas; agricultural and municipal waste; liquor hazardous wastes (sewage sludge, biomass); and materials that would be difficult to dispose of as waste (such as plastic and tires). Many of the petrochemical materials used in gasification have little or no commercial value because of factors such as refinery economics or stringent environmental regulations.

The proposed new gasification facilities target multifuels, with issues such as ash fusion temperature, gasifier temperature, and the type of gas feed desired for the turbine being important considerations.[5] With any carbon feedstock, the gasifier must be strategically placed because of transportation costs and raw material availability. Other factors, such as those associated with carbon feedstock processing (grinding and beneficiation), must also be considered because of equipment costs. A general flow sheet of petroleum refining in relation to gasification needs by a refinery, including the production of H_2, is shown in figure 1.2.*

Gasification of some industrial by-products, such as biomass, has been accomplished commercially with limited success. In the U.S., Weyerhaeuser* has gasified black liquor from pulp and paper processing, with issues of refractory liner service life impacting the on-line availability of the gasifier.[7] Corrosive alkali and sulfate compounds contained in the black liquor attack the working face of the refractory lining, causing short service life and limiting the widespread use of gasification to replace the black liquor recovery boilers. Gasification has an advantage over other processes used for black liquor and biomass processing—it occurs at temperatures high enough to break down and eliminate dioxins.[8] In the future, the gasification of biomass or agricultural/forestry residue may be appropriate in remote geographic areas that have small power demands and where biomass feedstock is available. When using biomass feedstock such as bulk fiber materials, the added cost associated with processing coarse biomass into the fine materials required as gasifier feed

* Use of commercial names such as Weyerhaeuser in this text does not constitute endorsement.

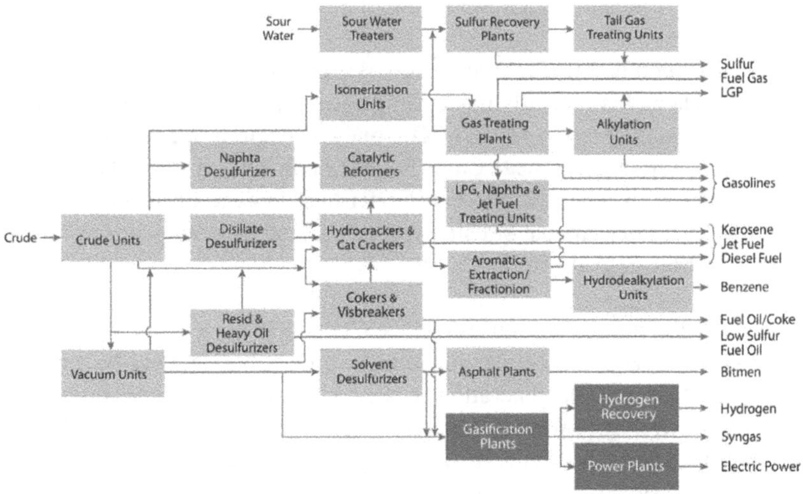

FIGURE 1.2 Petrochemical processing of petroleum into different products. (From Shell Gasification Process, available at www.uhde.biz, August 10, 2005. With permission.)

may be a cost-limiting factor. One gasification plant recently built in the U.K. has successfully used willow tree coppice as a carbon feedstock.

1.3 GASIFICATION PRODUCTS

The targeted use of syngas from gasification varies, with most syngas devoted to the production of single products such as power or H_2, and others devoted to producing a variety of products dictated by market demand for chemical feedstock, power, or other materials. Some products/outputs of gasification facilities are as follows:

- **Syngas (H_2 and CO)**—Used to generate heat, chemicals, electricity, or "town gas."
- **Chemicals**—H_2, CO, or both are used as chemical feedstock for the production of ammonia, oxo-chemicals, methanol, acetic acid, hydrogen, fertilizer, or synthetic hydrocarbon fuels (zero-sulfur diesel and other transportation fuels) manufactured using Fischer–Tropsch processing, and other chemicals.
- **Electricity**—Produced from the combustion of syngas or from gasification steam.
- **Steam**—Gasification or combustion by-product used in plant applications, power generation, and "over the fence" needs of nearby companies.
- **Gasification by-products**—CO_2 (used for enhanced oil and methane recovery, the food industry, urea fertilizer, possible enhanced greenhouse production, proposed geological disposal), S, N_2, Ar, Ni, and V.
- **Iron metal production**[8]—CO, H_2, or both are used to reduce iron oxide into metallic iron for steel production by the following reactions:

$$Fe_2O_3 + 3\ CO \rightarrow 2\ Fe + 3\ CO_2$$

$$Fe_2O_3 + 3\ H_2 \rightarrow 2\ Fe + 3\ H_2O$$

- Iron produced by the direct reduction of iron oxide is called direct reduced iron (DRI) and is typically made using natural gas. Two gasification facilities are designed to use H_2 and CO—one has been in operation since May 1999 (Saldanha Steel, near Cape Town, South Africa); the other is under construction in South Korea. These DRI facilities are designed to be syngas fuel flexible, capable of using syngas CO and H_2 combinations ranging from 100% CO to 100% H_2.

In general, the main syngas uses include CO and H_2 for chemical production or energy production; H_2 for chemical and refinery processing; and H_2, N_2, and CO_2 for fertilizer manufacture. The generation of chemicals is the predominant application for syngas, followed by power applications. Syngas CO and H_2 are considered good precursor materials for petrochemicals and agricultural products. Syngas produced from natural gas or coal is used in the manufacture of acetic acid, oxy-alcohols, isocyanates, plastics, and fibers. As with other carbon feedstock, heavier hydrocarbon material generated as a by-product or as a bottom material from petroleum refining (environmentally sensitive materials that are difficult to find applications for) is easily and economically processed by gasification into CO and H_2 used in the manufacture of high-value chemicals and energy.[9] Varying amounts of V and Ni heavy metals in the petroleum by-products or heavy fractions are converted during gasification into slag or high-value marketable products. Petroleum refineries have generated increasing amounts of these materials as the crude they process tends toward heavier oil and oil of higher sulfur content. Refineries have also had greater demand for H_2, a key material used by hydrocrackers to make lighter and cleaner fuels from low-quality oils, and a necessary material in fuel cells or a H_2 economy. This demand has been met through gasification, not through catalytic reformers. The trend worldwide is for more fuel that is lighter and cleaner, a demand driven by environmental stewardship and stricter emission regulations.

Some gasification facilities planned for the future are designed for syngas product flexibility, so gasification output can shift between syngas, H_2, CO, or combinations of them to meet changing industrial demand for power, steam, and chemicals. As with any chemical, specifications for syngas quality and purity vary with each application, necessitating different gasifier or chemical processes to treat the syngas. Because of transportation costs, most gasification facilities process (clean) syngas on site. Sulfur originating from the carbon feedstock, for instance, is removed at the gasification facility and marketed. Those gasification facilities based on IGCC designs produce some of the lowest NO_x, SO_x, particulate, solid, and hazardous air pollutants of any liquid or solid fuel technology used in power generation.[4,8,10] In general, near-zero-sulfur pollutants are desired in many syngas applications, necessitating sulfur cleanup in the ppm level for power generation (because of gas turbine and emission requirements) and in the ppb level for fuel cell applications.[11] Regarding CO_2 emissions, gasification has an advantage over other energy processes because it involves a closed loop, allowing for the possible collection, use, or disposal of CO_2 in deep-well injection to enhance oil or coal bed methane recovery, or "disposal" through mineral sequestration. At those gasification facilities devoted to making ammonia, CO_2 can also be recovered from the syngas and used to make urea (ammonia + CO_2 combine to produce urea fertilizer).

As with any chemical facility, process economics and transportation costs are critical factors in determining whether gasification syngas and the recovery of by-products will be profitable. Environmental factors such as the existence of proven technology for the recovery of SO_x, particulates, and mercury has made gasification attractive. When coal is used as a feedstock at Eastman Chemical, for instance, over 90% of the mercury contained in the coal is routinely collected.[10]

1.4 ENVIRONMENTAL ADVANTAGES

Gasification has many advantages that have led to its increased usage in chemical production and power generation, which are summarized below:

- Gaseous emissions:
  - Very low emissions compared to other processes—NO_x, SO_x and particulate emissions below current Environmental Protection Agency (EPA) standards.
  - Organic compound emissions are below environmental limits.
  - Mercury emissions can be reduced to acceptable environmental levels.
- SO_x can be processed into a marketable by-product.
- Ash can be liquefied into a slag that passes toxicity characteristic leaching procedure[12] (also known as TCLP) testing in most instances.
- CO_2 can be contained and recovered in the closed loops of gasifiers for remediation/reuse.
- Low-value carbon materials with environmental issues are easily utilized as a carbon feedstock.
- Gasifiers have product flexibility that allows output to be market driven.
- Gasification is a thermally efficient process.

1.5 HYDROGEN GENERATION BY GASIFICATION

In the U.S., the total H_2 consumption during 2003 was about 3.2 trillion cubic feet, with most utilized in petroleum refining and ammonia production.[10] Demand for H_2 is expected to grow, with worldwide needs projected to increase by 10 to 15% annually. In the U.S., most H_2 is generated by steam methane reforming, which constitutes about 85% of the total production. Gasification of hydrocarbon materials like coal, petcoke, and heavy oil, however, is starting to play a larger role in the production of H_2. This role is expected to increase as future natural gas supply and demand issues make the cost of generating H_2 from this feedstock too expensive or unreliable, and as refineries are forced to use lower-quality, heavy sour crude and as they produce cleaner-burning fuels. Refineries already collect by-product H_2 from off-gases generated during petroleum processing for reuse and cannot increase H_2 production by this route. The need for H_2 in petroleum refining is to hydrotreat crude, upgrading heavier hydrocarbon materials into higher-value fuels through hydrocracking or hydrodesulfurization (see figure 1.2). When gasification at a petrochemical facility is used to generate H_2, the gasifier is typically designed to give syngas flexibility, with excess syngas not needed internally used for power and steam or marketed.

Consideration must be given to purchasing H_2 as an over-the-fence raw material vs. building an on-site gasification plant. This decision should be based on gasification building, operation, and maintenance economics, and should consider if in-house expertise exists or can be assembled to operate the facility. Other factors, such as the consistency and availability of gasifier feedstock and the quantity, purity, pressure, and frequency of need for the H_2 output, will also dictate the technology used in H_2 generation. Another point to consider is the cost of H_2 transportation, storage, and dispensing, which are projected to be higher than the cost of production. In the U.S., H_2 transportation via pipeline is limited to about 500 miles.[10]

The commercial production of H_2 typically involves one of the following processes: (1) steam reforming, (2) water shift gas reaction, (3) partial oxidation, or (4) autothermal reforming. Electrolysis of water could be used to make H_2, but process economics are high when compared to the others processes listed; for that reason, electrolysis of water is not included. Currently, the production of H_2 by steam reforming has the lowest production cost of any process and is the most widely used, but as mentioned earlier, the cost and availability of the carbon feedstock may change that production cost in the future.[13]

Steam reforming, also known as steam methane reforming, involves reacting a hydrocarbon with steam at high temperature (700 to 1,100°C) in the presence of a metal catalyst, yielding CO and H_2. Of the processes used to make H_2, steam reforming is the most widely practiced by industry and can utilize a variety of carbon feedstocks, ranging from natural gas to naphtha, liquid petroleum gas (LPG), or refinery off-gas. Steam reforming, in its simplest form using methane as a feedstock, follows the general reaction

$$CH_4 + H_2O\ (gas) \rightarrow CO + 3\ H_2 \qquad (1.3)$$

Water shift gas reactions form CO_2 and H_2 using water and CO at elevated temperature, as shown in equation 1.4. The reaction may be used with catalysts, which can become poisoned by S if concentrations are high in the feed gas. The water shift gas reaction is used as a secondary means of processing syngas when greater amounts of H_2 are desired from gasification.

$$CO + H_2O\ (gas) \rightarrow H_2 + CO_2 \qquad (1.4)$$

Partial oxidation is the basic gasification reaction, breaking down a hydrogenated carbon feedstock (typically coal or petroleum coke) using heat in a reducing environment, producing CO and H_2 (equation 1.2). A number of techniques are utilized to separate H_2 from the CO in syngas or to enrich the H_2 content of the syngas. These include H_2 membranes, liquid adsorption of CO_2 or other gas impurities, and the water shift gas reaction (equation 1.4).

$$C_xH_y + x/2\ O_2 \rightarrow xCO + y/2\ H_2$$

Autothermal reforming is a term used to describe the combination of steam reforming (equation 1.3) and partial oxidation (equation 1.2) in a chemical reaction. It occurs when there is no physical wall separating the steam reforming and catalytic partial oxidation reactions. In autothermal reforming, a catalyst controls the relative extent of the partial oxidation and steam reforming reactions. Advantages of autothermal reforming are that it operates at lower temperatures than the partial oxidation reaction and results in higher H_2 concentration.[14]

In the above reactions, the partial oxidation process is the basis for producing H_2 and CO by gasification. As mentioned, depending on the amount of H_2 desired, other processes such as the water shift gas reaction may be used at the gasification facility to produce higher H_2 levels. It is important to remember that the ratio of H_2 to CO in gasification varies depending on the carbon feedstock, O_2 level, gasification temperature, and type of gasification process, in addition to other variables.

1.6 TYPES OF COMMERCIAL GASIFIERS

Different types of high-temperature gasifiers are commercially used to produce syngas, several of which are shown in figure 1.3. These gasifiers are known as (1) the General Electric slagging gasifier (figure 1.3a), (2) the ConocoPhillips slagging gasifier (figure 1.3b), (3) the Shell slagging gasifier (figure 1.3c and d), and (4) the Sasol–Lurgi fixed-bed dry-bottom gasifier (figure 1.3e). Several of the gasifier types, such as the General Electric (GE) and ConocoPhillips designs, were developed by other corporations and might be known by different names. All gasifiers and the support equipment are designed around a specific customer's needs, syngas application, carbon feedstock, or product requirements. Of the four types of gasifiers shown in figure 1.3, three types—the General Electric, ConocoPhillips, and Shell designs—can operate at temperatures high enough to form molten slag from solid impurities (ash) in the carbon feedstock. The fourth gasifier design, the Sasol–Lurgi fixed-bed dry-bottom gasifier (figure 1.3e), is designed to keep the ash as a free-flowing particulate, not a molten slag. Two of the gasifier designs (GE and Shell) are the dominant types of gasifiers used in the chemical production of H_2.

The *General Electric (GE) gasifier* (figure 1.3a) is a single-stage, downward-firing, entrained flow gasifier in which a carbon feedstock/water slurry (60 to 70% carbon, 40 to 30% water) and O_2 (95% pure) are feed into a reaction chamber (gasifier) under high pressure using a proprietary injector. The technology used in the GE gasifier was originally developed by Texaco in the 1950s to treat high-sulfur, heavy crude oil. In the gasifier, carbon, water, and O_2 combine according to equation 1.2, producing raw fuel gas (syngas) and molten ash. The GE design typically utilizes carbon feedstocks that include natural gas, heavy oil, coal, and petroleum coke, although a number of different feedstocks have been evaluated. GE gasifiers typically operate at pressures above 300 psi and at temperatures between 2,200 and 2,800°F. If a solid, the carbon feedstock must be fine enough to make pumpable slurry that can pass through the feed injector mounted at the top of the refractory-lined gasifier (less than 100 microns[15]). Ash in the carbon feedstock melts at the elevated gasification temperature, flowing down the gasifier sidewalls into a quench chamber, where it is collected and removed periodically through a lockhopper at

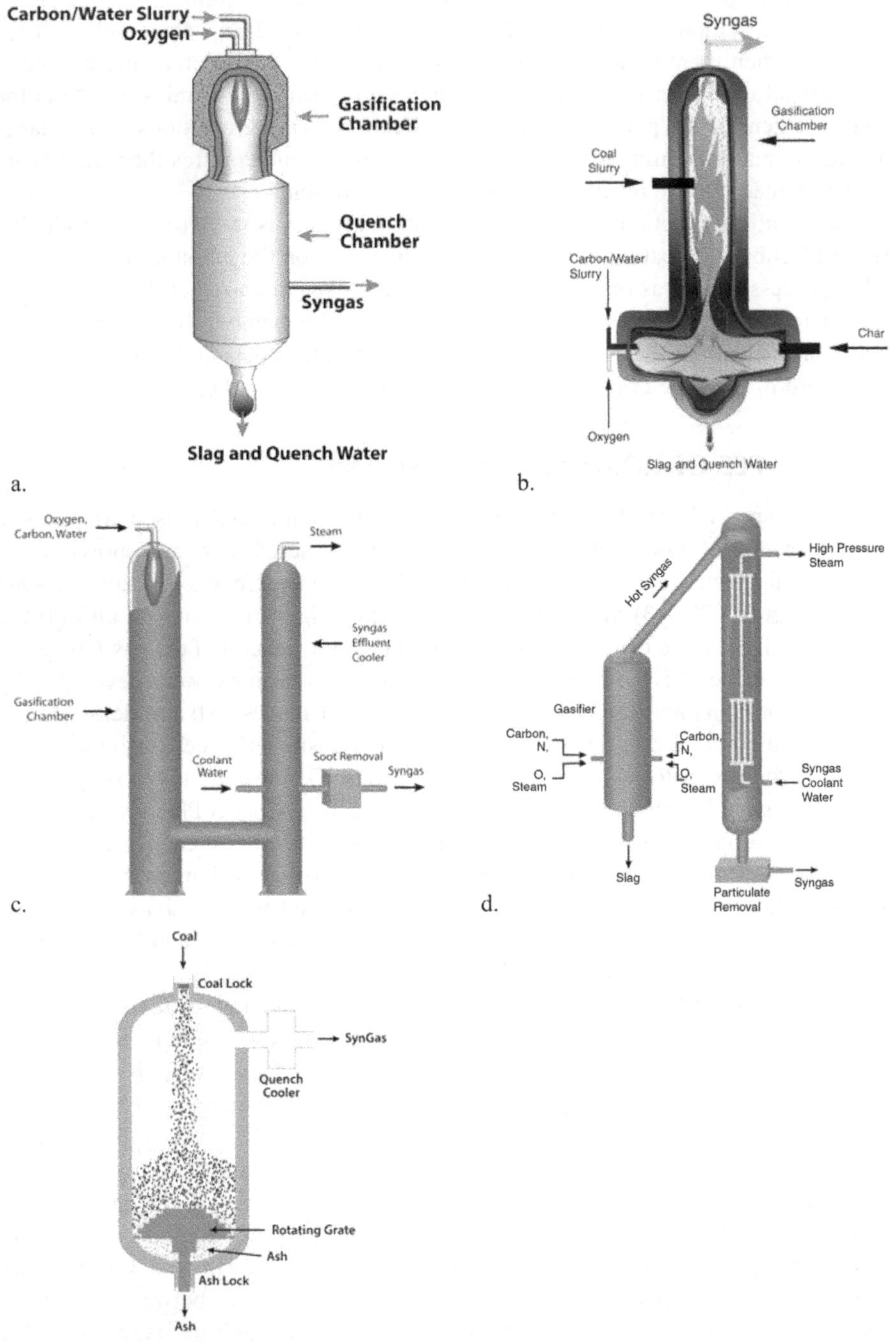

FIGURE 1.3 Several designs of commercially used gasifiers: (a) General Electric, (b) ConocoPhillips, (c) Shell—gas and liquid feedstock, (d) Shell—solid feedstock, and (e) Sasol–Lurgi fixed-bed dry-bottom gasifier.

the base of the quench chamber. One of two techniques is used to cool the syngas: a syngas cooler with heat exchangers or a water quench system. A scrubber then cleans and cools the syngas for processing or use elsewhere. Depending on the amount of scrubber fines and the carbon content of them, the fine particulate may be recycled to the gasifier. The raw syngas product from the gasifier consists primarily of H_2 and CO, along with a lower level of CO_2. The GE gasifier typically produces no hydrocarbons heavier than methane.[16] For the most part, metal oxides and other impurities present as solid material in the carbon feedstock become part of the glassy slag. Sulfur from the carbon feedstock forms H_2S during gasification, which is first chemically removed from the syngas during processing, and then is converted to commercially marketable elemental S. The refractory liner in the gasification chamber of the GE gasifier is air cooled. Corrosion/erosion from ash in the carbon feedstock leads to refractory liner replacement between 3 and 24 months in the gasifier, although in some special cases involving use of very low ash feedstock, liner life is approaching 3 years. The GE gasifier designed for coal can handle feedstock containing up to 4 wt% S on a dry basis, although higher S content carbon materials can be processed with an increase in the acid gas removal and sulfur recovery equipment.[5] It can handle a mix of up to 30% petroleum coke/coal blend, and with minor equipment modifications, mixed blends up to 70% petcoke. Limits exist on the allowable chloride content in the carbon feedstock because of the corrosion resistance of vessel construction material. Chlorides in the carbon feedstock are converted into HCl during gasification.[16]

The *ConocoPhillips gasifier* (figure 1.3b) is a two-stage pressurized, entrained flow slagging gasifier with an upward gas flow. The gasifier uses a carbon feedstock that is finely ground (less than 100 microns[15]) and mixed with water to make a pumpable feed ranging from 50 to 70 wt% carbon.[17] About 75% of this carbon–water mixture is then fed to a proprietary burner on one side of the gasification chamber base. On the opposite side of the gasifier base, recycled carbon char is fed to the second burner. The remaining carbon feedstock (25% of the total carbon feedstock slurry fed to the gasifier) is fed into the hot gases of the gasifier using a second stage injector located above the two opposing injectors (see figure 1.3b). Of the total carbon (carbon feedstock and char) introduced to the gasifier, about 80% is introduced in the two lower burners of the gasifier, and this is considered the first stage of gasification. Oxygen is fed only to the gasifier burner of the first stage with the slurry. Gasification of the carbon feedstock takes place at temperatures between 1,310 and 1,430°C, with ash becoming molten and flowing down the refractory sidewalls of the gasifier. The liquefied slag is then removed at the base of the gasifier. The carbon feedstock injected into the hot gas in the second stage becomes char that is later recycled as feed into the lower injector. In the second stage, where the balance of the carbon feedstock is introduced into the gasifier, an endothermic gasification reaction takes place, resulting in a gas temperature of about 1,040°C. This results in some hydrocarbons forming in the syngas. The hot gas leaving the gasifier is cooled in a fire-tube gas cooler to about 590°C, generating steam. Because the ConocoPhillips gasifier is air cooled and ash in the carbon feedstock liquefies, corrosion/erosion of the refractory liner of the gasifier occurs, with replacement necessary within 18 to 24 months.

Two *Shell gasifiers* (figure 1.3c and d) are used commercially, one for gas and liquid carbon feedstock (figure 1.3c) and one for solid carbon feedstock such as pulverized coal or petcoke (figure 1.3d). The first Shell gasifier (figure 1.3c) was developed in the 1950s to process fuel oil and bunker C-oil for the petrochemical industry.[18] Over time, heavier carbon source feedstocks with higher viscosity and higher levels of impurities (sulfur and heavy metal) were used, with feedstocks of short vacuum residue utilized in the 1970s and visbreaker and asphalt residues in the 1980s. Since the 1980s, petrochemical materials of even lower commercial value and quality have been used as carbon feedstock. Shell has recently developed a second type of gasifier (figure 1.3d), single-stage upflow, to utilize solid carbon feedstock, especially that with high ash content. Regardless of the feedstock, the Shell gasifier combines it with oxygen and steam (H_2O) to form syngas in a carefully controlled, reducing gasifier environment. Solid carbon feedstock such as coal must be ground to a fine particle size (less than 100 microns) and dried to less than 2% moisture. The dry ground material is then combined with a transport gas, usually N_2, and injected in the gasifier. It has been found that coal with an ash content as high as 40% can be utilized as a carbon feedstock in a Shell gasifier.[18] Oxygen for gasification is combined with steam before being fed into the gasifier, with both the carbon feedstock and the oxygen–steam gas combination preheated prior to injection in the gasifier. As in the GE and ConocoPhillips designs, injection of the carbon feedstock and the oxygen–steam into the pressurized gasifier uses a proprietary burner. A minimum temperature of 1.300°C is used to gasify feedstocks such as coal, which causes ash impurities to liquefy into molten slag that flows down the gasifier sidewall. Flux may be added to the carbon feedstock to control the ash melting and viscosity (flow) characteristics. The sidewall of the Shell gasifier is water cooled, which causes the molten slag to form a thin solid film over the gasifier refractory liner, protecting it from corrosive wear.[17] Because a water-cooled liner is used, refractory replacement occurs within 5 to 8 years, and thermocouple replacement about once a year. Typically, a Shell gasifier operates in a reducing atmosphere between 1.300 and 1.350°C, causing eutectics to form between the heavy metals and the slag, but also resulting in the formation of small quantities of soot (from carbon).[19] Under ideal operating conditions, a Shell gasifier has low O_2 consumption, creating high CO and minimal CO_2 in the syngas. After gasification, the syngas is passed through an effluent cooler, producing high-pressure steam and lowering the syngas temperature. Unburnt carbon (soot) and fine particulate ash remaining in the syngas are removed through quenching, forming oxides, sulfides, and carbonates of heavy metals and alkaline earth. The syngas also becomes saturated with water during quenching, which is used in the water shift reaction (Equation 1.4). If a syngas effluent cooler is used instead of a water quench cooler, gasification efficiency can increase by 5 percentage points.

The *Sasol–Lurgi gasifier* (figure 1.3e) was developed and put into service during the 1950s, and was used to process coal with ash content ranging from 10 to 35% and moisture up to 30%.[20] Approximately 100 individual units are in use throughout the world and produce over 25% of the total syngas produced worldwide. The type of Sasol–Lurgi gasifier in figure 1.3e is air cooled, with ash remaining as discrete particles. Because of this design, carbon feedstock must be a solid. that is. noncaking. and does not form a liquid or "sticky" ash that can agglomerate at the gasification temperature. Other important coal feedstock variables for a Sasol–Lurgi gasifier include burn rate, particle sizing, thermal fragmentation of the carbon feedstock, ash fusion temperature, bed void tendency, gas channeling, and the theoretical carbon yield. Coal used for the carbon feedstock is ground to a specific size range (between 6- and 50-mm particles[15]), permitting gas passage during gasification. Coal is introduced through a lockhopper at the top of the gasifier, with oxygen and steam (H_2O) introduced in the gasifier base. Gas is pulled from the base to the top of the gasifier, with the hot ash at the base of the gasifier preheating incoming steam and O_2 gases. The countercurrent gas flow of the syngas at the top of the gasifier preheats the incoming carbon feedstock and cools the syngas. The carbon feed passes through the gasifier by gravity. In the top of the gasifier, where the coal is preheated, moisture is driven off. Toward the base of the gasifier, pyrolysis of the carbon feedstock takes place, followed by gasification with oxygen and steam. At the very base of the gasifier, carbon has been depleted from the feedstock and only ash remains, which is removed by a rotating screen. The ash is kept below the fusion temperature so it remains as particles. Gasification occurs in stages, with a process pressure of about 430 psi. Critical reactions occur at about 1,000°C, with a crude syngas composition produced in the gasifier consisting primarily of H_2, CO, and methane. Gasification using the Sasol–Lurgi process is known to be reliable and tolerant of carbon feedstock changes. Condensates from the Sasol–Lurgi process are used to produce tars, oils, nitrogen compounds, phenolic compounds, and sulfur. Once the syngas is cleaned, it can be used as a town gas (a substitute for natural gas), in power generation, or as a chemical feedstock. Chemical processes built in conjunction with Sasol–Lurgi gasifiers include high-temperature Fischer–Tropsch conversion processes used to produce acetone, acetic acid, and ketones, and low-temperature Fischer–Tropsch conversion processes used to produce specialty waxes, high-quality diesel, kerosene, and ammonia.

1.7 GASIFIER/FEEDSTOCK EFFECT ON SYNGAS COMPOSITION

In general, depending on the type of gasifier used, the H-to-C ratio in the carbon feedstock, the feed rate in a gasifier, and the amount of oxygen introduced during gasification, syngas produced by gasification can have a range of H_2/CO ratios.[20] The H-to-C mole ratios for some carbon feedstock materials are about 0.1 for wood, 1 for coal, 2 for oil, and 4 for methane.[10] In general, the higher the hydrocarbon content in a carbon feedstock, the lower the ratio of H_2 to CO after gasification. This trend is shown in table 1.1, where the approximate H_2/CO ratios obtainable by gasifying a number of different carbon materials in a slagging gasifier are listed.[9]

Typical carbon feedstock properties and the resulting H_2/CO content in wt% after gasification in a Shell slagging gasifier are shown in table 1.2.[21] In contrast, H_2/CO ratios between 1.7 and 2.0 are produced in a Sasol–Lurgi fixed-bed dry-ash gasifier using coal as a carbon feedstock.[20] Specific ratios of H_2/CO are more important in petrochemical production than in the refinery, fertilizer, or power applications. This is because the H_2 and CO generated by gasification are used as the basic building blocks for chemicals such as methanol, phosgene, oxo-alcohols, and acetic acid. Syngas H_2/CO needs can range from a 2:1 ratio for methanol production to 100% CO

TABLE 1.1
Ratios of H_2/CO Produced by the Gasification of Different Carbon Feedstocks Using a Slagging Gasifier

Feedstock	H_2/CO Ratio
Natural gas	1.75
Naphtha	0.94
Heavy oil	0.90
Vacuum residue	0.83
Coal	0.80
Petroleum coke	0.61

TABLE 1.2
H_2 and CO Properties for Different Carbon Feedstocks Gasified Using a Shell Gasifier

Feedstock	Natural Gas	Liquefied Waste	Vacuum Residue	Liquefied Coke
C/H ratio (wt%)	3.35	9.2	9.7	11.9
S (wt%)	—	3.1	6.8	8.0
Ash (wt%)	—	0.01	0.08	0.16
H_2, CO in product (vol%)	95.3	94.0	92.9	92.8
H_2/CO ratio in syngas product (mole/mole)	1.69	0.89	0.88	0.78

in acetic acid production. It is important to remember that when heavier hydrocarbon by-products from petroleum refining are used in gasification, ash content is usually greater, plus the syngas will require more rigorous processing to remove unwanted materials, increasing production costs. The economic viability of any gasification process depends upon the carbon feedstock cost and the ability to optimize the syngas for the consuming industries.[10]

In practice, petroleum refineries have found it economical to gasify bottom materials (such as petroleum coke or heavy oils) for H_2 syngas production, with excess syngas used for power and steam generation. Fertilizer producers use H_2 from syngas to produce ammonia, and can use the CO_2 by-product for urea production. At most gasification facilities, the syngas receives additional physical or chemical processing on site to alter/optimize the H_2/CO ratio for the desired application. It is important to remember that beneficiation costs associated with the carbon feedstock and the syngas are limited by market demand and alternative source for chemicals. Sulfur, a common syngas impurity, is routinely removed because of environmental regulations or because of the negative impact it can have on materials it contacts during use, such as catalysts or turbine blades.

1.8 COMMERCIAL GASIFICATION

Different gasifier feedstocks in use or planned throughout the world are listed in table 1.3.[4] The carbon feedstock for the majority of these gasifiers originates from petroleum or coal, with some units designed to use both. Although petcoke is listed as a separate carbon feedstock in table 1.3, it is a by-product of petroleum processing and could be listed in that category. Those gasifiers using biomass or organic waste as a carbon feedstock require special gasifier linings and operate at lower gasification temperatures than petcoke or coal gasifiers (biomass/waste gasifiers listed are manufactured by Foster Wheeler). Regardless of the carbon feedstock, the location and size of a gasification complex is dictated by feedstock availability, transportation cost, and product demand.

Current or planned applications for syngas throughout the world are listed in table 1.4. Of the 155 syngas applications listed in table 1.4 for gasifiers, 105 facilities are used in chemical synthesis, 28 in power generation, and 11 in gaseous fuel production. Because a gasification facility is designed and built based on a targeted carbon feedstock and syngas application, limited flexibility exists in changing or modifying a facility without incurring high costs. Some syngas plants designed for future syngas production are considering the need for feedstock or product flexibility, and are designing this flexibility into the plant so the input/output can be driven by market forces.[11] This is particularly important in the petrochemical industry, where carbon feedstock can vary and demand for the chemical feedstock produced by gasification can be cyclic. A breakdown of the chemical applications for syngas in the chemical industry is listed in table 1.5, with the majority used in ammonia, oxo-chemicals, methanol, H_2, and CO synthesis.[4]

TABLE 1.3
Carbon Feedstock in Different Types of Gasifiers Used or Planned throughout the World

Gasifier Type	Carbon Feedstock Type (Number of Gasifiers Utilizing)				
	Petroleum	Coal	Gas	Petcoke	Biomass/Waste
GE	32	16	18	4	None
Shell	25	20	4	1	None
ConocoPhillips	None	3	None	4	None
Sasol–Lurgi	None	6	None	None	None
Foster Wheeler	None	None	None	None	6
Others[a]	2	6	1	None	7

Source: Information available at www.gasification.org, August 10, 2005.

[a] Other types of gasifiers, with the number of them in use in parentheses, are as follows: GTI U-Gas (2), GSP (2), Lurgi dry ash (2), Lurgi circulating fluidized bed (2), Lurgi multipurpose (1), low-pressure Winkler (1), BGL (1), Foster Wheeler pressurized circulating fluidized bed (1), Thermo Select (1), TSP (1), Krupp Kroppers PRGNFLO (1), and Koppers-Totzak(1).

TABLE 1.4
Syngas Applications Output for Gasifiers That Are Operating or Are Planned

Gasifier Type	Gasification End Product Use (Number of Gasifiers Dedicated to That Purpose)					
	Chemicals	Power	Gaseous Fuels	FT Liquids	Multiple	Not Specified
GE	60	8	2	None	None	None
Shell	41	3	1	2	None	3
ConocoPhillips	None	4	None	1	2	None
Sasol–Lurgi	2	None	1	3	None	None
Foster Wheeler	None	1	5	None	None	None
Others[a]	2	12	2	None	None	None

Source: Information available at www.gasification.org, August 10, 2005.

[a] Other types of gasifiers are as follows: GTI U-Gas (2), GSP (2), Lurgi dry ash (2), Lurgi circulating fluidized bed (2), Lurgi multipurpose (1), low-pressure Winkler (1), BGL (1), Foster Wheeler pressurized circulating fluidized bed (1), Thermo Select (1), TSP (1), Krupp Kroppers PRGNFLO (1), and Koppers-Totzak(1).

TABLE 1.5
Industrial Applications for Syngas from Those Gasifiers Whose Output Is Classified as Chemical in Table 1.4

Primary Syngas Application	Number of Gasifiers	Percent of Total
Ammonia	37	35
Oxo-chemical	19	18
Methanol	17	16
Hydrogen	11	10
Carbon monoxide	7	7
Syngas	4	4
Acetic anhydride	1	1
Acetyls	1	1
Unknown	8	8

Source: Information available at www.gasification.org, August 10, 2005.

1.9 GASIFICATION FOR H_2 PRODUCTION

Worldwide, it is estimated that approximately 50 million tons of H_2 is produced and consumed annually.[22] Of this total, approximately 90% is produced by steam reforming (equation 1.3) and 10% by gasification (equation 1.2).[11] Regardless of the process used (steam reforming or gasification), the primary products are H_2 and CO, along with by-products that include CO_2, S, and other gaseous impurities. To raise the H_2 output from steam reforming or gasification, the water shift reaction (equation 1.4) is used to convert CO to H_2.[11] Of the total amount of H_2 originating from syngas and used in chemical synthesis, approximately 20% is consumed by refineries and 80% in ammonia, methanol, CO, and oxo-chemical production. The quantities of H_2 gas utilized are large, as indicated by the amount consumed by Shell, which uses more than 2.6 million tons/year of H_2 gas that it produces from steam methane reforming, coal gasification, oil residue gasification, and platforming (naphtha type carbon feedstock processed over a platinum-containing catalyst to produce a reformate and hydrogen).[22]

Data on several gasification facilities producing H_2 as the primary or secondary product are listed in table 1.6.[4,23] This table is not complete because it does not include all on-site facilities devoted to the production of H_2 converted directly into chemicals like ammonia or urea. In table 1.5, 37 plants were listed as producing ammonia from syngas. H_2 used in ammonia production is first separated from the syngas, then reacted with N_2 separated from the air (equation 1.5). A catalyst is used in the reaction chamber where the reaction of equation 1.5 occurs. Some gasification facilities take fertilizer production an additional step, reacting ammonia with CO_2 to produce a liquid urea fertilizer.[24] It is of interest to note in table 1.6 that only two types of gasifiers are predominantly used in H_2 production, the GE and Shell designs (figure 1.2a and c).

$$3\,H_2 + N_2 \rightarrow NH_3 \tag{1.5}$$

When the GE gasifier is used to produce H_2, two types of syngas cooling systems are used (table 1.6): the direct water quench and the radiant syngas cooler. In the direct water quench, hot syngas from the gasifier is fed into a water quench ring, cooling the syngas through direct water contact. This is considered thermally inefficient, but introduces water into the syngas necessary for the water gas shift reaction used to produce H_2 from CO (equation 1.4).[9] When a syngas radiant cooler is used, hot syngas passes directly from the gasification chamber to another chamber containing a syngas cooler, producing high-pressure saturated steam. Use of the radiant syngas cooler maximizes heat recovery, which can be important depending on the value of steam in the gasification facility or as a marketable product.

When a feedstock such as natural gas is used in the production of H_2, it is often pressurized to match downstream process requirements before entering the gasifier.[9] No specifications exist on the most efficient pressures to use in a gasifier or in the production facility. Operating pressure must be determined based on reaction rate efficiencies, equipment sizing issues, product requirements (quality, quantity, and frequency of gas need), available feedstock, gasification and syngas process technology, and projected consumption demand. Catalysts, absorbents, and operating temperature are other factors to consider for each processing stage because of cost and process limitations. In many instances, multiple stages of some processes are necessary to produce the quantity and purity necessary in gases like H_2. Each plant stage impacts production cost and efficiencies. Examples of some syngas plants used to produce H_2 are as follows:

Coffeyville, KS[24]—This facility uses a GE gasifier to convert petcoke into high-purity H_2, which is subsequently converted to ammonia. In addition to petcoke as a carbon feedstock, the gasifier is able to use low-value refinery

TABLE 1.6
Commercial Gasification Facilities Dedicated to Hydrogen Production

Plant Name	Location	Gasifier Type	Date Built	Feedstock	Syngas Cooler	Syngas Output (106 Nm3/d)	Syngas Application
AGIP IGCC	Sannazzaro, Italy	Shell	2005	Cracked residue (1200 mt/d)	Fire-tube boiler	3.34	Power, H2
Brisbane H2	Brisbane, Queensland, Australia	GE	October 2000	Natural gas (15 MMscf/d), refinery off-gases	Direct water quench	0.8	H2 (30 MMscf/d)
Coffeyville Nitrogen Plant	Coffeyville, KS	GE	July 2000	Petcoke (1100 mt/d)	Direct water quench	2.14	Ammonia (1000 mt/d), H2
Convent H2	Convent, LA	GE	1984	H-oil bottoms (650 mt/d)	Direct water quench	1.88	H2 (62.5 MMscf/d)
Gela Ragusa H2	Gela Ragusa, Italy	GE	1963	Natural gas (16.8 MMscf/d)	Direct water quench	1.15	H2
Kaohsuing Syngas	Kaohsuing, Taiwan	GE	August 1984	Bitumen(1000 mt/d)	Direct water quench/syngas cooler	2.14	H2 (158,200 Nm3/h)
LaPort Syngas	LaPorte, TX	GE	August 1996	Natural gas(28.7 MMscf/d)	Radiant syngas	1.85	H2 (50 MMscf/d), power, steam
Leuna Methanol Anlage	Leuna, Germany	Shell	1985	Visbreaker residue (2400 mt/d)	Fire-tube boiler	7.2	H2 (42.4 MMscf/d)
Ludwigshafen H2	Ludwigshafen, Germany	GE	1968	Fuel oil (345.5 mt/d)	Unknown	0.98	H2 (35 MMscf/d)
Most Gasification Plant	Most, Czech Republic	Shell	January 1971	Cracked residue (1250 mt/d)	Fire-tube boiler	3.6	H2, methanol, power, steam

Opit/Nexen	Alberta, Canada	Shell	2006	Asphalt (3100 mt/d)	Unknown	7.5	H2, steam
Paradip Gasification H2/Power	Naapattinam, Orissa, India	Shell	2006	Petcoke (3500 mt/d)	Fire-tube boiler	6.5	H2 (160 MMscf/d)
Pernis Shell IGCC/H2	Rotterdam, Netherlands	Shell	October 1997	Visbreaker residue (1650 mt/d)	Fire-tube boiler	4.7	H2 (100 MMscf/d), power, steam
Rafineria Gdariska SA	Gdansk, Poland	Shell	2008	Asphalt (1650 mt/d)	Unknown	4.5	H2, power
Singapore Syngas	Jurong Island, Singapore	GE	June 2000	Refinery residue (572 mt/d)	Direct water quench	1.6	H2 (25 MMscf/d)
Texas City Syngas	Texas City, TX	GE	April 1996	Natural gas (31.6 MMscf/d)	Radiant syngas	1.92	H2 (40 MMscf/d)

Source: Information available at www.gasification.org, August 10, 2005; Zuideveld, P. and de Graaf, J., Overview of Shell Global Solutions, Worldwide Gasification Developments, paper presented at the Proceedings of Gasification Technologies 2003, San Francisco, October 12–15, 2003.

materials as supplemental carbon, and can supply H_2 to a nearby refinery when its economic value exceeds the commercial value of ammonia. The gasification facility is also capable of processing ammonia into urea–ammonium–nitrate (a liquid fertilizer) by capturing CO_2 from gasification and reacting it with the ammonia. The general gasification process mixes petcoke and water with a flux (if needed) to form a high-solids-concentration slurry that is fed to a gasifier burner, where it is mixed with pure oxygen and injected into the gasifier. In the gasification chamber, syngas (H_2, CO, CO_2, H_2S, and minor amounts of other compounds) is formed at temperatures between 1,320 and 1,480°C. Mineral impurities in the petcoke are melted at these temperatures, forming a slag, which flows down the gasifier sidewall into the quench chamber. The quench chamber serves two purposes, cooling the syngas and quenching the slag. Periodically, a lockhopper at the bottom of the gasifier allows solidified slag to exit, while the syngas product continuously exits the gasifier to a water scrubber used to remove solid particulates. The scrubber also saturates the syngas with moisture, which reacts CO in the water shift unit (in the presence of a catalyst) to form H_2 and CO_2 (equation 1.4). When the syngas exits the shift unit, it is over 40% CO_2. Heat from the shift reaction produces steam, which is used in the ammonia unit and in the refinery. The cooled syngas is next passed to an acid gas removal (AGR) unit based on the Selexol process. This unit concentrates H_2S to about 44%, which is sent to a Claus unit for CO_2 and sulfur removal. At Coffeyville, the bulk of the CO_2 is removed from the syngas, with a portion of it reused in urea production. In the future, if other applications for CO_2 are identified, it can be purified and sold. Syngas exiting the AGR unit is about 96 mol% H_2. This high-H_2 feedstock is sent to a pressure swing adsorption (PSA) unit where remaining impurities are extracted, resulting in a H_2 gas of 99.3% purity. After PSA processing, the main impurity remaining in the H_2 is N_2. The purified H_2 is fed to the ammonia unit, where ammonia is manufactured using N_2 from the air separation unit (equation 1.5). The tail gas from the PSA unit is about 75% H_2 and CO, which are compressed and recycled back to the water shift unit.

Eni SpA, AGIP, Sannazzaro refinery, Italy[23]—This gasifier was built because of Italian legislation with the goal of reducing emissions from power stations, and is based on a Shell gasifier design. The gasification facility produces syngas for a new 1,000-MW gas–syngas power plant, with some H_2 recovered for refinery needs using membrane separation technology.

Liuzhou Chemical Industry Corporation, Siuzhou, Guangxi, PRC[23]—This gasifier uses a Shell design based on coal feedstock and converts 1,200 t/d of coal into 2.1×10^6 Nm^3/d of syngas. The syngas is used to manufacture ammonia-based fertilizer and oxo-alcohols. Some process CO_2 is recovered and used to make urea fertilizer.

OPTI Canada Inc.–Nexen Petroleum Inc., Long Lake, Alberta, Canada[23]—This gasification facility was designed using a Shell gasifier to process heavy asphaltene by-products generated from processing oil sands for bitumen. When placed in operation, this gasifier will process approximately 3,100 t/d of carbon feedstock, producing H_2 for hydrocracking of oil sand bitumen into high-quality synthetic crude and producing steam for the extraction of the bitumen from the sands.

Shell Nederland refinery, Pernis, Rotterdam, the Netherlands[23]—This gasification facility was built in 1997 using a Shell gasifier and was part of a refinery upgrade. The gasification facility produces H_2 for petroleum refining. It contains three gasification trains to meet H_2 refinery needs, with two of the three gasifiers on-line and the third targeted for repair at any given time. The gasifiers process a carbon feedstock of vacuum flashed cracked residue from the thermal cracking unit of the refinery or a mixture of straight-run vacuum residue and propane asphalt. The gasification facility has a feed rate of approximately 1,650 t/d and a higher H_2 capacity than needed by the refinery. Excess syngas capacity is used to produce electricity using gas turbines. The gasifiers operate at about 1,300°C and a pressure of 940 psi. After gasification, the syngas is cooled below 400°C, with the liberated heat used to produce high-pressure steam. A low-temperature CO shift reaction is used to increase the amount of H_2 obtained from the syngas. An integrated gas treatment unit removes H_2S and CO_2 from the syngas, with about 3,000 t/d of CO_2 released to air. Future use of the CO_2 by nearby greenhouses to enhance plant growth is being considered. The CO level in the CO_2 is reduced to about 1 vol% during high- and low-temperature shift reactions.[21] About two-thirds of the syngas produced by gasification is used in the production of H_2 for hydrocracking (up to 285 t/d), with the remainder used in power generation. H_2 generated at the Shell Nederland gasification facility is about 98% pure, with a pressure of 680 psi. Soot/ash recovered in the quench unit is marketed because of the high percentage of V and Ni present in the ash, which can contain up to 65% vanadium oxide. Because of many factors, H_2 produced at this facility is at a lower cost than that produced by steam–methane reforming.

Rafineria Gdanska SA, Gdansk, Poland[23]—This gasification facility uses a Shell gasifier and was built as part of a refinery upgrade. The goals of the upgrade were to reduce refinery emissions while processing a higher sulfur carbon feedstock, produce higher-value products from the refinery, and produce a lower-emission gasoline. Carbon feedstock for the gasifier will be about 1,600 t/d of asphaltenes, with most of the syngas targeted for H_2 manufacture and use in the hydrocracking unit. Excess syngas will be used for power generation. Steam generated from the gasification process will be used by the refinery.

Sinopec, Zhijiang, Hubei, and Anqing, Anhui, PRC[23]—These gasification facilities are scheduled to be built based on Shell gasifiers and will use coal feedstock to manufacture fertilizer.

Texas City Gasification Project Texaco Gasifier[9]—This gasification facility uses a GE gasifier and began operation in June 1996. It markets H_2 and supplies CO as a feedstock to a chemical company that uses it to manufacture chemicals such as acetic acid and special alcohols. The carbon source is natural gas, which is preheated before entry in the gasifier. Gasification syngas is cooled in a syngas cooler, which generates high-pressure saturated steam. The syngas is then processed in an AGR unit, which also removes and recycles CO_2 back to the gasifier. A cold box separates CO from H_2, with the H_2 further processed by a PSA unit. After beneficiation, the H_2 can be up to 99.9% pure.

1.10 SYNGAS FOR CHEMICAL PROCESSING

As noted previously, syngas must be cleaned or processed to reduce or remove impurities such as H_2S, COS, HCN, CO_2, N_2, and carbon/soot, and may be processed to concentrate or increase the quantity of gases like H_2, CO, or CO_2. The level of processing is determined by the application, with most having limits on S that originates from the gasification carbon feedstock. After gasification, S is typically present as H_2S and must be removed because of emission controls, the corrosive effect of sulfur, or the "poisoning" effect it has on the catalytic materials used in many chemical or gasification processes. Two of the many processes used to remove S at a gasification facility are (1) the glycol-based absorber–stripper process, which uses a mixture of tetraethylene glycol dimethyl ether ($C_{10}H_{22}O_5$) to capture more than 98% of the H_2S,[25] and (2) the sodium hydroxide reaction, which uses an aqueous solution of sodium hydroxide to react with H_2S and produce sodium sulfide, removing from 85 to 95% of the H_2S from the sour gas. CO_2 in the syngas is also removed by the sodium hydroxide, forming sodium bicarbonate.[24] Multiple stages of these or other chemical beneficiation processes are often used to reach purification levels higher than can be achieved from a single gas pass. Although high-purity levels are obtained, the process redundancy creates high processing costs associated with the additional equipment setup and maintenance and with energy/efficiency losses. Some of the chemical processes used in a gasification facility to produce specific gases or desired purity levels can include the following:

1. **Shift unit**—Reacts syngas CO and moisture (H_2O) at a low temperature in the presence of a catalyst using the shift gas reaction (equation 1.4), forming H_2 and CO_2.
2. **Catalytic hydrolysis reactor**—Hydrolyzes the syngas COS to CO_2 and H_2S, and HCN to NH_3 and CO for ensuring that environmental emission limits are met in the syngas.
3. **AGR unit**—Separates and concentrates H_2S in the syngas for feed to a Claus unit.
4. **Claus unit**—Produces S from syngas H_2S concentrate.
5. **PSA unit**—Used to purify a H_2 syngas stream of specific desired impurities through the use of absorbents. The gas purity of the H_2 produced is determined by factors such as which absorbents are used.
6. **Fischer–Tropsch synthesis**[8]—Uses a catalyst to react syngas at high temperatures (330 to 350°C) and pressures (25 bars), or low temperatures (180 to 250°C) and high pressures (45 bars), producing different straight-chain hydrocarbons that range from methane to high molecular weight waxes, according to the reaction $CO + 2\,H_2 \rightarrow -[CH_2]- + H_2O$. Fischer–Tropsch

synthesis processing occurs in the presence of a catalyst and produces carbon chain lengths from 1 to 15.

7. **Sasol advanced synthol process**—A high-temperature process (approximately 340°C) that uses syngas to produce gasoline and olefins.

1.11 MATERIALS OF CONSTRUCTION

The gasification chamber used to contain the chemical reaction among the carbon feedstock, water, and oxygen at high temperature is lined with a number of different refractory materials. These materials are determined by the type of carbon feedstock (gas/liquid feed or solid materials like coal or petcoke), whether the gasification chamber is designed to liquefy feedstock ash or keep it as discrete particles, the quantity of ash generated, and if the gasifier sidewalls are air or water cooled. The refractory lining is designed to provide protection of the steel shell for sustained, uninterrupted periods of time so the gasifier can operate effectively and economically. A steel shell thickness is determined by internal pressure, shell temperature, and the vessel diameter and can range up to 3 inches or more. A gasifier's operational temperature has a large influence on liner materials, with gasifiers such as the Sasol–Lurgi (figure 1.3e) operating at temperatures below the ash liquification (ash impurities remain as a dry particulate). It uses a refractory liner designed to withstand abrasive wear, the gasification atmosphere, and the high operating temperature. Gasifiers that form molten slag focus on chemical wear and corrosion. Most gasifiers produce a molten ash (slag) that is highly corrosive, causing high refractory wear that negatively impacts refractory service life. Gasifiers that form molten slag include the GE (figure 1.3a), ConocoPhillips (figure 1.3b), and Shell (figure 1.3c and d) designs. Common slag elements in a coal feedstock that become part of the ash or slag include Si, Al, Fe, and Ca, while a petcoke feedstock can contain V and Ni in addition to the other elements. Other elements such as Na or K may be present, depending on the feedstock source.

High alumina brick or monolithic refractory materials are predominantly used as hot face liners in gas or liquid feedstock gasifiers, and high chrome oxide (air-cooled gasifier) or high thermal conductivity (water-cooled gasifier) materials with coal or petcoke feedstock. The service life of different refractory linings varies, with those using oil as a carbon source typically lasting 4 to 5 years (high-wear areas may need replacement in as short as 2 years), gas up to 10 years (various areas of the gasifier requiring replacement between 6 and 10 years), and solid feedstock (coal or petcoke feedstock) up to 2 years (high-wear areas may require replacement in as little as 3 months). Refractory liners used in a gas or liquid gasifier are typically high alumina materials (94 to 99% Al_2O_3) that are low in SiO_2 and FeO, while material used in solid-fuel air-cooled slagging gasifiers are high in chrome oxide (up to 95% Cr_2O_3).

Gasifiers that use biomass feedstock have unique requirements due to the high alkali and alkaline earth oxides, components that vigorously attack any refractory linings. Biomass from black liquor gasification, for instance, produces salts of $NaCO_3$, Na_2S, and NaOH, with minor components of SiO_2, NaCl, and KCl,[26] which have been found to melt between 400 and 780°C depending on composition, and has a viscosity lower than coal slag by a factor of 1,000. Use of water-cooled SiC linings has not shown promise because of constant attack of the lining, with no stable equilibrium established, while samples without water cooling showed rapid and complete dissolution. Volkmann and Just[26] mention that a refractory is needed with the following qualities:

1. Resistance of molten black liquor slats containing $NaCO_3$, Na_2S, and NaOH
2. Resistance to gaseous reaction products such as Na and NaOH vapors as well as gasification syngas
3. Resistance to thermal shock from temperature and gas pressure variations
4. The ability to bond to cooling coils or screen
5. Good thermal conductivity

For these and other reasons, biomass liners are still being researched, with MgO and $MgAl_2O_4$ spinel refractories holding some promise.[27] High alumina refractories or refractories of a mullite base are not used in biomass gasifiers because of the large volume change that occurs when Na_2O (from the biomass) interacts with high alumina refractories to form beta alumina, or in mullite refractories, to form phases such as beta alumina and nepheline. These are compounds with large volume changes that disrupt the brick structure by causing cracking and spalling. Besides refractory damage, stresses created by the volume change can be so large as to cause containment shell damage. Because of these and other material issues, biomass gasification thus far is very limited.

The refractory lining in gas, liquid, or solid feedstock gasifiers can be between two and six layers in thickness,[28] with a typical sidewall lining shown in figure 1.4. The hot face or working lining is designed for direct contact with the gasification environment, followed by a backup refractory material and an insulating refractory lining. The refractory materials in each layer depend on the gasifier design, location in the gasifier, gasification temperature and atmosphere, and carbon feedstock. The main purpose of a lining is to protect the high-pressure steel shell from syngas and other gasification gases, elevated temperature, particulate abrasion, and slag corrosion. Any material in a gasifier must be thermodynamically stable to hot gases such as H_2, CO, CO_2, H_2O, and H_2S. If ash is liquefied in a gasifier, varying ash chemistry, the quantity of slag generated, and the number of components in it makes the use of phase diagrams to determine refractory stability of limited value. In practice, few phase diagrams of relevance beyond three components exist. Other factors limiting the use of phase diagrams include high gasification pressure, an oxidizing reducing/reducing gasifier environment, and an atmosphere that includes O_2, CO, H_2, or sulfur compounds. Besides corrosion/phase stability, limited information also exists on the ability of a refractory material to withstand the high-temperature particulate impact/abrasive associated with the burner and its carbon feedstock. Another requirement of the refractory lining is to reduce the gasifier shell temperature to an acceptable temperature, yet keep the shell interior above acid dew point condensation temperature (e.g., 290°C for H_2SO_4 at a 95 wt% concentration, preventing steel shell corrosion, which could result in catastrophic shell failure). In certain locations, dew point condensation may also be a concern on the shell exterior, requiring higher shell temperatures.

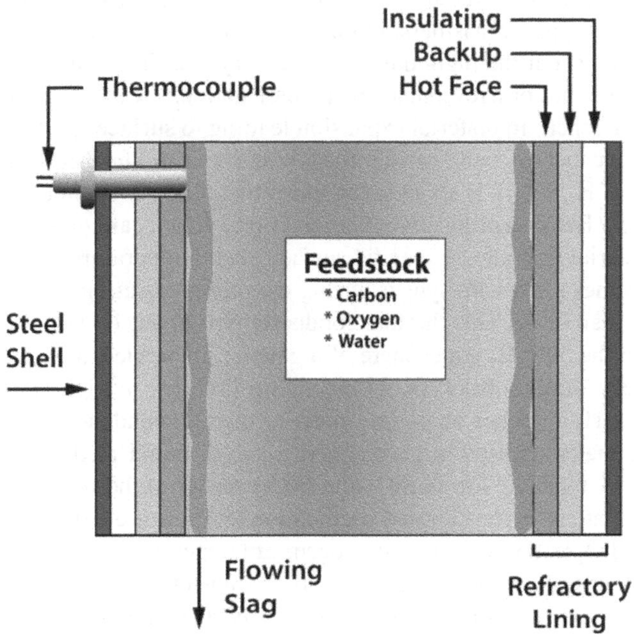

FIGURE 1.4 Cross-section of a gasification chamber.

Regardless of the gasifier type, expansion and movement of the individual refractory layers in a gasifier are important design considerations, in both the circumference and vertical dimensions. Fiber insulation is often used to allow for expansion, and is typically placed at the wall or on top of the vessel to account for refractory expansion and permanent growth. Usually a 70 to 75% compression is allowed in fiber material to permit flexibility, yet is not allowed for hot spots because of excessive fiber space or permanent shaping of the fiber from overcompression. Differences in movement between layers of refractory can cause shear of items like thermocouples that extend through the gasifier shell and refractory to monitor gasification reaction temperatures. When all refractory issues are combined, they have a significant impact on gasifier operation and are viewed by gasifier users and designers, in a recent survey, as the leading issue contributing to the low on-line availability of commercial gasifiers and to their low commercial acceptance.[29] Industry desires a gasifier availability of over 90% if greater use of this technology is to occur. Material needs in two types of gasification systems—gas and liquid feedstock gasifiers, and solid feedstock gasifiers—are as follows.

1.11.1 Liners for Gas or Liquid Feedstock

In a gasifier utilizing gas or liquid as a carbon source, refractory failure is primarily caused by thermal, chemical, and structural wear of the refractory liner. The hot face refractory lining in the gasifier is typically a dense high-alumina material (low in SiO_2 and FeO), followed by a porous layer composed of a material such as bubbled alumina, and backed up by an insulating refractory layer. Since slag is not an issue in these types of gasifiers, failure mechanisms other than by slag corrosion occur. Gasifier heat at the high-alumina refractory working lining can lead to thermal expansion issues or irreversible creep deformation.[30] Rapid temperature cycling can cause differences in material expansion leading to surface spalling or joint failure. An added concern in any gasifier (gas, liquid, or solid feedstock) is the thermal conductivity of H_2, which is about seven times that of air, even though it has a much lower density.[30] Because of the high thermal conductivity, care must also be taken in choosing material liners for a specific surface shell temperature. Porous refractory material becomes filled with syngas during operation, resulting in a higher thermal conductivity than in air. This thermal conductivity is about 1.5 to 2 times that of air when used in the 50% H_2 atmosphere of a gasifier. In service, a backup lining may be a superduty (mullite) brick vs. an insulating firebrick or high-alumina material because of the H_2 influence on porous material thermal conductivity.[28]

Hot face materials used in gasifiers with gas or liquid feedstock are high-alumina refractory typically low in SiO_2 and FeO because of thermodynamic concerns with chemical attack at the elevated gasification temperatures. If SiO_2 is present, H_2 reduction of SiO_2 in the refractory can occur at temperatures above 980°C, causing it to be reduced to SiO vapor and removed or transferred elsewhere in the refractory lining.[30] This reaction becomes very likely at temperatures over 1,200°C.[28] Research[30] has indicated that the removal of SiO_2 from a refractory is impacted by refractory porosity, gasifier temperature, gasifier pressure, and feedstock throughput. It has been found to occur mainly at the surface of a material, where material strength can be adversely impacted. The thermodynamic concern with iron is due to the Boudouard reaction ($2\ CO \rightarrow CO_2 + C$),[30] which Fe catalyzes. This reaction occurs as low as 510°C, maximizes by 570°C, and nearly disappears by 730°C.[31] The C from the reaction builds up and can cause structural weakening or free layers of C (or Fe_2C) to form at joints, pores, voids, or cracks. This buildup leads to thermal expansion mismatches between materials and disruption of the refractory structure, forming voids or compressive stress that can break a refractory microstructure or cause it to weaken from thermal cycling and the expansion differences. Carbon buildup can also result in increased or nonuniform head transfer to the shell.

Refractory flaws can also be caused by the frequency of temperature cycling, the rate of temperature drop, and the amount of temperature drop, which can lead to thermal shock or structural flaws from thermal expansion differences. Where monolithic linings are used, anchors attach the refractory material to the shell. Those anchors can experience failure from mechanical stresses, metal fatigue, or corrosion, leading to gaps between the refractory shell and the lining. Any gap impacts heat transfer and can initiate other types of refractory failure, such as slag corrosion because of heat buildup at those sites.

A concern common to both liquid and solid gasifier feedstock centers on vanadium, which is present as an impurity. Vanadium can attack an alumina lining depending on its valance state, decreasing refractory service life. When vanadium is present as V_2O_3 (stable phase present in the reducing environment of a gasifier), it has a melting temperature of about 1,970°C, while V_2O_5 (stable phase present in an oxygen-rich environment) melts at about 660°C. Because of the lower melting temperature and its phase interactions with Al_2O_3, V_2O_5 will produce low melting point liquids in an Al_2O_3 lining, leading to rapid and excessive refractory wear. In practice, the valance of vanadium depends on the oxygen partial pressure of the gasifier and should not be an issue except during gasifier preheat or cooldown. Behavior similar to that between vanadium and alumina exists in refractories that are high in chrome oxide.

In a gas or liquid gasifier, firebrick linings are predominantly used because of their material strength and historical usage. Monolithic materials such as castables, however, are seeing increased usage[30] for many reasons, including fewer bonds and increased speed of installation. Low cement castables with calcium aluminate bonds have been found to give monolithics adequate strength for gasifier applications. When repair work is necessary in a gasifier, they are also easier and quicker to repair than bricked areas. In general, a monolithic structure is thought to give a uniform and predictable shell temperature.

1.11.2 Liners for Solid Feedstock

A slagging gasifier operates in a temperature range where ash in the carbon feedstock melts and flows down the gasifier sidewalls, as shown in figure 1.4. Two types of slagging gasifiers are commercially used: air cooled (GE and ConocoPhillips designs) and water cooled (Shell design). Air-cooled slagging gasifiers are lined with refractory materials that contain between 60 and 95% chrome oxide. These materials evolved from research in the 1970s to the 1980s that indicated a chrome oxide content of about 75%[32] was necessary to provide the best chemical resistance to gasifier slag corrosion. Since that time, three types of high–chrome oxide refractory materials have been or are currently used in gasifiers. These types are listed in table 1.7. Of these, chrome oxide–alumina and chrome oxide–alumina–zirconia (brick types A and B) are used in the majority of air-cooled slagging gasifiers as hot face liners, while the chrome oxide–magnesia material (brick type C) usage has

TABLE 1.7

Chemical Composition of Three Classes of High Chrome Oxide Refractories Used in Air-Cooled Slagging Gasifiers (wt%)

Material (wt%)	Brick Type		
	A	B	C[a]
Cr_2O_3	90.3	87.3	81.0
Al_2O_3	7.0	2.5	0.4
MgO	0.28	0.12	17.0
ZrO_2	0.01	5.2	NA
SiO_2	0.3	0.2	0.1
$Fe^{+2 \text{ and } +3}$	0.23	0.28	0.3
CaO	0.28	0.03	0.3

[a] Data from manufacturer's technical data sheet.

become more historical because of concerns over hexavalent chrome. The formation of hexavalent chrome during a coal or petcoke slagging gasifier operation, however, has not been reported and is not known to occur.

In an air-cooled slagging gasifier, zoning (the use of different refractory materials at different locations in a furnace) is practiced because of different wear mechanisms and wear rates at different locations in the gasifier, and because of high material costs. In general, chrome oxide content ranging from 60 to 95% is used to line the working face of a gasifier (figure 1.4), with lower chrome oxide content found in the low-wear areas and higher chrome oxide content (approaching 95%) found in the higher-wear locations. The backup lining is typically a high-alumina (approximately 90%)/low-chrome-oxide (approximately 10%) refractory, which serves as an emergency lining to contain the gasifier environment in case of failure of the hot face lining. A third refractory layer, an insulating refractory, often backs up the hot face and backup linings, reducing thermal loss and controlling shell temperature. The insulating lining can be up to 90% Al_2O_3 that is low in silica (under 1%) or may be insulating firebrick or superduty (mullite) brick,[28] depending on the location. The backup lining reduces shell temperature of the gasifier, but must keep it above acid and liquid dew point condensation temperatures.

As viewed from the interior of a gasifier on the hot face, an example of wear dominated by chemical corrosion in a high-chrome-oxide refractory used in an air-cooled slagging gasifier is shown in figure 1.5. Chemical corrosion involves the dissolution of the refractory in the slag as it flows over or penetrates within the refractory pores, and is thought to be one of the main causes of wear. Corrosion can also lead to the removal of large refractory particles or grains as the bond phase is weakened or removed. Chrome oxide can make up to 95 wt% of the hot face refractories (due to zoning) and is considered to be an excellent refractory material because it interacts with several

FIGURE 1.5 Refractory surface wear dominated by chemical corrosive (dissolution).

components of the gasifier slag, forming high melting point phases (solid solutions or spinels). It is also highly insoluble in the molten slag during normal gasifier operation. The refractories used to line a gasifier are not fully dense, in part because of thermal shock associated with the refractory material in the application. Because of the porous nature of chrome oxide refractories and the small thermal gradient across them during service, slag can penetrate deeply within the refractory, setting up the basis for refractory wear by spalling and chemical corrosion.

Refractory wear by several types of spalling in a gasifier is shown in figure 1.6. The pinch spalling (a) probably originates from compressive hoop stress due to the vessel steel shell or improperly manufactured or installed refractory. Thermal spalling (b), visible on the brick edge in this example, is caused by rapid temperature fluctuation in the vessel. Structural spalling (c) may be due to a complex combination of factors, such as shell stress loading, slag infiltration, thermal cycling, or long-term creep. As shown, pinch spalling and thermal spalling occur in isolated areas of the gasifier. Structural spalling and corrosion can occur throughout the gasifier and are the predominant wear mechanisms in air-cooled slagging gasifiers.

Bakker[32] discussed how spalling can incrementally remove large portions of a gasifier refractory, rapidly shortening refractory service life as large pieces of material are physically removed vs. a slow material dissolution in slag. The effects of both chemical dissolution and spalling on the wear of a refractory are shown in figure 1.7. As mentioned, factors such as the gasifier operational temperature, thermal cycling of the gasifier, and slag infiltration into a refractory have a pronounced influence on refractory spalling.

The combined wear mechanisms of high–chrome oxide gasifier linings in air-cooled slagging gasifiers are shown in figure 1.8. Refractory wear or failure is influenced by a number of factors, including gasifier design (air vs. water quench in the lower cone area), how the gasifier is operated (material throughput, temperature, number of cycles per campaign), the composition of the refractory and how it withstands chemical corrosion/physical wear, the quality of the refractory (internal flaws or exterior dimensions), and how well the material is installed. In most slagging gasifiers, the most common means of refractory failure is by chemical corrosion and spalling, although wear caused by burner misalignment can be problematic.

Water-cooled slagging gasifier linings, such as the Shell gasifier, have a different refractory lining makeup than the chrome oxide materials used in the GE or ConocoPhillips designs. This type of gasifier uses a thermally conductive hot face refractory lining to solidify (freeze) gasifier slag on the refractory surface, preventing refractory corrosion. The refractory linings can be high in SiC, a conductive ceramic material, and may contain alumina with a bonding material such as calcium aluminate cement if a monolithic vs. brick lining is used. In use, a lining wears to some equilibrium thickness that is a balance of refractory surface temperature and frozen slag thickness. If the refractory lining is installed as a monolithic material, it is held in place by steel anchors fastened to the gasifier steel shell. Failure of the gasifier refractory is typically by separation of the refractory lining from the steel shell, leading to a poor heat transfer and fluidization of slag on the refractory surface, followed by refractory corrosion. Refractory failure can also occur by a reaction between SiC in the refractory and FeO from the slag, forming SiO_2, CO, and metallic iron. Separation

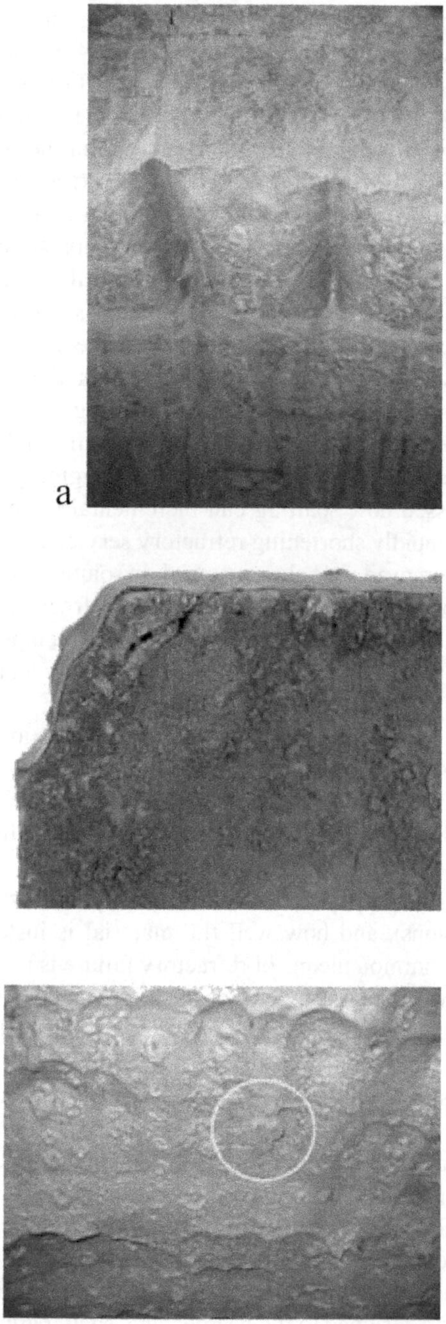

a

b

c

FIGURE 1.6 Spalling examples in gasifier refractory: (a) pinch spall along refractory joint, (b) thermal spall, and (c) structural spall (circled material).

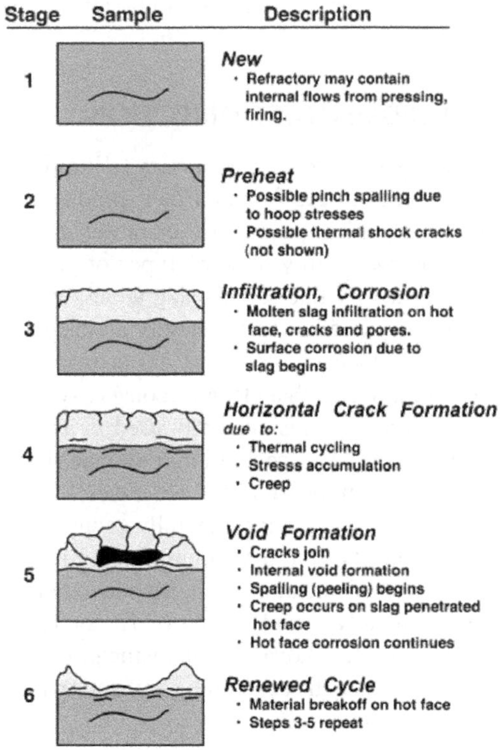

FIGURE 1.7 Stages of refractory wear.

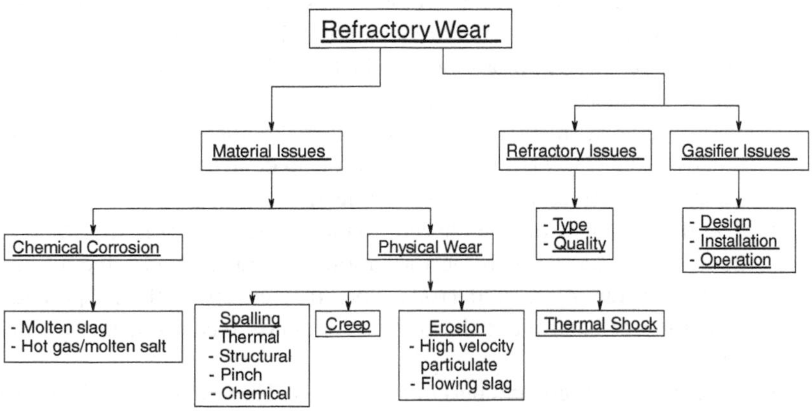

FIGURE 1.8 Causes of refractory failure in a slagging gasifier.

of the steel shell from the refractory is caused by steel anchor failure due to hot gas or slag corrosion, metal fatigue, or metal failure. In practice, a water-cooled gasifier lining can last upwards of 8 years.

1.12 RESEARCH NEEDS/FUTURE DIRECTION

Technological advances in gasification are necessary to improve the efficiency, reliability, and on-line availability of gasifiers and their product quality if gasification is to play a larger role in chemical production or power generation. Limits exist, for instance, in the on-line availability of some types of gasifier liners because of slag attack and wear of refractory liners. This slag attack/wear can result in a gasifier cycling down for repair or reline as frequently as once every 2 to 3 months. Other material issues exist associated with wear or corrosion of gasification valves or downstream process components that clean or convert syngas into raw materials used in other processes. Gasification is currently used in a variety of applications, including supplying syngas for power generation, raw materials (primarily CO and H_2) for chemical processes, steam for industrial processes, and H_2 for petrochemical refining. It has also proven to be an environmentally sound technology for converting low-value carbon by-products from petroleum refining into higher-value chemicals used throughout society. Gasification holds tremendous potential as a reliable, environmental, and economically sound source of H_2 for future applications that include fuel cells and automobiles. Advances in gasification and syngas processing technology would greatly enhance the economic justification for H_2 technology. Key areas of gasification research are as follows:

1. Gasification:[11]
    - Plant designs should be optimized to allow for flexibility in carbon feedstock, differences in manufacturers' equipment, and differing process technology.
    - Future gasification facilities should be designed with syngas output flexibility to meet varying market demands. This may include co-production capability so two or more products, such as H_2, syngas, or power, can be produced and marketed.
    - Improved instrumentation and control technology of the gasification process (especially temperature monitoring devices).
    - Longer-life refractory liners for gasifiers.
    - The development of non-chrome oxide refractories (because of real/perceived concerns with the formation of hexavalent chrome and the associated environmental risks. Note that except with the potential formation of hexavalent chrome with biomass feedstock, this is not a known issue.
    - Longer-life burners and burner assemblies able to handle a wider variety of feedstock.
    - Increased efficiency gasification systems.
    - Improved wear and corrosion-resistant materials for downstream use in gasification plants.

- Measurement techniques to actively monitor refractory lining thickness during gasifier operation.

2. Syngas processing:
   - Advanced cleaning and processing technology is necessary for raw syngas, with an emphasis on gas throughput, composition, and purification.
   - Advanced syngas cleaning and conditioning processes are needed that efficiently remove contaminants at high temperatures, producing ultraclean syngas.
   - Advances in syngas processing are needed that capture and sequester CO_2 efficiently and at a low cost.
   - More outlets for syngas products and by-products are necessary.
   - Sulfur-tolerant catalysts for water–gas shift reactors are necessary to lower syngas processing costs.
   - The development of novel membranes for advanced, low-cost separation/concentration of specific syngases is necessary. This would include H_2, CO_2, CO, and gaseous impurities (such as H_2S). Membranes to separate other gases from air, such as O_2 or N_2, are also needed. Improvements could include a reduction in the number of steps or stages in a process or in processing speed.
   - Advanced technology concepts/processes are necessary that could combine processing stages such as H_2 separation and the water shift reaction into one step, with the goal of reducing syngas processing costs.[11,33]

Future trends in gasification are clearly toward increasing fuel output flexibility, feedstock flexibility, gasification efficiency, and on-line availability. Liner material research is split into three types: materials for units that generate dry ash, slagging units that are water cooled, and slagging units that are air cooled. Research in air-cooled slagging gasifiers is directed at improved chrome oxide and non-chrome oxide linings. Currently used chrome oxide liner materials do not meet the service life needs of industry and will probably only show incremental performance improvements in the future unless changes are made in the design and operation of gasifiers. As the cost of gas and liquid feedstock increases and the availability of high-quality coal/petcoke decreases, gasifiers will be utilized for lower-quality feedstock such as coal with high ash and sulfur impurities. Gasifiers must be designed to operate efficiently for long periods with these feedstocks. With increases in the quantity and potential hazardous materials in the ash, slagging gasification will become more important because of its ability to fuse ash (so it occupies less volume) and transform it into an environmentally acceptable glass phase. Because gasification is a closed system, the capture of CO_2 greenhouse gas emissions may become an important factor in the future consideration of this technology. The increased recovery of trace impurities in the carbon feedstock ash will also become important for gasification by-product recovery. The rising cost and instability of some petroleum-based carbon feedstock will also act as a driving force for increased use of gasification in countries with large deposits of fossil fuel. Other feedstocks, such as biomass, require the development of a reliable and efficient gasification process. In general, gasification must compete on a cost basis with other technology if it is to be widely utilized by industry.

1.13 SUMMARY

Gasification has been reliably used by industry since the 1950s and 1960s to produce syngas (H_2 and CO) for use in power generation or as a feedstock in chemical synthesis. The process of gasification occurs when water, oxygen, and a carbon source are reacted at elevated temperatures and pressures in a reducing gasifier environment. The general chemical equation for gasification is $C_xH_y + x/2\ O_2 \rightarrow x\ CO + y/2\ H_2 + (heat, C)$. Gasification uses carbon from a variety of sources, the most common being coal or petroleum based. It is considered an excellent way to process by-products from petroleum processing, materials that have environmental liabilities limiting their use. The gasification process typically occurs at temperatures between 1,250 and 1,575°C and pressures from 300 to 1,200 psi—temperatures and pressures that break down the carbon feedstock into CO and H_2 (called syngas). A number of different types of gasifiers are used commercially, with two varieties (General Electric and Shell) comprising 120 of the 155 facilities planned or in use throughout the world. By-products or waste materials generated in the gasification process (such as H_2S, COS, HCN, CO_2, N_2, and carbon/soot) are typically removed at the gasification site by beneficiation. Syngas is also processed for H_2 that is used in the petrochemical industry or to manufacture fertilizer. To produce a higher H_2 yield, syngas usually has the CO present converted into CO_2 and H_2 through the water shift gas reaction [$CO + H_2O\ (gas) \rightarrow H_2 + CO_2$]. Currently about 10% of all H_2 produced originates from syngas. Syngas used in applications like power generation or as a source of H_2 for fuel cells must have the S reduced to very low levels because it reacts with other attacks of materials. The removal of S can involve multiple stages of chemical processing, raising production costs. Because of the rising cost of natural gas used to produce H_2 by steam methane reforming [$CH_4 + H_2O\ (gas) \rightarrow CO + 3\ H_2$], gasification is expected to play a larger role in future gas production. The refractory liners of gasifiers have been identified as one of the major limitations to greater on-line availability of commercial gasifiers. Air-cooled slagging gasifiers use high—chrome oxide prefired shapes that fail by chemical corrosion and spalling. Water-cooled gasifier liners fail because of separation of the monolithic refractory material from the steel shell, leading to refractory failure by chemical corrosion or thermodynamic interaction with the slag. Gasification research or syngas processing research is needed if the gasification process is to have the cost competitiveness and on-line reliability of other chemical processes.

REFERENCES

1. Richter, N., Introduction to Gasification (ChevronTexaco), available at www.gasification.org/Docs/02Richter.pdf, August 2, 2005.
2. Information, Shell Gasification Process, available at www.uhde.biz, August 10, 2005.
3. General information, available at www.shell.com, August 1, 2005.
4. Information, available at www.gasification.org, August 10, 2005.
5. Jaeger, H., 630 MW IGCC Targets Cost Parity at $1600/kw and 38.5 % Efficiency, *Gas Turbine World*, 35, 18–24, 2005.

6. Marruffo, F., Chirinos, M.L., Sarmiento, W.B.O., Bitumenes, S.A., Orinoco, and Hernandez-Carstens, E., Orimulsion®: A Clean and Abundant Energy Source, paper presented at the 17th Congress of the World Energy Council, Houston, TX, September 14, 1998, available at www.worldenergy.org.
7. Information, Shell Gasification Process, available at www.uhde.biz, August 10, 2005.
8. Rezaie, A., Headrick, W.L., Fahrenholtz, W.G., Moore, R.E., Velez, M., and Davis, W.A., Interaction of Refractories and Alkaline Containing Corrodants, *Refractories Applications and News*, 9, 26–31, 2004.
9. Phillips, G., Gasification Offers Integration Opportunities and Refinery Modernization, paper presented at Petrotech 2001, October 29–30, 2001, available at www.fwc.com/publications/tech_papers/env/pdfs/GP1109.pdf.
10. Philcox, J.E., and Fenner, G.W., Gasification: An Attraction for Chemical Reactions, available at www.praxair.com/Praxair.nsf, August 8, 2005.
11. O'Keefe, L.F., Hydrogen Production, Coal Gasification and "FutureGen" Initiative, paper presented at Accelerating Deployment of Hydrogen Fuel Cell Vehicles, The Keystone Center, Washington, DC, June 5, 2003.
12. Stiege, G.J., Gasification Technologies Program: Overview of Program, Focus on H_2 Production, paper presented at the Hydrogen Workshop, September 19, 2000, available at www.netl.doe.gov/publications.
13. Austgen, D., Transitioning to a Hydrogen Fuel Infrastructure, paper presented at the 3rd API Conference on the Oil and Gas Industry's Voluntary Actions to Address Climate Change, September 29, 2004, information available at www.api-ec.api.org.
14. *The Fuel Cell Handbook*, 6th ed., produced by E.G. and G. Services under contract for the U.S. Department of Energy, Morgantown, W. VA., Publication DOE/NETL-2002/2270, available from National Technical Information Services, chapter 8.1.1, 2002.
15. Donaldson, A.M. and Mukherjee, K.K., Gas Turbine "Refueling" via IGCC, *Power*, March 2006, pp. 34, 36–38.
16. U.S Environmental Protection Agency, *Texaco Gasification Process Innovative Technology Evaluation Report*, EPA/540/R-94/514, July 1995.
17. U.S. Department of Energy, National Energy Technology Laboratory, Gasification: Gasifier Technologies, www.netl.doe.gov/coal/Gasification/description/gasifiers, August 9, 2005.
18. Rich, J.W., et al., WMPI: Waste Coal to Clean Liquid Fuels, paper presented at the Proceedings of Gasification Technologies 2003, San Francisco, October 12–15, 2003.
19. Information, available at www.shell_Gasification_Process, August 3, 2005.
20. van Dyk, J.C., Keyser, M.J., and Coertzen, M., Sasol's Unique Position in Syngas Production from South African Coal Sources Using Sasol-Lurgi Fixed Bed Dry Bottom Gasifiers, paper presented at the Proceedings of the Gasification Technologies 2004, Washington, DC, October 3–6, 2004.
21. de Graaf, J.D., van den Berg, R., and Zuideveld, P.L., Information, Shell Gasification Process, www.shell_Gasification_Process, August 3, 2005.
22. Bentham, J., Moving Research into Reality: The Next Stretch on the Hydrogen Journey, paper presented at the World Hydrogen Energy Conference, Yokohama, Japan, July 1, 2004, available at www.shell.com/static/hydrogen-en/downloads/speeches/speech_whec.pdf.
23. Zuideveld, P. and J. de Graaf, P., Overview of Shell Global Solutions, Worldwide Gasification Developments, paper presented at the Proceedings of Gasification Technologies 2003, San Francisco, October 12–15, 2003.

24. Ferguson, C.R., Falsetti, J.S., and Volk, W.P., Refining Gasification: Petroleum Coke to Fertilizer at Farmland's Coffeyville, Kansas Refinery, Paper AM-99-13, paper presented at the NPRA 1999 Annual Meeting, San Antonio, TX, March 21–23, 1999.
25. Doctor, R.O., Molburg, J.C., Brockmeier, N.F., and Stiegel, G.J., Designing for Hydrogen, Electricity and CO_2 Recovery from a Shell Gasification-Based System, paper presented at the Proceedings of the 18th Annual International Pittsburgh Coal Conference, Newcastle, New South Wales, Australia, December 4–7, 2001.
26. Volkmann, D. and Just, T., Refractories for Gasification Reactors: A Gasification Technology Supplier's Point of View, *Refractories Applications and News*, 9, 11–16, 2004.
27. Rezaie, A., Headrick, W.L., and Fahrenholtz, W.G., Identification of refractories for high temperature black liquor gasifiers, in *Proceedings of the Unified International Technical Conference on Refractories, UNITECR '05*, Orlando, FL, November 2005, 4 pp.
28. Taber, W.A., Refractories for Gasification, *Refractories Applications and News*, 8, 18–22, 2003.
29. U.S. Department of Energy, *Gasification Markets and Technologies — Present and Future: An Industry Perspective*, Report 0447, July 2002, pp. 1–53.
30. Johnson, R.C. and Crowley, M.S., State of the art refractory linings for hydrogen reformer vessels, in *Proceedings of the Unified International Technical Conference on Refractories, UNITECR '05*, Orlando, FL, November 2005, 4 pp.
31. Raymon, N.S. and Saddler, L.Y., III, Refractory Linings Materials for Coal Gasifiers: A Literature Review of Reactions Involving High-Temperature Gas and Alkali Metal Vapors, USBM Information Circular 8721, 22, 1976.
32. Bakker, W.T., Refractories for Present and Future Electric Power Plants, *Key Engineering Materials*, 88, 41–70, 1993.
33. U.S. Department of Energy, Fossil Energy: DOE's Hydrogen from Coal R+D Program, available at www.fe.doe.gov/programs/fuels/hydrogen/Hydrogen_ from_coal_R+D, August 1, 2005.

2 Materials for Water Electrolysis Cells

Paul A. Lessing

CONTENTS

2.1 BACKGROUND OF HYDROGEN GENERATION VIA ELECTROLYSIS

Hydrogen generation can be accomplished via traditional DC electrolysis of aqueous solutions at temperatures less than about 100°C. However, electrolysis of steam can also be accomplished at higher temperatures at the cathode of electrolytic cells utilizing solid membranes. The solid membranes typically are electronic insulators and need to be gas-tight (hermetic), but have the special property of being able to conduct ions via fast diffusion through the solid. Generally the cells (cathode/electrolyte/anode) are known by the chemical name of their solid electrolytes. It has been found for some operating hydrogen fuel cell anode/electrolyte/cathode systems that the fuel cell reactions at the electrodes are reversible and can be operated in an electrolysis mode. However, reversibility has not been demonstrated for all cathode/electrolyte/anode combinations.

Hydrogen production via conventional electrolysis largely depends upon the availability of cheap electricity (e.g., from hydroelectric generators). Consequently, only about 5% of the world hydrogen production is via electrolysis. The only complete hydrogen production process that is free of CO_2 emissions is water electrolysis (if the electricity is derived from nuclear or renewable fuels). However, 97% of the hydrogen currently produced is ultimately derived from fossil energy. Currently, the most widely used and economical process is steam reforming of natural gas, a process that results in CO_2 emissions.

2.2 LOW-TEMPERATURE ELECTROLYSIS OF WATER SOLUTIONS

The reversible electrical potential ($\Delta G/nF = E_{rev}$) to split the O–H bond in water is 1.229 V. In addition, heat is needed for the operation of an electrolysis cell. If the heat energy is supplied in the form of electrical energy, then the thermal potential is 0.252 V (at standard conditions), and this voltage must be added to E_{rev} (i.e., add entropic term $T\Delta S$ to ΔG). The (theoretical) decomposition potential for water at standard conditions (for $\Delta H \cong \Delta H°$) is then 1.480 V. This is shown in figure 2.1. Anode and cathode reactions for electrolysis (see figure 2.1) are:

$$\text{Anode: } 2\ OH^- \rightarrow 1/2\ O_2 + H_2O + 2\ e^- \qquad (2.1)$$

$$\text{Cathode: } 2\ H_2O + 2\ e^- \rightarrow H_2 + 2\ OH^- \qquad (2.2)$$

For alkaline electrolysis, OH^- ions must be able to move through the membrane (under influence of the electric field) from the cathode chamber into the anode chamber to supply OH^- to participate in the reaction (equation 2.1) at the anode.

Irreversible processes that occur at the anode and cathode and the electrical resistance of the cells cause the actual decomposition potential (voltage) to increase to about 1.85 to 2.05 V. This means that the electrolysis efficiency will be between 72 and 80%. The total electrical resistance of the cell is dependent upon the conductivity of the electrolyte, the ionic permeability of the gas-tight diaphragm that separates the anodic region from the cathodic region, and the current density (normally in the fairly moderate range of 0.1 to 0.3 A cm^{-2}). Higher KOH concentrations (up to 47%) yield higher conductivity, but this usually greatly increases the corrosion of various cell components.

Common aqueous electrolytes are better conductors at slightly elevated temperatures (70 to 90°C), so the electrolysis cells are operated at these conditions. The original discovery of electrolytic water splitting used acidic (diluted H_2SO_4) water, but in industrial plants an alkaline (e.g., 25 wt% KOH) medium is preferred because corrosion is more easily controlled and cheaper materials can be utilized. Diaphragms (see figure 2.1) are made either of polymers (polysulfonate type) or from porous ceramics (e.g., asbestos or barium titanate). In some configurations, the electrodes are placed directly at the surface of the diaphragm to reduce the voltage drop and minimize heat losses. The cathode material has historically been made from steel and the anode material from nickel or nickel-coated steel. The cell walls have been made from carbon steel. The heat generated in the electrolyte must be removed by water cooling. Pure water has to be added to the cell to replace the water that is dissociated to hydrogen and oxygen gases.

In order to reduce the actual cell voltage downward toward the 1.48 value (reduce energy consumption), many different catalytic materials have been examined for use as anodes or cathodes (or coatings on underlying electrodes). Research was conducted in Germany in the 1980s and 1990s on advanced materials and designs for alkaline water electrolysis cells.[1] Electronically conductive, metal oxides (e.g., $La_{0.5}Sr_{0.5}Co_3$, $LaNi_{0.2}Co_{0.8}O_3$, or RuO_2) were investigated for use as anodes and various metal alloys (e.g., Ni/Co, fine Raney iron, Raney Ni/Co, Pt-black platinized Ni) were evaluated by Wendt et al.[2] for possible use as cathodes. Raney nickel is a highly porous nickel coated onto supporting nickel or stainless steel electrodes, and can be produced by a number of different methods.[3] Many times these activated electrodes provide enhanced performance, but they can have a short lifetime.

In recent years (1999 to present), experiments on metal coatings (e.g., Ni-Fe-Mo, Ni-Fe, or Ni-Co alloys) as catalysts for the cathode (in order to reduce polarization) have been conducted.[4] Mild steel is often used as the underlying substrates. Materials evaluated for catalysts have included hydrogen storage alloys (Mm = Misch metal; $Ni_{3.6}Co_{0.75}Mn_{0.4}Al_{0.27}$, $LaNi_{4.9}Si_{0.1}$, and Ti_2Ni). These alloys were layered on top of a nickel-molybdenum coating with an underlying nickel foam substrate and seem to show promise for both electrocatalytic activity and stability.[5] Work on mixed-metal oxide catalysts (in order to reduce anode polarizations) has included deposition (e.g., sol-gel method) of spinel ($NiCo_2O4$) on substrates[6] of mild steel, nickel, or titanium. These layered structures demonstrated a high (compared to Ni) and stable activity (during 200 h of operation).

There has been some recent interest in selective electrolysis of seawater (e.g., electrolytes of 0.5 *M* NaCl @ pH 12) in desert coastal areas (no freshwater) to produce hydrogen (for possible use with carbon dioxide to produce methane) and oxygen (not chlorine). In a study by Abdel Ghany et al.,[7] anodes of $Mn_{1-x}Mo_xO_{2+x}$ (on IrO_2/Ti substrates) were prepared using anodic deposition from $MnSO_4$-Na_2MoO_4 solutions. When running at a current density of 1,000 Am^{-2} at 30°C, an increase in solution temperature resulted in dissolution of the oxides as molybdate and permanganate ions. Additions of iron to the oxides greatly aided in the chemical stability (30 to 90°C range) and also enhanced the oxygen evolution efficiency.

The fluorinated polymer polytetrafluorethylene (PTFE) diaphragm is stable in hot KOH; however, membranes made with this material tend to become gas clogged and are not suitable as diaphragm materials. Wendt and Hofmann[8] conducted a study to replace the conventional asbestos diaphragm (that dissolves in caustic KOH at temperatures above 90°C) with polymer-bonded (PTFE-type) composites. These composites included an inorganic material (ZrO_2, Ca- or Ba-titanate, or K-hexatitanate). The polymer-bonded materials showed too high of an electrical resistance for "sandwich" cell designs, so they were not pursued. The polymer-bonded materials might, however, be used as gaskets.

The electrolysis diaphragm generally is fabricated to include fine pores (vs. being an ionic (OH^-) conductor) such that it passes electrolytes. But, it must prevent unhindered intermixing of the catholyte and anolyte since these liquids are really a two-phase mixture of electrolyte with a dispersion of gas bubbles (hydrogen and oxygen, respectively) and hydrogen gas cannot be mixed with oxygen gas. In order to operate efficiently, the diaphragm must not be clogged by gas bubbles that may intrude into the pore mouths or that may precipitate out within the pores from supersaturated (high-pressure operation) electrolyte solutions. The diaphragm must also offer sufficiently high hydrodynamic resistance to retard intermixing of oxygen-saturated

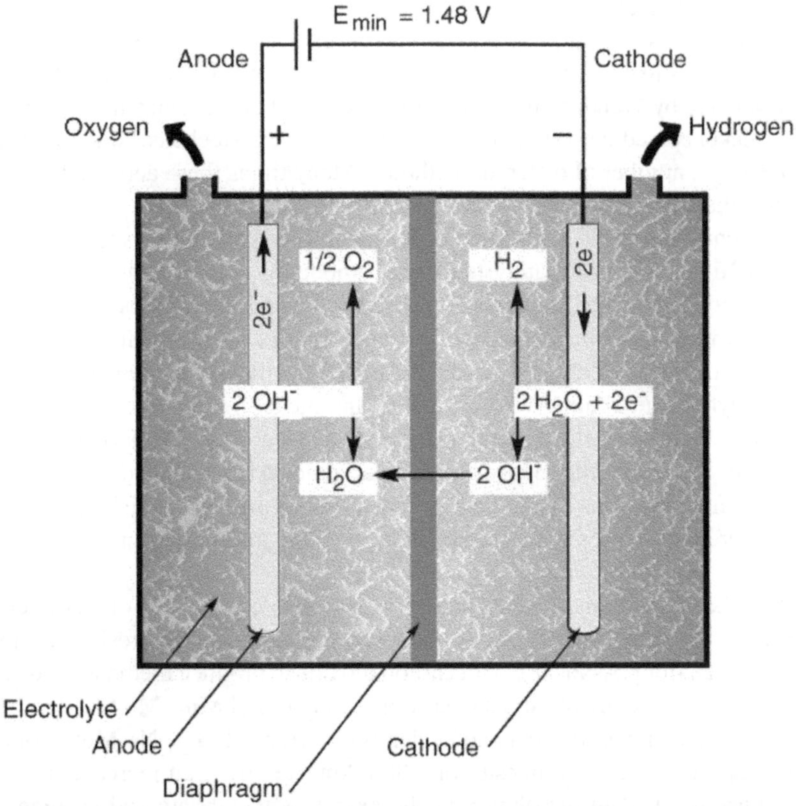

FIGURE 2.1 (a) Schematic of water (alkaline) electrolysis. (b) Two large (200 Nm^3/h) atmospheric, alkaline, multicell electrolysis stacks generating hydrogen at the Norsk Hydro Company.

anolyte with hydrogen-saturated catholyte due to any pressure differences between the two chambers and also prevent diffusion of gas molecules.

Diaphragms made of sintered metals are not easily incorporated into a bipolar-type cell and do not permit zero-gap cell geometries. Therefore, Wendt and Hofmann[8] further investigated metal-ceramic cermets. Nickel (low carbon, low sulfur) was the most stable against corrosion (220°C) of the various other metals that were evaluated (titanium, zirconium). Nickel mesh-supported, sintered nickel cermets utilizing $NiTiO_3$ or $BaTiO_3$ were fabricated and showed good in-cell performance.

Norsk Hydro Electrolyzers (NHE) in Norway is a leading producer of alkaline electrolyzers (see figure 2.1b, where individual cells are linked in series electrically and geometrically in a bipolar filter press configuration). Kreuter and Hofmann[9] discuss the efficiency, operability, safety, and economics of scaling up alkaline-type electrolysis cells to large plants, including the advanced pressurized (30-bar) alkaline prototype built by Gesellschaft fur Hochleistungselektrolyseure zur Wasserstofferzeugung (GHW) in Germany.

2.3 LOW-TEMPERATURE PEM-TYPE ELECTROLYZERS

Proton exchange membrane or PEM-type water electrolyzers utilize thin films (e.g., 0.25 mm) of a proton-conducting ion exchange material instead of a liquid electrolyte. When a reverse polarity is applied to a PEM fuel cell, the fuel cell reactions are reversed and become water electrolysis reactions (see equations 2.6 to 2.8). PEM fuel cells have been the subject of research and development for decades. In the 1960s NASA used PEM cells for their Hope, Gemini, and Biosatellite missions. After a lull in the 1980s, a rush of development began in the early 1990s for transportation applications. This was initiated by improvements in bonded electrodes, which enabled much higher current densities. These improvements can be advantageous to PEM cells used as electrolyzers.

The PEM cells typically use sulfonated polymer (e.g., Nafion™) electrolytes that conduct the protons away from the anode to the cathode (in electrolysis mode). For smaller generators, the solid polymer can be more attractive than a dangerous, caustic electrolyte. A complicating factor is that the solid-state conduction of the protons is accompanied by multiple water molecules $(H_2O)_nH^+$. Also, the membrane must be kept hydrated to sustain the conduction mechanism. Therefore, water recycling becomes a large consideration since water is constantly removed from the anode and reappears at the cathode (mixed with the hydrogen). At temperatures less than 100°C, gaseous hydrogen is easily removed from liquid water, but the hydrogen still contains water vapor that most likely requires dehumidification (e.g., pressure swing adsorption dryer). Electrodes generally have utilized finely divided platinum black or, more recently, IrO_2 or RuO_2 (for increased electronic conductivity) as catalysts.[10] Research is currently being conducted into PEM-type membranes that have better kinetics, yet are chemically stable at elevated temperatures such that they could operate in steam.[11]

PEM water electrolysis cells have a potential advantage over traditional low-temperature electrolysis cells (e.g., KOH in water electrolytes with palladium, titanium, or alternative metal or ceramic electrodes[12,13]) because PEM devices have been shown to be reversible. They can "load level" by generating electricity from hydrogen (and oxygen) operating as a fuel cell when needed (peak) and reverse to operate as an electrolyzer by consuming electricity to produce hydrogen (and oxygen). This is convenient if excess electricity is available during low periods of consumption (off-peak).[14] PEM electrolysis cells could also be used in hybrid systems utilizing solar energy.[15]

Because of all the developments in PEM fuel cell technology, small PEM electrolysis plants are becoming available. Small (up to 240 SCF/h = 6 Nm^3/h) PEM electrolysis units are now available commercially from Proton Energy Systems,[16] and efforts are being made to reduce their production cost.[17] Hamilton Sundstrand[18] has been manufacturing SPE™ electrolysis systems (PEM) for a number of years for the U.S. Navy. Treadwell Corp.[19] has recently developed PEM generators (20 to 170 standard liters per minute or SLPM) at pressures of up to 1,100 psi. Hydrogenics Corp.[20] is manufacturing two units (1.1 and 30 Nm^3/h), and Giner Electrochemical Systems[21] is developing a PEM electrolysis unit. Some critical attention to cell stack lifetime must be paid in light of the degradation and thinning of Nafion 117 PEM electrolytes identified in long-term tests in Switzerland[22] (two 100-kW PEM water electrolyzer plants). The thinning process proceeded via dissolution of the membrane from the interface between the cathode and the membrane. The degradation rate depended upon the position within an individual cell as well as the position of the cell in the electrolyzer stack.

Ando and Tanaka[23] have recently used a Nafion electrolyte in electrolysis mode to decompose two water molecules to simultaneously generate one molecule of hydrogen and one of hydrogen peroxide (used in paper/pulp and chemical industries). They do this by using a high applied voltage (1.77 to 2.00 V) in a two-electron transfer process (cathode, $2\ e^- + 2\ H^+ \rightarrow H_2$; anode, $2\ H_2O \rightarrow HOOH + 2\ H^+ + 2\ e^-$) and a NaOH anolyte collection solution. No oxygen is generated.

2.4 LOW-TEMPERATURE INORGANIC MEMBRANE ELECTROLYZERS

Electrolyzers operated at low temperatures do not take full advantage of thermodynamic efficiency advantages. The required cell voltage drops considerably (to $E_0°$ = 0.9 V at 927°C) because of the positive entropy value ($\Delta G° = \Delta H° - T\Delta S°$) when operating at high temperatures. However, sealing bipolar plate devices should be easier at low temperatures since thermal cycling would not result in high stresses due to thermal expansion mismatches between cell components and sealing material. Also, inorganic membranes will be more chemically stable in the 200 to 300°C temperature range than most organic proton-conducting membranes. A typical pressurized-water nuclear reactor[24] heats water from 285°C to 306°C (at 2150 psia) in its core and might be a heat source (heat-exchanged steam at temperatures significantly lower than the core temperature) for a low-temperature electrolysis device.

Solid inorganic materials exhibiting fast proton conduction at low temperatures seem to be more prevalent than fast oxygen ion conductors. Some proton-conducting glasses achieve high proton mobility due to incorporation of water (bonded to POH groups). These glasses can be fabricated by sol-gel techniques at low temperatures.

However, the gels are deliquescent and also are easily fractured into pieces when heated.[25] This limits the practical application of these glasses to very low temperatures, and therefore limits the flux values of hydrogen that can be achieved. Fabrication of proton-exchanged β"-alumina compositions is difficult because waters of hydration are lost during firing, and therefore the crystal structure is irreversibly destroyed.[26] One approach used to solve this problem, for β"-alumina, has been to fabricate a potassium ion crystal structure by firing to high temperatures. Then, at room temperature, protons can be electrochemically ion exchanged into the crystals from a mineral acid.[27,28] Since the potassium ion is larger than the sodium ion, using the potassium composition lessens lattice strain during the proton exchange process. In these oxide ceramics, two protonic species can exist. The first type is a H_2O molecule associated with a proton as a hydronium ion (H_3O^+). The second type is a proton bound to an oxygen ion of the crystal lattice ($=OH^+$).

Ion exchange techniques have also been applied to compositions of the family of three-dimensional sodium ion-conducting "NASICON." NASICON is a three-dimensional conductor, whereas β"-alumina is a two-dimensional conductor. NASICON membranes have primarily been used for efficiently producing caustic (NaOH) from concentrated sodium salts dissolved in water.[29] NASICON is a family of compositions; the original NASICONs were solid solutions derived from $NaZr_2P_3O_{12}$ by partial replacement of P by Si with Na excess to balance the negative charges to generate the formula $Na_{1+x}Zr_2P_{3-x}Si_xO_{12}$ ($0 \leq x \leq 3$). NASICON compositions have been prepared by a sol-gel route, and then the membranes ion exchanged with hydronium ions.[30] However, severe difficulties with cracking of dense membranes occur during the ion exchange.[31] Recently, a sintered proton-exchanged NASICON-type composition known as PRONAS™ has become available in experimental quantities from a commercial supplier.[32] This material was designed for use in liquid systems, but reportedly has been tested as a membrane for hydrogen gas separation. Presumably, the PRONAS composition was sintered and then proton exchanged at room temperature; however, no chemical composition or processing details are available at this date.

Historically, there have been only a few articles regarding materials (including various phosphates) exhibiting fast proton conduction at low temperatures. These include early reviews by Farrington and Briant[33] and McGeehin and Hooper.[34] McGeehin concludes that slow proton conduction is associated with the instability of the hydride (H^-) ion in oxidizing environments and the ease with which the small proton (H^+) is trapped. Problems associated with fabricating dense, polycrystalline membranes of these materials should be parallel to those of NASICON. The low-temperature proton conductivity of materials such as $CsHSO_4$,[35] $M_3H(XO_4)_2$ (M = K, Rb, Cs, and X = S, Se),[36] CsH_2PO_4,[37,38] $H_5GeMo_{11}VO_4$ 0.24 H_2O,[39] H_xMoO_3 ($0 < x < 2$)[40] (hydrogen molybdenum bronze), or the similar $H_{0.46}$ WO_3 (hydrogen tungsten bronze) have been studied. However, no work seems to be extant related to fabrication of these materials into membranes for fuel cell or steam electrolysis applications.

2.5 MODERATE-TEMPERATURE INORGANIC MEMBRANE ELECTROLYZERS

Steam electrolysis is feasible at moderate temperatures using cells constructed with solid inorganic (ceramic) membranes. These temperatures could range from approximately 500 to 800°C using ceramic membranes that are either oxygen ion or proton conductors. This temperature regime is a good match to approximate coolant outlet temperatures that would be generated by various experimental nuclear reactor concepts,[41] such as Gas-Cooled Fast Reactor System (GFR) at 850°C, Lead-Cooled Fast Reactor System (LFR) at 550°C (perhaps up to 800°C), Molten Salt Reactor (MSR) at 700°C, Sodium-Cooled Fast Reactor System (SFR) at 550°C, and Supercritical-Water-Cooled Reactor System (SCWR) at 550°C. Of course the steam temperature in a secondary cooling loop would be somewhat less than a reactor's coolant outlet temperature due to heat exchanger inefficiencies.

One approach to enable operation at lower temperatures while using traditional materials like cubic phase zirconia is to reduce the thickness of zirconia electrolyte using any one of a number of diverse fabrication techniques, such as tape calendaring,[42] vacuum plasma spraying,[43,44] reactive sputtering,[45] pulsed-laser plasma evaporation,[46] or chemical vapor deposition (CVD).[47] Very thin electrolytes generally have to be supported by a thicker, porous electrode. Wang[45] mentions the problem of microporosity that is normally observed in zirconia electrolytes when using the evaporative-type deposition techniques, whereas CVD-type coatings are generally much more hermetic. INL has performed experiments with Liquid Injected Plasma Deposition (LIPD; see figure 2.2) where mixed cation salts (e.g., metal nitrates) are dissolved in water or alcohol and pumped to be misted into a plasma plume. The metal nitrates are decomposed in the plasma to form very fine mixed-metal oxide particles. These particles are melted in the plasma and are concurrently deposited on a substrate. Porous layers that can be used as electrodes are easily formed. Efforts are ongoing to produce thin, dense/hermetic layers that would be an inexpensive substitute for CVD coatings. A mock-up of the experimental apparatus in use at INL is shown in figure 2.3. For illustration, it does not show the plasma torch, but it does show the programmable syringe pump to control the injection rate of liquid solution (left), liquid/air injection nozzles (red tips), holder with injection ports (including nozzle shroud), and sample to be coated (in holder at right).

The other approach to operating at lower temperatures is to develop new electrolyte compositions with higher ionic conductivities (for a given temperature range). Even though these electrolytes have higher ionic conductivities than zirconia at temperatures in the 600 to 800°C range, they generally have not been applied at higher temperatures for a variety of reasons: (1) low activation energy for diffusion such that, while ionic conductivity is higher than zirconia at moderate temperatures, it can be lower than zirconia at high temperatures; (2) chemical instabilities, interdiffusion, or reactions with other cell components (electrodes, bipolar plate, sealants); (3) poor high-temperature mechanical or creep properties; or (4) a desire to use the electrolyte in cell stacks in conjunction with low-cost metal bipolar plates that operate best at low to moderate temperatures (due to problems with low-conductivity oxidation layers formed at high temperatures).

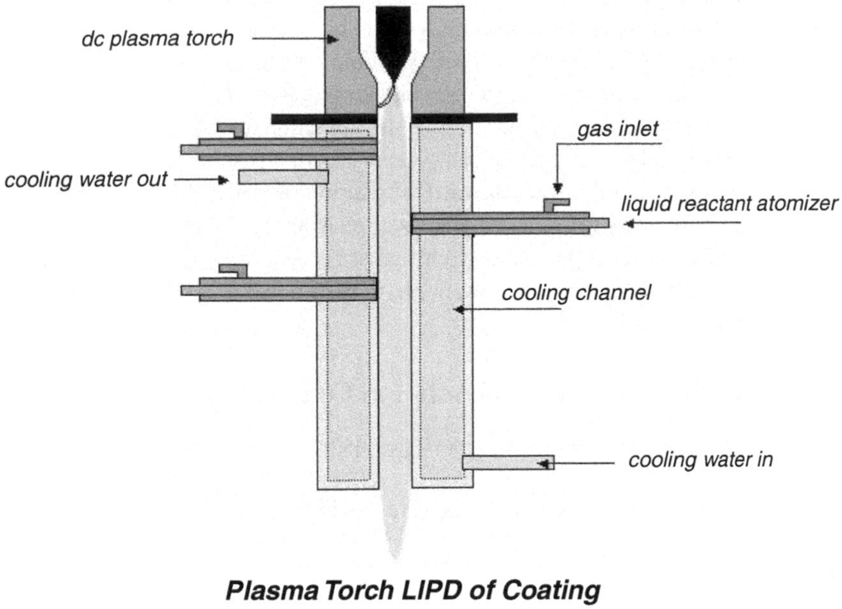

FIGURE 2.2 Schematic of Liquid Injected Plasma Deposition technique.

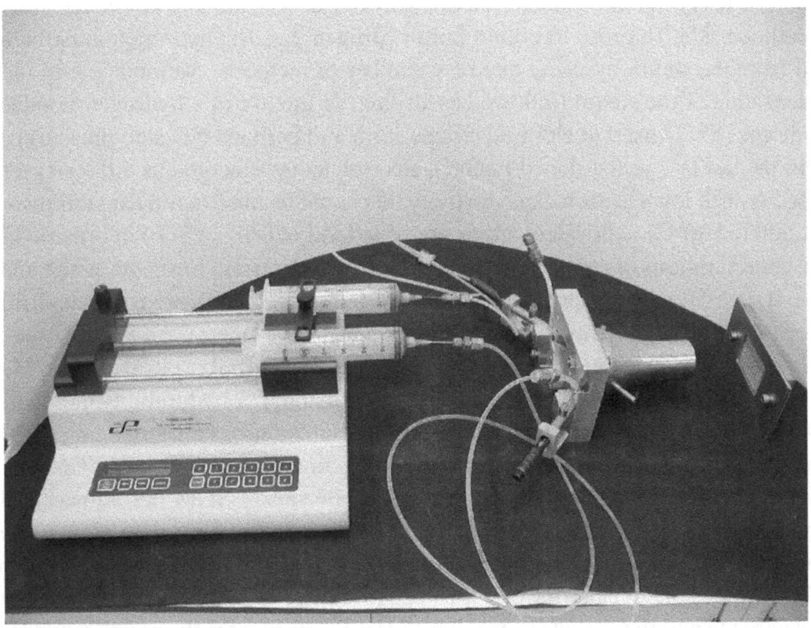

FIGURE 2.3 Equipment in use at INL for Liquid Injected Plasma Deposition.

Over the last decade there has been significant R&D to reduce the operating temperature of solid-oxide fuel cells (SOFCs). This primarily is intended to enable the use of cheaper and higher-conductivity (compared to electronically conductive ceramics like doped lanthanum chromite) bipolar plates made from metal alloys and at the same time minimizing formation of the low-conductivity metal oxide layers that greatly increase IR (current × resistance) losses in a bipolar plate stack configuration. This has spurred the trend toward fabrication of much thinner (e.g., films in range of 1 to 50 microns for reduced electrical resistance) electrolytes that are electrode supported. Porous support electrodes must be very smooth, such that the thin electrolyte layers that are deposited do not have thru-holes or voids that cause a loss of gas-tightness.

2.5.1 Moderate-Temperature Oxygen Ion Conductors

The electrolysis reactions to produce hydrogen using oxygen ion conductors are:

$$\text{Cathode: } H_2O + 2\,e^- \rightarrow H_2 + O^{-2} \qquad (2.3)$$

$$\text{Anode: } O^{-2} \rightarrow \tfrac{1}{2}\,O_2 + 2\,e^- \qquad (2.4)$$

$$\text{Overall: } H_2O \rightarrow H_2 + \tfrac{1}{2}\,O_2 \qquad (2.5)$$

During the electrolysis reaction, oxygen is removed from the reaction site via the membrane (oxygen ion conductor), leaving hydrogen gas and any unreacted steam on the cathode side. In order to obtain pure hydrogen gas, the hydrogen must be separated from the steam by using one of a number of methods. Methods could include condensation of the steam (followed by drying) or the use of a hydrogen-conducting membrane (likely used at elevated temperature and perhaps elevated pressure).

In the last few years, doped $LaGaO_3$ electrolyte has emerged as a fast oxygen ion conductor with low electronic conductivity that could be used at reduced temperatures (e.g., 600 to 800°C). Aliovalent atoms are added to $LaGaO_3$ (ABO_3) in order to create large concentrations of oxygen vacancies. Typical dopants are Sr on the A site and Mg on the B site[48,49] known as strontium and magnesium doped lanthanum gallate (LSGM), or occasionally Ba on the A site.[50] Other studies have been conducted to measure doped $LaGaO_3$'s electronic conductivity[51–53] and develop suitable electrodes.[54–57] Questions regarding $LaGO_3$'s high-temperature strength, toughness/durability (compared to ZrO_2), and long-term interactions with electrode combinations are still being answered by single-cell fuel cell tests.[58,59] Single cells utilizing plasma-sprayed LSGM electrolytes have been recently reported by Ma et al.[60] Because LSGM has a lower melting point than zirconia, it may be easier to plasma spray gas-tight films than when using zirconia. LSGM development has been slowed by its chemical reaction with nickel in the fuel electrode.[61,62] Recently a CeO_2 (Sm-doped) buffer layer has been added between the electrolyte and the fuel electrode, which largely eliminates the reaction.[63,64] Huang et al.[65] notes greatly improved performance with $La_{0.6}Sr_{0.4}CoO_{3-\delta}$ (LSMCo) cathodes compared to LSM cathodes. An LSGM (strontium- and magnesium-doped $LaGaO_3$) electrolyte (thin film, anode supported) single cell has been tested as an electrolyzer at

800°C. The cell exhibited a steady current density of 700 mA/cm^2 for 350 h.[66] Ishihara et al.[67] has also reported that doped $PrGaO_3$ is a fast oxygen ion conductor, but it does not seem to hold any advantage over $LaGaO_3$.

Doped ceria (CeO_2) has been a longtime oxygen ion-conducting SOFC electrolyte candidate.[68] Its ionic conductivity is about one order of magnitude greater than zirconia's in the 500 to 600°C range. Ceria has not been viewed as viable at high temperatures because of excess electronic conductivity. However, if the operating range is below 700°C, then its ionic transference number is greater than about 0.9, and it could be considered a candidate electrolyte for a moderate-temperature electrolyzer. Typical dopants for CeO_2 are Gd (10 to 20% substitution for Ce),[69] Y,[70,71] and Sm.[72] The materials cost for doped ceria electrolyte is significantly lower than that for doped $LaGaO_3$ electrolytes.[73]

Bismuth oxide (Bi_2O_3) is a much better oxygen ion conductor than doped CeO_2 at intermediate temperatures and always has held promise as a high-performance electrolyte. However, despite over 30 years of studies, Bi_2O_3 is still plagued with crystallographic and chemical stability problems that have prevented implementation in practical long-lived cells. As reviewed by Azad et al.,[74] α-Bi_2O_3 (monoclinic) is stable below 730°C, while the very high conductivity δ-Bi_2O_3 (cubic, CaF_2 type) is only stable between 730°C and its melting temperature of 825°C. This is much too narrow of a range and is too close to the Bi_2O_3 melting point. The δ-Bi_2O_3 contains 25% vacant oxygen sites, which results in the extremely high oxygen ion conductivity (approximately 1 Ω^{-1} cm^{-1} near the melting point). The δ-Bi_2O_3 also must be phase-stabilized by doping (e.g., Y_2O_3) in order to avoid the cracking that results from the volume change associated with the $\delta \rightarrow \alpha$ phase change. Even stabilized δ-Bi_2O_3 is prone to reduction into metallic bismuth (even at moderately low oxygen partial pressures). These features lead to the tentative conclusion that δ-Bi_2O_3 is not a good candidate to be an electrolysis cell's electrolyte. However, because of the promise of high conductivity at low to moderate temperatures, researchers in the 1990s studied a wide variety of bismuth oxide–containing compounds. Because yttria-stabilized Bi_2O_3 will transform to a rhombohedral phase (via diffusion) when annealed at less than 700°C,[75] some research was conducted on rhombohedral phase Bi_2O_3 stabilized by alkaline–earth oxide dopants (e.g., CaO-Bi_2O_3, SrO-Bi_2O_3, or BaO-Bi_2O_3)[76] or Nb_2O_5 -Bi_2O_3,[77] which appeared to be more stable (remained as cubic phases) than Y_2O_3-Bi_2O_3.

During the 1990s a new group of low-temperature oxygen ion-conducting compounds based on bismuth vanadate ($Bi_4V_2O_{11}$) were studied.[78] Crystal structures were studied into the mid-1990s, and it was found that $Bi_4V_2O_{11}$ exhibits three phases (δ β, γ) between room temperature and 800°C. The γ phase is the high-temperature, highest oxygen conductivity phase due to anion vacancies and a disordering of the anion vacancies. The γ structure can be stabilized to room temperature by partial substitution of various metal ions for vanadium. These compounds were termed BIMEVOX. Investigations of fabrication with possible application as an electrolyte, with particular interest in copper substituted material (BICUVOX, e.g., $Be_2V_{0.9}Cu_{0.1}O_{5.35}$),[79] followed. There is some electrical conductivity data measured on BICUVOX "cells,"[80,81] but no actual fuel cell data seem to be available. This may be an indication of increased electronic conductivity[82] (electronic shorting of cells) or the material's dilation when this type of material is reduced in a hydrogen-containing atmosphere. For a depleted steam electrolysis gas stream, the H_2/H_2O ratio could be in the 0.85 to 0.90 range, which could cause reduction at the fuel electrode (cathode). At this time, BIMEVOX electrolytes could not be considered good candidates for moderate-temperature electrolytes for steam electrolysis cells.

New electrode compositions need to be considered for use with moderate-temperature electrolytes. Platinum's coefficient of thermal expansion (CTE) is a good match to those of zirconia and doped CeO_2. Porous platinum is known to have excellent catalytic activity, but due to high cost, platinum is usually used only in the developmental testing of some single cells. Traditional conducting perovskite electrodes (air) have been developed with thermal expansion coefficients (CTEs) to approximate those of zirconia. Since ceria interacts too much with strontium-doped lanthanum manganites, other perovksite compositions have been proposed for air electrodes ($La_{0.8}Sr_{0.2}Fe_{0.8}Co_{0.2}O_{3-\delta}$ and $LaFe_{0.5}Ni_{0.5}O_{3-\delta}$).[83] A strong need for alternative lower-temperature SOFC anodes to replace nickel cermets has not been clearly identified (although copper has been used to prevent carbon deposition when using hydrocarbon fuels). Ni has been shown[84] to exhibit the highest electrochemical activity for H_2 oxidation (and assuming reversibility, for H_2 reduction in an electrolyzer) of the group: Ni, Co, Fe, Pt, Mn, and Ru. For operating fuel cells, overvoltages (polarizations) of Ni/samaria-doped ceria (SDC) and Pt/SDC anodes were very small compared with those of Ni/YSZ and Pt/YSZ cermet anodes. Electrode polarization generally is not a problem when operating at 950 to 1,000°C; however, polarization becomes a very significant problem at intermediate temperatures, especially for the air electrode. A recent review of SOFC anodes by Jiang and Chan[85] is a good source for Ni/ZrO_2 information as well as for information on various other cermets or conducting oxides, such as gadolinium- or samarium-doped ceria, titanate-based oxides, and lanthanum chromite-based materials. The Jiang and Chan article also reviews thick, anode-supported and porous metal-supported thin-film electrolytes, where the porous support material provides the structural strength. Because of improved performance from the thin electrolytes, these type cells are being considered for operation in the 600 to 800°C range. Since the porous support can have a significant thickness (e.g., in the 500- to 2,000-μm range), polarization losses due to gas diffusion can become significant. Therefore, a graded pore-size structure would become important with large-pore channels to enable easy diffusion of gases in most of the electrode, yet have a high surface area to enable the reaction near the electrolyte interface.

2.5.2 Moderate-Temperature Proton Conductors

Using proton-conducting ceramics as an electrolyte for a steam electrolyzer involves the same reactions as for a low-temperature proton-conducting polymer membrane:

$$\text{Anode: } H_2O \rightarrow 2\,H^+ + \tfrac{1}{2}\,O_2 + 2\,e^- \qquad (2.6)$$

$$\text{Cathode: } 2\,H^+ + 2\,e^- \rightarrow H_2 \qquad (2.7)$$

$$\text{Overall: } H_2O \rightarrow H_2 + \tfrac{1}{2}\,O_2 \qquad (2.8)$$

Therefore, the proton-conducting ceramics represent a significantly different technology than the oxygen ion-conducting ceramics, for example, zirconia, ceria, or lanthanum gallate. For fuel cell operations,[86] the proton-conducting cells have a thermodynamic advantage over oxygen ion-conducting cells (due to product water being swept from the cathode by excess air required for cell cooling). Applications that are driven by maximizing efficiency at the expense of power density favor proton cells. Proton conductors like the cerates ($BaCeO_3$ and $SrCeO_3$) have been studied for a number of years, while doped barium zirconate ($BaZrO_3$) has been advancing strongly in the last couple of years due to reports of high conductivity and good chemical resistance to CO_2 (not relevant for steam electrolysis). The aliovalent doping creates oxygen vacancies; an incorporation example is given by equation 2.9:

$$2\ BaO + Gd_2O_3\ (\text{into } BaCeO_3 \text{ lattice}) \rightarrow 2\ Ba^X_{Ba} + 2\ Gd^{/}_{Ce} + 5\ O^X_O + V^{\bullet\bullet}_O \quad (2.9)$$

Water vapor in the cell can react with the oxygen vacancies to form protons per equation 2.10:

$$H_2O + V^{\bullet\bullet}_O + O^X_O \rightarrow 2\ OH^{\circ}_O \quad (2.10)$$

The OH°_O species is a proton bound to an oxygen ion in the lattice. However, the proton can hop from one oxygen ion to another, giving rise to proton conductivity.

Twenty years ago, Iwahara et al.[87] introduced doped (Y, Yb, Sc) $SrCeO_3$ as a proton-conducting electrolyte with tests using platinum electrodes. He later reported[88] cell tests in both fuel cell and steam electrolysis mode (for hydrogen production) using both platinum and nickel fuel electrodes. A small electrolyzer was fabricated using $SrCe_{0.95}Yb_{0.05}O_{3-\delta}$ electrolyte, and pure, very dry hydrogen gas was produced[89] at 750°C at the rate of about 3 l/h. Emphasis later shifted to doped (Gd or Nd) $BaCeO_3$ because of increased proton conductivity.[90,91] The temperature range of application for electrolyzers was anticipated by Iwahara to be 600 to 800°C. There was some concern about the chemical stability of $BaCeO_3$ in CO_2 and H_2O. However, even though $BaCeO_3$ dissolves in boiling water, it is relatively stable as a dense electrolyte at high temperatures in high water vapor atmospheres.[92]

There has been considerable interest in developing proton-conducting perovskite ceramics in Germany. $BaZrO_3$ is a newly considered compound originally proposed by K. D. Kreuer[93] for use in the 500 to 800°C range. It is very refractory (good thermodynamic phase stability) and has good[94] proton conductivity if it is doped with acceptors (e.g., Y). Proton conductivity has been increased in $BaZrO_3$ grain boundaries by forming solid solutions with small amounts of $BaCeO_3$.[95] Recently, electrical and mechanical properties were measured and fabrication techniques developed for barium calcium niobate ($Ba_3Ca_{1+x}Nb_{2-x}O_{0-\delta}$),[96,97] but cell performance data are not yet available. Kreuer recently published a careful review of the considerations and problems involved with fabricating SOFCs utilizing proton-conducting perovskites.[98] The electrolyte thickness and electrodes have not been optimized for maximum performance. However, these materials have not shown sufficient conductivity to compete (in fuel cell or electrolyzer applications) with the best oxygen ion conductors until the temperature is less than about 700°C.

Kobayashi et al.[99] conducted steam electrolysis experiments using $SrZr_{0.9}Yb_{0.1}O_{3-\delta}$ tubular electrolytes (2-mm walls) with platinum electrodes (cermet with the electrolyte powder) at low temperatures (460 to 600°C) and was successful in generating hydrogen and oxygen. They used the low temperatures in an attempt to avoid excessive electronic (hole) conductivity in the electrolyte.

2.5.3 Moderate-Temperature Bipolar Plates (Interconnects)

At low to moderate temperatures new possibilities arise for using various metals as bipolar plates (for series connected cells in a bipolar stack arrangement). Most metals have too high (e.g., 15 E-6 $°C^{-1}$) of thermal expansion to match that of zirconia (10.5 E-6 $°C^{-1}$). In order to get a lower thermal expansion metal (to match zirconia), SOFC developers originally tried to use special high-chromium alloys like 95 Cr_4–5 Fe (Plansee alloy) or 94 Cr–5 Fe–1 Y_2O_3. However, they ran into the problem of high temperature Cr oxidation. The problem is primarily found on the cathode (air) side of a SOFC. The reaction is $Cr_2O_3 + ½\ O_2 \rightarrow 2\ CrO_3$ (high vapor pressure gas). The Cr must diffuse through the Cr_2O_3 protective coating such that Cr can continually evaporate as CrO_3 from the outer (exposed to air) surface at temperatures (some literature) beginning as low as 200°C. Once in the vapor state, Cr oxide condenses in the LSM cathode and at the LSM–electrolyte interface. One proposed mechanism is for Mn^{+2} ion to remove the oxygen from the CrO_3, resulting in precipitation of Cr crystallites.[100] Kofstad and Bredesen[101] point out that a Cr problem may also exist at the anode (fuel) side of a SOFC if high water vapor partial pressures spur the formation and evaporation of chromium oxyhydroxides (e.g., CrO_2OH). This could be a problem for the cathode during operation at high temperatures as an electrolyzer because of the high water content.

The presence of alloying elements in the interconnect tends to minimize the tendency for the Cr oxidation to take place (especially after oxide scale formation). Alloy elements like Y, Ce, Hf, Zr, and Al are reported to slow scale growth. However, these elements tend to form scales with low electronic conductivity, whereas Cr_2O_3 scales are semiconductors. Yang et al.[102] have reviewed the alloys being considered for SOFC bipolar plates. They present an evaluation of oxidation behavior that indicates chromia scales on chromia-forming alloys, especially the ferritic stainless steels, can grow to microns or even tens of microns thick after exposure for thousands of hours in the SOFC environment (even in the intermediate temperature range). They note that this scale growth will lead to an area-specific resistance (ASR) that is likely to be unacceptable. Nonetheless, iron-based ferritic steels (body centered cubic or BCC structure) are generally recommended because they have a reasonable CTE match to zirconia, and are less expensive and more easily fabricated than chromium-based alloys. Operating at the lower temperatures may help by slowing the evaporation and diffusion kinetics. The Cr issue is one of the primary reasons why SOFC developers are beginning to coat the air side of the interconnect with various conducting-oxide diffusion barriers.[103] One issue is maintaining a thin but protective conductive scale (Cr_2O_3) on the air side; the other issue is preventing the Cr evaporation and subsequent condensation reactions. In order to limit the growth rate of Cr_2O_3 scale on the metal interconnect (minimize the electrical resistance at the surface), various ceramic (conductive) coatings have been applied on the air side of the metal interconnects. However, some interdiffusion of elements between the protective coating and metallic interconnect has been observed to lead to nondesirable phases.[104] Most SOFC generator designs have noncell components, such as gas inlet chambers or electrical leads, that will be exposed to high-temperature air where CrO_3 formation could be problematic. The use of alloys like Hastelloy S (67% Ni, 15.5% Cr, 15.5% Mo, 1% Fe, 0.02% La) and Haynes alloy 214 could solve these problems. Haynes alloy 214 is specifically designed for service in high-temperature air at 900°C and above. It is an alumina former that displaces Cr_2O_3 on the metal surface. The total Cr in the 214 alloy is only 16%, which could also reduce the CrO_3 vaporization issue for nonstack structural elements.

Oxidation in H_2–H_2O mixtures could be a long-term problem for uncoated metallic bipolar electrolyzer plates with low H_2 content gas. Horita et al.[105] documents oxidation in Fe-Cr alloys using 1% H_2-Ar (balance) bubbled through water at 50°C (approximately 10% H_2 content). A higher H_2 content and the use of coatings would greatly lessen this problem.

One solution to the interconnect oxidation problem is being developed at INL. It is to form a thin, strontium-doped, lanthanum chromite (LSC) coating (for low electrical resistance) on a porous NiAl plate.[106,107] The NiAl is exposed to the fuel gas in a SOFC or hydrogen plus steam in an electrolyzer. There is some concern that the NiAl structural component will be slowly oxidized in a steam/hydrogen mixture. Oxidation tests are being conducted at INL using a 85% H_2O/15% H_2 (minimum) mixture at high temperatures. One oxidation reaction possibility is 2 NiAl + 3/2 O_2 $\rightarrow$ 2 Ni + Al_2O_3. However, this probably will not cause significant conductivity problems because of the formation of metallic Ni. Another possibility is a thin adherent coating of amorphous alumina within the open pores of the NiAl structure, but not a continuous coating. A noncontinuous alumina layer should not pose much of a problem. The other reaction possibility would be 2 NiAl + 2 O_2 $\rightarrow$ Ni + $NiAl_2O_4$ (spinel); this may present a problem, but there could be sufficient leftover nickel to preserve some electrical conductivity.

Other proposed solutions to interconnect oxidation can be found by searching patents. A ceramic plate (e.g., zirconia) with metal filled via holes extended through the thickness has been proposed by Hartvigsen et al.,[108] which is similar to a patent application by Badding et al.[109] For application at intermediate temperatures, the "via" filler material could be silver (m.p. = 962°C) since silver oxide is not stable at high temperatures and silver is tremendously less expensive than platinum or palladium. A metallic interconnect plate with gas-tight, silver-filled holes is described by Meulenberg et al.[110] as providing lowered contact resistance at temperatures up to 800°C. Wang et al.[111] describe sputter-deposited silver/yttria-stabilized zirconia cermets for electrodes as stable at temperatures up to 750°C. To reduce scale formation on the fuel cell interconnect, coating FeCrAl and FeCrMn(LaTi) alloys with nickel foils (dense, hot laminated) has been studied at 800°C in a 4% H_2–3% H_2O–remainder Ar atmosphere.[112] These nickel foils seemed to be helpful in preventing oxide scales. In some cases a stable nickel aluminide layer was formed at the interface between the alloy and the Ni foil.

2.6 HIGH-TEMPERATURE INORGANIC MEMBRANE ELECTROLYZERS

2.6.1 High-Temperature Oxygen Ion Conductors

The most common high-temperature cells being investigated are solid-oxide fuel cells (SOFCs) using yttria- or scandia-stabilized zirconia (cubic phase) electrolytes that are rapid oxygen conductors. Over many years, yttrium and scandium have been used to substitute on the zirconium lattice site to stabilize the cubic structure and increase oxygen ion diffusion by creating oxygen vacancies to compensate for their aliovalent (Y^{+3} or Sc^{+3} on Zr^{+4} site) charges.[113] Yttria provides excellent structural stabilization and good ionic conductivity. Scandia has been long known to provide higher ionic conductivity,[114] but at significant additional material cost.[115] Loss of conductivity for scandia-stabilized zirconia has been reported[116] due to phase changes upon aging at high temperatures (i.e., 1,000°C). This instability certainly would be less of a problem for cells operated at lower temperatures (e.g., 800°C). For long-life operation at high temperatures, it is very important to use suitable electrodes that do not interact (e.g., interdiffuse) unduly with the electrolyte or lose their activity (e.g., sintering). Fuel cells using zirconia electrolytes have traditionally used Ni-ZrO_2 and doped $LaMnO_3$ electrodes. These combinations have proven to be structurally and chemically stable at high temperatures for long periods with fuel cells operating for up to 25,000 h with performance degradation of less than 0.1% per 1,000 h.[117] Some interdiffusion and formation of nonconductive compounds (e.g., $La_2Zr_2O_7$) has been reported.[118] These interactions are more severe at high temperatures[119] and long times.

Early testing of electrolysis cells utilizing tubular yttria-stabilized zirconia electrolytes was reported by Donitz and Erdle[120] at Dornier System GmbH (Friedrichshafen, Germany) and Hino and Miyamoto[121] at JAERI (Japan). The German work was part of the high-temperature steam electrolysis Project "HOT ELLY" that began in about 1980. There has been recent successful testing in the U.S. at the Idaho National Laboratory (INL) and Ceramatec, Inc., of planar-design, zirconia electrolyte, solid-oxide fuel cells as steam electrolyzers.[122,123] Single cells and cell stacks utilizing yttria- and scandia-stabilized zirconia electrolytes were tested over a range of operating temperatures (700 to 850°C) and steam/H_2 input compositions. No activation polarization was observed near open-circuit voltages. There was a linear and symmetric behavior in the current-voltage (I-V) characteristics from the fuel cell mode to the electrolyzer mode of operation (up to the point where steam is largely depleted). Cell degradation characteristics were at least as good in the electrolysis mode as in the fuel cell mode.

The operating temperature of most zirconia membranes has been within the 800 to 1,000°C range. These temperatures may be consistent with utilization of heat from a new generation of proposed high-temperature gas-cooled reactors.[124,125] The Very High Temperature Reactor (VHTR) reference concept has been described as a helium-cooled, graphite-moderated, thermal neutron spectrum reactor with an outlet temperature of 1,000°C or higher.[126] In the U.S. there are investigations to combine a nuclear reactor with a high-temperature steam electrolysis plant to

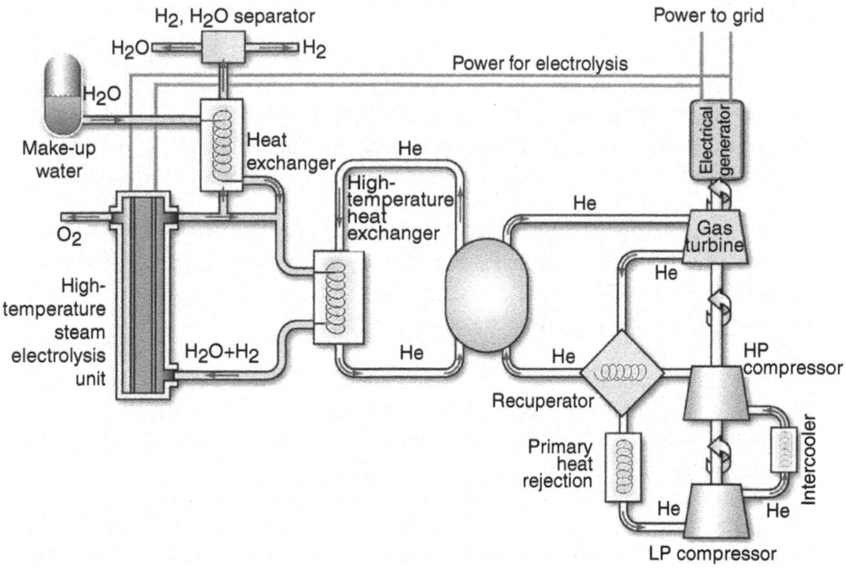

FIGURE 2.4 High-temperature steam electrolysis to generate hydrogen using heat and electricity from a high-temperature gas-cooled nuclear reactor.

generate hydrogen. Materials concerns have recently caused the initial outlet temperature goal for the U.S. design to be lowered to 900 to 950°C.[127] A schematic diagram of a combined nuclear–steam electrolysis plant is shown in figure 2.4. Process heat would be available for generating electricity and heating steam, after heat exchanging of the helium coolant. The temperature of the steam available to the electrolysis process will depend upon the heat exchanger efficiencies and certainly will be significantly lower than the latest proposed outlet temperature of 900 to 950°C. Some additional heat that would increase the cell temperature may be derived from IR losses within the electrolysis cells. The high-temperature electrolysis process will utilize both heat and electricity generated by the reactor. Another VHTR design, the Pebble Bed Modular Reactor (PBMR), is being developed in South Africa through a worldwide international collaborative effort led by South Africa's Electricity Supply Commission (ESKOM; supplies approximately 95% of that country's electricity). The PBMR currently has an average helium coolant exit temperature of 900°C under normal operating conditions.[128]

ACKNOWLEDGMENTS

This work was supported by the U.S. Department of Energy's Office of Nuclear Energy Science and Technology, under DOE-NE Idaho Operations Office Contract DE-AC07-05ID14517.

REFERENCES

1. Wendt, H. and Imarisio, G., Nine years of research and development on advanced water electrolysis. A review of the research programme of the Commission of the European Communities, *J. Appl. Electrochem.*, 18, 1–14 (1988).
2. Wendt, H., Hofmann, H., and Plzak, V. Anode and cathode-activation, diaphragm-construction and electrolyzer configuration in advanced alkaline water electrolysis, *Int. J. Hydrogen Energy*, 9, 297–302 (1984).
3. Wendt, H., Hofmann, H., and Plzak, V., Materials research and development of electrocatalysts for alkaline water electrolysis, *Mater. Chem. Physics*, 22, 27–49 (1989).
4. Ramesh, L. et al., Electrolytic preparation and characterization of Ni-Fe-Mo- alloys: cathode materials for alkaline water electrolysis, *Int. J. Energy Res.*, 23, 919–924 (1999).
5. Hu, W., Electrocatalytic properties of new electrocatalysts for hydrogen evolution in alkaline water electrolysis, *Int. J. Hydrogen Energy*, 25, 111–118 (2000).
6. Suffredini, H.B. et al., Recent developments in electrode materials for water electrolysis, *Int. J. Hydrogen Energy*, 25, 415–423 (2000).
7. Abdel Ghany, N.A. et al., Oxygen evolution anodes composed of anodically deposited Mn-Mo-Fe oxides for seawater electrolysis, *Electrochem. Acta*, 48, 21–28 (2002).
8. Wendt, H. and Hofmann, H., Cermet diaphragms and integrated electrode-diaphragm units for advanced alkaline water electrolysis, *Int. J. Hydrogen Energy*, 10, 375–381 (1985).
9. Kreuter, W. and Hofmann, H., Electrolysis: the important energy transformer in a world of sustainable energy, *Int. J. Hydrogen Energy*, 23, 661–666 (1998).
10. Rasten, E., Hagen, G., and Tunold, R., Anode catalyst materials for PEM-electrolysis, in *New Materials for Electrochemical Systems IV. Extended Abstracts of the Fourth International Symposium on New Materials for Electrochemical Systems*, Montreal, Quebec, Canada, July 9–13, 2001, pp. 278–280.
11. Linkous, C.A. et al. Development of new proton exchange membrane electrolytes for water electrolysis at higher temperatures, *Int. J. Hydrogen Energy*, 23, 525–529 (1998).
12. Weikang, H. et al., A novel cathode for alkaline water electrolysis, *Int. J. Hydrogen Energy*, 22, 621–623 (1997).
13. Weikang, H., Electrocatalytic properties of new electrocatalysts for hydrogen evolution in alkaline water electrolysis, *Int. J. Hydrogen Energy*, 25, 111–118 (2000).
14. Oi, T. and Sakaki, Y., Optimum hydrogen generation capacity and current density of the PEM-type water electrolyzer operating only during the off-peak period of electricity demand, *J. Power Sources*, 129, 229–237 (2004).
15. Morizonoa, T., Watanabe, K., and Ohstsuka, K., Production of hydrogen by electrolysis with proton exchange membrane (PEM) using sea water and fundamental study of hybrid system with PV-ED-FC, *Mem. Fac. Eng.*, 31, 213–218 (2002).
16. HOGEN™ Hydrogen Generators from Proton Energy Systems, 50 Inwood Rd., Rocky Hill, CT 06067, 860-571-6533, www.protonenergy.com.
17. Friedland, R.J. and Speranza, A.J., Hydrogen Production through Electrolysis, paper presented at the Proceedings of the 2001 DOE Hydrogen Program Review, NREL/CP-570-30535, Golden, CO, June 4-5, 2001.
18. Hamilton Sundstrand (A United Technologies Company), One Hamilton Rd., Windsor Locks, CT 06096, 860-654-6000.
19. Treadwell Corp., 341 Railroad St., Thomaston, CT 06787, 860-283-8251.
20. Hydrogenics Corp., 5985 McLaughlin Rd., Mississauga, Ontario, Canada, L5R 1 B8, 905-361-3660.

21. Giner Electrochemical Systems, LLC, 89 Rumford Ave., Newton, MA 02466, 781-529-0500.
22. Stucki, S. et al., PEM water electrolysers: evidence for membrane failure in 100 kW demonstration plants, *J. Appl. Electrochem.*, 28, 1041–1049 (1998).
23. Ando, U. and Tanaka, T., Proposal for a new system for simultaneous production of hydrogen and hydrogen peroxide by water electrolysis, *Int. J. Hydrogen Energy*, 29, 11349–11354 (2004).
24. Foster, R. and Wright, R.L. Jr., *Basic Nuclear Engineering*, Boston, MA: A. Allyn and Bacon, 1968.
25. Nogami, M., Matsushita, H., Kasuga, T., and Hayakawa, T., Hydrogen gas sensing by sol-gel-derived proton-conducting glass membranes, *Electrochem. Solid-State Lett.*, 2, 415–417 (1999).
26. Kuo, C.K., Tan, A., Sarkar, P., and Nicholson, P.S., Water partial pressure-dependent conductance and humidity effects on hydronium-β''-Al_2O_3 ceramics, *Solid State Ionics*, 58, 311–314 (1992).
27. Schafer, G., Zyl, A.V., and Weppner, W., Process for the Production of K- or Rb-β''- or -β-Aluminum Oxide Ion Conductors, U.S. Patent 5,474,959, December 12, 1995.
28. Schafer, G.W., Kim, H.J., and Aldinger, F., Protonated β''-aluminas, correlation of ion-exchange rates, chemical composition and resulting lattice constants, *Solid State Ionics*, 97, 285–289 (1997).
29. Joshi, A.V., Liu, M., Bjorseth, A., and Renberg, L., NaOH Production from Ceramic Electrolytic Cell, U.S. Patent 5,290,405, March 1, 1994.
30. Damasceno, O., Siebert, E., Khireddine, H., and Fabry, P., Ionic exchange and selectivity of NASICON sensitive membranes, *Sensors Actuators B*, 8, 245–248 (1992).
31. Slade, R.C.T. and Young, K.E., Hydronium and ammonium NASICONs: investigations of conductivity and conduction mechanism, *Solid State Ionics*, 46, 83–88 (1991).
32. "PRONAS" available from Ceramatec, Inc., 2425 South 900 West, Salt Lake City, UT 84119, 801-978-2152.
33. Farrington, G.C. and Briant, J.L., Fast ionic transport in solids, *Science*, 204, 1371–1379 (1979).
34. McGeehin, P. and Hooper, A., Fast ion conduction materials [review], *J. Mater. Sci.*, 12, 1–27 (1977).
35. Mizuno, M. and Hayashi, S., Proton dynamics in phase II of $CsHSO_4$ studied by H-1 NMR, *Solid State Ionics*, 167, 317–323 (2004).
36. Kaminura, H. et al., On the mechanism of superionic conduction in the zero-dimensional hydrogen-bonded crystals $M_3H(XO_4)_2$; (M = K, Rb, Cs, and X = S, Se), *Physica Status Solidi C*, 1, 8 (2004).
37. Park, J.-H., Possible origin of the proton conduction mechanism of CsH_2PO_4 crystals at high temperatures, *Phys. Rev. B*, 69, 54104-1-6 (2004).
38. Yaroslavtsev, A.B. and Kotov, V.Y., Proton mobility in hydrates of inorganic acids and acid salts, *Russ. Chem. Bull.*, 54, 555–568 (2002).
39. Wu, Q.Y. and Meng, G.Y., Preparation and conductibility of solid high-proton conductor molybdovana dogermanic heteropoly acid, *Mater. Res. Bull.*, 35, 85–91 (2000).
40. Adams, S., CDW superstructures in hydrogen molybdenum bronzes $HxMoO_3$, *J. Solid State Chem.*, 149, 75–87 (2000).
41. U.S. Department of Energy, *A Technology Roadmap for Generation IV Nuclear Energy Systems*, GIF-002-00, Nuclear Energy Research Advisory Committee and the Generation IV International Forum, December 2002.
42. Guan, J. et al., Ceramic oxygen generators with thin-film zirconia electrolytes, *J. Am. Ceram. Soc.*, 85, 2651–2654 (2002).
43. Rambert, S. et al., Composite ceramic fuel cell fabricated by vacuum plasma spraying, *J. Eur. Ceram. Soc.*, 19, 921–923 (1999).

44. Henne, R.H. et al., Light-Weight SOFCs for Automotive Auxiliary Power Units, paper presented at the 2nd International Conference on Fuel Cell Science, Engineering and Technology, Rochester, NY, June 14–16, 2004.
45. Wang, L.S. et al., Sputter deposition of yttria-stabilized zirconia and silver cermet electrodes for SOFC applications, *Solid State Ionics*, 52, 261–267 (1992).
46. Chu, W.F., Thin- and thick-film solid ionic devices, *Solid State Ionics*, 52, 243–248 (1992).
47. Windes, W.E. and Lessing, P.A., Plasma spray coatings for SOFC, in *2002 Fuel Cell Seminar Abstracts*, Courtesy Associates, Washington, DC, 471–474, 2002.
48. Chen, T.Y. and Fung, K.Z., A and B-site substitution of the solid electrolyte $LaGaO_3$ and $LaAlO_3$ with the alkaline-earth oxides MgO and SrO, *J. Alloys Compounds*, 368, 106–115 (2004).
49. Kurumada, M., Ito, A., and Fujie, Y., Preparation of $La_{2-x}Sr_xGa_{0.8}Mg_{0.2}O_{3-\delta}$ electrolyte for solid oxide fuel cell by citrate method using industrial raw materials, *J. Ceram. Soc. Jpn.*, 111, 200–204 (2003).
50. Choi, S.M., et al., Oxygen ion conductivity and cell performance of $La_{0.9}Ba_{0.1}Ga_{1-x}Mg_xO_{3-\delta}$ electrolyte, *Solid State Ionics*, 131, 221–228 (2000).
51. Kharton, V.V. et al., Ionic and p-type electronic conduction in $LaGa(Mg,Nb)O_{3-\delta}$ perovksites, *Solid State Ionics*, 128, 79–90 (2000).
52. Maffei, N. and de Silveira, G., Interfacial layers in tape cast anoe-supported doped lanthanum gallate SOFC elements, *Solid State Ionics*, 159, 209–216 (2003).
53. Kim, J.H. and Yoo, H.I., Partial electronic conductivity and electrolytic domain of $La_{0.9}Sr_{0.1}Ga_{0.8}Mg_{0.2}O_{3-\delta}$, *Solid State Ionics*, 140, 105–113 (2001).
54. Zhang, X.G. et al., Interactions of $La_{0.9}Sf_{0.1}Ga_{0.8}Mg_{0.2}O_{3-\delta}$ electrolyte with Fe_2O_3, Co_2O_3 and NiO anode materials, *Solid State Ionics*, 139, 145–152 (2001).
55. Majkic, G. et al., High-temperature deformation of $La_{0.2}Sr_{0.8}Fe_{0.8}Cr_{0.2}O_{3-\delta}$ mixed ionic-electronic conductor, *Solid State Ionics*, 146, 393–404 (2002).
56. Kostogloudis, G.C. et al., Chemical compatibility of alternative perovskite oxide SOFC cathodes with doped lanthanum gallate solid electrolyte, *Solid State Ionics*, 134, 127–138 (2000).
57. Zhang, X.G. et al., Interface reactions in the NiO-SDC-LSGM system, *Solid State Ionics*, 133, 153–160 (2000).
58. Wang, S.Z., High performance fuel cells based on $LaGaO_3$ electrolytes, *Acta Physico-Chim. Sin.*, 20, 43–46 (2004).
59. Kuroda, K. et al., Characterization of solid oxide fuel cell using doped lanthanum gallate, *Solid State Ionics*, 132, 199–208 (2000).
60. Ma, X. et al., The power of plasma, *Ceramic Industry*, June 2004, pp. 25–28.
61. Pengnian, H. et al., Interfacial reaction between nickel oxide and lanthanum gallate during sintering and its effect on conductivity, *J. Am. Ceram. Soc.*, 82, 2402–2406 (1999).
62. Maffei, N. and de Silveira, G., Interfacial layers in tape cast anode-supported doped lanthanum gallate SOFC elements, *Solid State Ionics*, 159, 209–216 (2003).
63. Huang, K.G. et al., Increasing power density of LSGM-based solid oxide fuel cells using new anode materials, J. *Electrochem. Soc.*, 148, A788–A794 (2001).
64. Wang, S.Z. and Tatsumi, I., Improvement of the performance of fuel cells anodes with Sm^{+3} doped CeO_2, *Acta Physico-Chim. Sin.*, 19, 844–848 (2003).
65. Huang, K.Q. et al., Electrode performance test on single ceramic fuel cells using as electrolyte Sr- and Mg-doped $LaGaO_3$, *J. Electrochem. Soc.*, 144, 3620–3624 (1997).
66. Elangovan, S., Hartvigsen, J.J., O'Brien, J.E., Stoots, C.E., Herring, J.S., and Lessing, P.A., Operation and Analysis of Solid Oxide Fuel Cells in Steam Electrolysis Mode, paper presented at Session B07, 6th European SOFC Forum, Lucerne, Switzerland, June 28–July 2, 2004.

67. Ishihara, T. et al., Oxide ion conductivity in doubly doped $PrGaO_3$ perovskite-type oxide, *J. Electrochem. Soc.*, 146, 1643–1649 (1999).
68. Maricle, D.L. et al., Enhanced ceria: a low-temperature SOFC electrolyte, *Solid State Ionics*, 52, 173–182 (1992).
69. Kharton, V.V. et al., Ceria-based materials for solid oxide fuel cells, *J. Mater. Sci.*, 36, 1105–1117 (2001).
70. Hidenori, Y. et al., High temperature fuel cell with ceria-yttria solid electrolyte, *J. Electrochem. Soc. Solid-State Sci. Technol.*, 2077–2080 (1988).
71. Kirk, T.J. and Winnick, J., A hydrogen sulfide solid-oxide fuel cell using ceria-based electrolytes, *J. Electrochem. Soc.*, 140, 3494–3496 (1993).
72. Lu, C. et al., SOFCs for direct oxidation of hydrocarbon fuels with samaria-doped ceria electrolyte, *J. Electrochem. Soc.*, 150, A354–A358 (2003).
73. Alfa Aesar 2004 catalog prices: La_2O_3 (99.99%) $108/kg, Ga_2O_3 (99.999%) $3400/kg, CeO_2 (99.9%) $84/kg, Y_2O_3 (99.99%) $212/kg, Gd_2O_3 (99.99%) $320/kg.
74. Azad, A.M., Larose, S., and Akbar, S.A., Review bismuth oxide-based solid electrolytes for fuel cells, *J. Mater. Sci.*, 29, 4135–4151 (1994).
75. Fung, K.Z. et al., Massive transformation in the Y_2O_e-Bi_2O_3 system, *J. Am. Ceram. Soc.*, 77, 1638–1648 (1994).
76. Fung, K.Z. et al., Thermodynamic and kinetic considerations for Bi_2O_3-based electrolytes, *Solid State Ionics*, 52, 199–211 (1992).
77. Joshi, A.V. et al., Phase stability and oxygen transport characteristics of yttria- and niobia-stabilized bismuth oxide, *J. Mater. Sci.*, 25, 1237–1245 (1990).
78. Abraham, F. et al., The BIMEVOX series: a new family of high performance oxide ion conductors, *Solid State Ionics Diffusion React.*, 40/41, 934–937 (1990).
79. Simner, S.P. et al., Synthesis, densification, and conductivity characteristics of BICUVOX oxygen-ion-conducting ceramics, *J. Am. Ceram. Soc.*, 80, 2563–2568 (1997).
80. Yaremchenko, A.A. et al., Physicochemical and transport properties of Bicuvox-based ceramics, *J. Electroceram.*, 4, 233–242 (2000).
81. Pasciak, G. et al., Solid electrolytes for gas sensors and fuel cells applications, *J. Eur. Ceram. Soc.*, 21, 1867–1870 (2001).
82. Priovano, C. et al., Characterisation of the electrode-electrolyte BIMEVOX system for oxygen separation. Part I. In situ synchrotron study, *Solid State Ionics*, 159, 167–179 (2003).
83. Kharton, V.V. et al., Ceria-based materials for solid oxide fuel cells, *J. Mater. Sci.*, 36, 1105–1117 (2001).
84. Setoguchi, T. et al., Effects of anode materials and fuel on anodic reaction of solid oxide fuel-cells, *J. Electrochem. Soc.*, 139, 2875–2880 (1993).
85. Jiang, S.P. and Chan, S.H., A review of anode materials development in solid oxide fuel cells, *J. Mater. Sci.*, 39, 4405–4439 (2004).
86. Hartvigsen, J., Elangovan, S., and Khandkar, A., A Comparison of Proton and Oxygen Ion Conducting Electrolytes for Fuel Cell Applications, paper presented at AIChE Annual Meeting Fuel Cells for Utility Applications and Transportation: Engineering & Design II, St. Louis, MO, November 11, 1993.
87. Iwahara, H. et al., High temperature type proton conductor based on $SrCeO_3$ and its application to solid electrolyte fuel cells, *Solid State Ionics*, 9/10, 1021–1026 (1983).
88. Iwahara, H., High temperature-type proton conductive solid oxide fuel cells using various fuels, *J. Appl. Electrochem.*, 16, 663–668 (1986).
89. Iwahara, H., Oxide-ionic and protonic conductors based on perovskite-type oxides and their possible applications, *Solid State Ionics*, 52, 99–104 (1992).
90. Iwahara, H. et al., Proton conduction in sintered oxides based on $BaCeO_3$, *J. Electrochem. Soc. Solid-State Sci. Technol.*, 135, 529–533 (1988).

91. Chen, F.L. et al., Preparation of Nd-doped $BaCeO_3$ proton-conducting ceramic and its electrical properties in different atmospheres, *J. Eur. Ceram. Soc.*, 18, 1389–1395 (1998).
92. Bhide, S.V. and Virkar, A.V., Stability of $BaCeO_3$-based proton conductors in water-containing atmospheres, *J. Electrochem. Soc.*, 146, 2038–2044 (1999).
93. Kreuer, K.D., Aspects of the formation and mobility of protonic charge carriers and the stability of perovskite-type oxides, *Solid State Ionics*, 125, 285–302 (1999).
94. Kreuer, K.D. et al., Proton conducting alkaline earth zirconates and titanates for high drain electrochemical applications, *Solid State Ionics*, 145, 295–306 (2001).
95. Kreuer, K.D., Proton-conducting oxides, *Annu. Rev. Mater. Res.*, 33, 333–359 (2003).
96. Hassan, D. et al., Proton-conducting ceramics as electrode/electrolyte materials for SOFC's. Part I. Preparation, mechanical and thermal properties of sintered bodies, *J. Eur. Ceram. Soc.*, 23, 221–228 (2003).
97. Fehringer, G. et al., Proton-conducting ceramics as electrode/electrolyte: materials for SOFCs: preparation, mechanical and thermal-mechanical properties of thermal sprayed coatings, material combination and stacks, *J. Eur. Ceram. Soc.*, 24, 705–715 (2004).
98. Kreuer, K.D., Proton-conducting oxides, *Annu. Rev. Mater. Res.*, 33, 333–359 (2003).
99. Kobayashi, T. et al., Study on current efficiency of steam electrolysis using a partial protonic conductor $SrZr_{0.9}Yb_{0.1}O_{3-\delta}$, *Solid State Ionics*, 138, 243–251 (2001).
100. Jian, S.P. et al., A comparative investigation of chromium deposition at air electrodes of solid oxide fuel cells, *J. Eur. Ceram. Soc.*, 22, 361–373 (2002).
101. Kofstad, P. and Bredesen, R., High temperature corrosion in SOFC environments, *Solid State Ionics*, 52, 69–75 (1992).
102. Yang, S.G. et al., Solid oxide fuel cells: materials for the bipolar plates are under development to reduce costs while maintaining performance and durability, *Adv. Mater. Processes*, 161, 34 (2003).
103. Larring, Y. and Norby, T., Spinel and perovskite functional layers between Plansee metallic interconnect (Cr-5 wt% Fe-1 wt% Y_2O_3) in ceramic $(La_{0.85}Sr_{0.15})_{0.91}MnO_3$ cathode materials for solid oxide fuel cells, *J. Electrochem. Soc.*, 147, 3251–3256 (2000).
104. Kung, S.C. et al., Performance of Metallic Interconnect in Solid-Oxide Fuel Cells, paper presented at the 2000 Fuel Cell Seminar, Portland, OR, October 30–November 2, 2000.
105. Horita, T. et al., Evaluation of Fe-Cr alloys as interconnects for reduced operation temperature SOFCs, *J. Electrochem. Soc.*, 150, A243–A248 (2003).
106. Windes, W.E. and Lessing, P.A., Fabrication Methods of a Leaky SOFC Design, paper presented at the Eighth International Symposium of Solid Oxide Fuel Cells (SOFC VIII), 203rd Meeting of the Electrochemical Society, Paris, April 27–May 2, 2003.
107. Windes, W.E. and Lessing, P.A., A Low CTE Intermetallic Bipolar Plate, paper presented at the Eighth International Symposium of Solid Oxide Fuel Cells (SOFC VIII), 203rd Meeting of the Electrochemical Society, Paris, April 27–May 2, 2003.
108. Hartvigsen, J.J. et al., Via Filled Interconnect for Solid Oxide Fuel Cells, U.S. Patent 6,183,897 B1, February 6, 2001.
109. Badding, M.E. et al., High Performance Solid Electrolyte Fuel Cells, U.S. Patent Application 20010044041, November 22, 2001.
110. Meulenberg, W.A. et al., Improved contacting by the use of silver in solid oxide fuel cells up to an operating temperature of 800°C, *J. Mater. Sci.*, 36, 3189–3195 (2001).
111. Wang, L.S. et al., Sputter deposition of yttria-stabilized zirconia and silver cermet electrodes for SOFC applications, *Solid State Ionics*, 52, 261–267 (1992).
112. Meulenberg, W.A. et al., Oxidation behaviour of ferrous alloys used as interconnecting material in solid oxide fuel cells, *J. Mater. Sci.*, 38, 507–513 (2003).

113. Subbarao, E. and Maiti, H.S., Solid electrolytes with oxygen ion conduction, *Solid State Ionics*, 11, 317–338 (1984).
114. Kilner, J.A. and Brook, R.J., A study of oxygen ion conductivity in doped non-stoichiometric oxides, *Solid State Ionics*, 6, 237–252 (1982).
115. USGS, Year 2000 data: Sc_2O_3 (99.99% pure) $3,000/kg, Sc_2O_3 (99.9% pure) $700/kg; Y_2O_3 (99.99% pure) $ 200/kg.
116. The system Y_2O_3-Sc_2O_3-ZrO_2: phase stability and ionic conductivity studies, *J. Eur. Ceram. Soc.*, 7, 197–206 (1991).
117. Singhal, S.C., Science and technology of solid-oxide fuel cells, *MRS Bull.*, 25, 16–21 (2000).
118. Anderson, H.U. and Nasrallah, M.M., Characterization of Oxide for Electrical Delivery Systems, paper presented at the EPRI/GRI Fuel Cell Workshop on Fuel Cell Technology Research and Development, New Orleans, April 13–14, 1993.
119. Misuyasu, H. et al., Microscopic analysis of lanthanum strontium manganite yttria-stabilized zirconia interface, *Solid State Ionics*, 100, 11–15 (1997).
120. Donitz, W. and Erdle, E., High-temperature electrolysis of water vapor: status of development and perspectives for application, *Int. J. Hydrogen Energy*, 10, 291–295 (1985).
121. Hino, R. and Miyamoto, Y., Hydrogen production by high-temperature electrolysis of steam, in *High Temperature Applications of Nuclear Energy*, International Atomic Energy Agency, 1994, pp. 119–124.
122. Herring, J.S., Anderson, R., Lessing, P.A., O'Brien, J.E., Stoots, C.M., Hartvigsen, J.J., and Elangovan, S., Hydrogen Production through High-Temperature Electrolysis in a Solid Oxide Cell, paper presented at the National Hydrogen Association 15th Annual Conference, Los Angeles, April 26–29, 2004.
123. O'Brien, J.E., Stoots, C.M., Herring, J.S., and Lessing, P.A., Characterization of Solid-Oxide Electrolysis Cells for Hydrogen Production via High-Temperature Steam Electrolysis, Paper 2474, paper presented at the 2nd International Conference on Fuel Cell Science, Engineering, and Technology, Rochester, NY, June 14–16, 2004.
124. Herring, J.S., O'Brien, J.E.,. Stoots, C.M,. Lessing, P.A, Anderson, R.P., Hartvigsen, J.J., and Elangovan, S., Hydrogen Production from Nuclear Energy via High-Temperature Electrolysis, Paper 4322, pape presented at the 2004 International Conference on Advances in Nuclear Power Plants (ICAPP'04), Pittsburgh, PA, June 13–17, 2004.
125. Herring, J.S., O'Brien, J.E., Stoots, C.M., Lessing, P.A., Anderson, R.P., Hartvigsen, J.J., and Elangovan, S., Hydrogen Production through High-Temperature Electrolysis Using Nuclear Power, paper presented at the AIChE Spring National Meeting, New Orleans, April 25–29, 2004.
126. *Very High Temperature Reactor (VHTR) Survey of Materials Research and Development Needs to Support Early Deployment*, INL/EXT-03-004-141, January 31, 2003.
127. *Design Features and Technology Uncertainties for the Next Generation Nuclear Plant*, INEEL/EXT-04-01816, Independent Technology Review Group, June 30, 2004.
128. Ion, S. et al., Pebble Bed Modular Reactor: The First Generation IV Reactor to Be Constructed, paper presented at the World Nuclear Association Annual Symposium, London, September 3–5.

3 High-Temperature Electrolysis

S. Elangovan and J. Hartvigsen

CONTENTS

3.1 BACKGROUND

Emphasis on energy security issues has brought much needed attention to economic production of hydrogen as the secondary energy carrier for nonelectrical markets. The recent focus on hydrogen comes from its environmentally benign aspect. However, much of the hydrogen currently produced is used near the production facility for chemical synthesis, such as ammonia and methanol production, and for upgrading as well as desulfurization of crude oil. While steam reforming of methane is the current method of production of hydrogen, the fossil fuel feed consumes nonrenewable fuel while emitting greenhouse gases. Thus, in the long run, efficient, environmentally friendly, and economic means of hydrogen production using renewable energy need to be developed. Additionally, when excess energy production capacity exists, for example, during off-peak hours, efficient generation of hydrogen may be an option to make an effective use of the investment in power generation infrastructure. Steam electrolysis, particularly using high-temperature ceramic membrane processes, provides an attractive option for efficient generation of ultra high purity (UHP) hydrogen.

The electrolysis reaction can be expressed as:

$$H_2O + 2e^- \rightarrow H_2 + O^= \text{ Cathode reaction} \tag{3.1}$$

$$O^= \rightarrow \frac{1}{2}O_2 + 2e^- \text{ Anode reaction} \tag{3.2}$$

$$H_2O \xrightarrow{Energy} H_2 + \frac{1}{2}O_2 \text{ Overall reaction} \tag{3.3}$$

The enthalpy of the overall reaction is ΔH = 242 KJ/mole at 298K and 248 KJ/mole at 1,000K. A schematic of an electrolysis cell using an oxygen ion conductor is shown in figure 3.1. The benefit of high-temperature electrolysis (HTE) stems from the fact that a portion of endothermic heat of reaction can be supplied by thermal energy instead of electric energy. Figure 3.2 shows the energy input required for electrolysis of steam. It can be seen that at higher temperatures substantial energy is provided as thermal energy, resulting in considerable reduction of primary (electrical) energy. The high temperature also allows high current density operation as both ohmic resistance losses from the electrolyte and electrode materials, and non-ohmic resistance losses from the electrode reaction processes are thermally activated. Hydrogen production via room temperature electrolysis of liquid water has the disadvantage of much lower overall thermal-to-hydrogen efficiencies of 24 to 32% (including power generation), while at higher temperatures practical efficiency can be as high as 50 to 60%.

3.2 MATERIALS AND DESIGN

The high operating temperature that is necessary for an efficient electrolysis process requires the use of materials that are stable at those temperatures. In general, the materials and fabrication technology that are used for high-temperature solid-oxide fuel cells (SOFCs) are directly applicable to high-temperature electrolysis devices. The high-temperature electrolysis cell is commonly referred to as the solid-oxide electrolysis cell (SOEC). In fact, regenerative fuel cells, where the same device can be used in both the fuel cell and electrolysis modes, are an interesting option for load leveling when the electricity demand varies widely during the day. Thus, much of the research work done in the SOFC area is directly applicable to SOEC technology. Thus, not only the materials set but also the cell and multicell stack designs of SOEC have followed the technology advances of SOFC development. The cell components, electrolyte, anode, and cathode in general are of similar chemistry among the various designs. The interconnect material used for connecting individual cells into a stack varies with the cell design. It should be noted that the hydrogen generation electrode and oxygen evolution electrode in an SOEC are called cathode and anode,

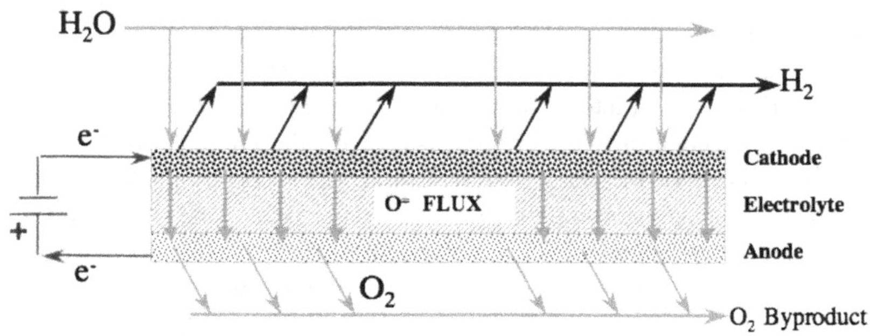

FIGURE 3.1 Schematic of an electrolysis cell.

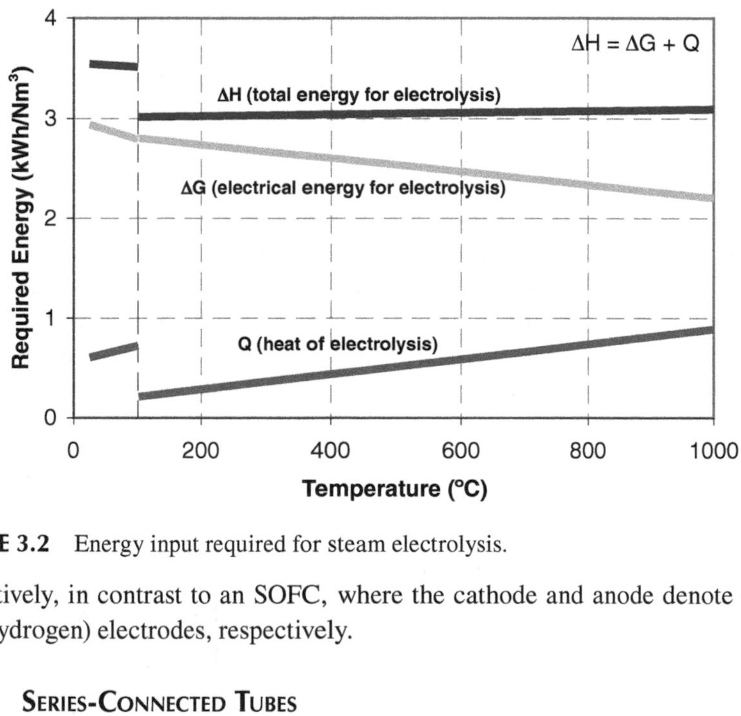

FIGURE 3.2 Energy input required for steam electrolysis.

respectively, in contrast to an SOFC, where the cathode and anode denote air and fuel (hydrogen) electrodes, respectively.

3.2.1 Series-Connected Tubes

The earliest large-scale high-temperature electrolysis work was done at Dornier System Gmbh in Germany.[1] The cell used traditional SOFC materials such as 9 mol% yttria-doped zirconia (YSZ) as the electrolyte, a cermet mixture of 50:50 wt% nickel and YSZ as the hydrogen electrode, and Ca-doped $LaMnO_3$ perovskite ($La_{0.5}Ca_{0.5}MnO_3$) as the air electrode.[2] An intermediate layer consisting of a mixture of perovskite and YSZ was applied between the electrolyte and air electrode.[3] The electrolyte was sintered to sufficient density to be gas impermeable, whereas the electrodes remained porous to allow gas diffusion to the electrode–electrolyte interface for the oxidation and reduction reactions. The electrolyte was about 300 microns thick, and the electrodes were about 250 microns thick. The typical cell diameter was 14 mm, with an active cell length of 10 mm.[4–6]

In order to increase the overall hydrogen output, several cells are connected in electrical series using an interconnect ring to form a tubular HTE stack, as shown in figure 3.3. Selection of material composition for the interconnect is one of the challenging aspects of the stack design. Unlike the electrodes, which face either oxidizing or reducing environments, the interconnects must be stable over a wide range of oxygen partial pressures and possess high electronic conductivity. An intimate mixture of a perovskite similar to the air electrode and V_2O_5-doped CeO_2 was reportedly used as the interconnect.[7] The perovskite exhibits high conductivity in the oxygen partial pressures ranging from 1 bar to 10^{-5} bar, while the doped ceria has high electronic conductivity at oxygen partial pressures lower than 10^{-5} bar at 1,000°C. A layered structure of the perovskite and doped ceria or titania was also suggested.[7] Cylindrical tubes of zirconia and short rings of interconnect were diffusion bonded to fabricate a series-connected stack of cells, as shown in figure 3.3. The hydrogen electrode is deposited on the inside and the air electrode on the outside using a spray process with appropriate masking to provide a series electrical connection (also shown in the figure).

At high current density operation, however, the interfaces between the interconnect ring and the adjacent electrolyte cylinders may act as an electrolytic cell causing oxygen migration from one interface to the other, resulting in evolution of molecular oxygen. This phenomenon may lead to disruption of the interfacial bond between the interconnect and the electrolyte. This necessitates an additional layer between the interconnect and the electrolyte that is electrically insulating. An annular insulating ring bonded to one or both sides of the interconnect ring has been suggested.[2] A

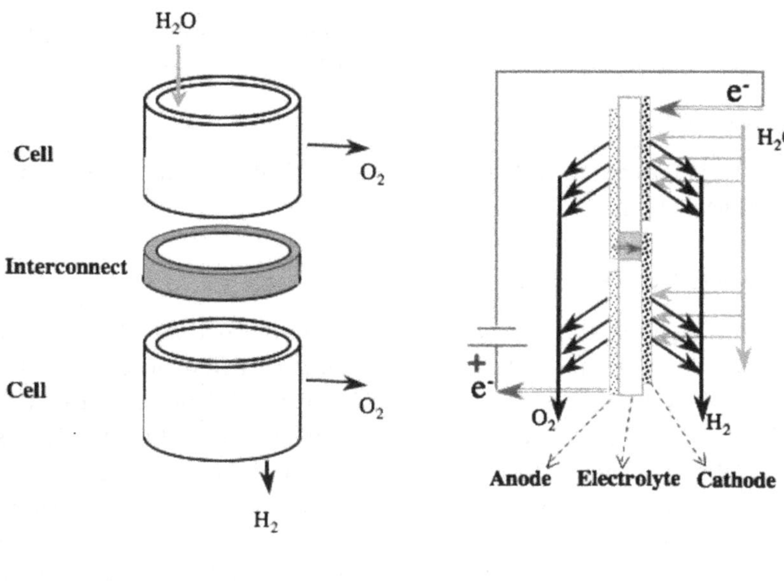

FIGURE 3.3 Series-connected tubular cell stack.

ceramic insulating material such as $La_{0.95}Mg_{0.05}Cr_{0.5}Al_{0.5}O_3$ is preferred, as it has a thermal expansion match with the electrolyte and interconnect ring.

Typical operating conditions of a 10-cell series-connected structure were 1,000°C at a stack voltage of 13.3 V and a current density of 0.37 A/cm^2. A hydrogen production rate of 113 Nml/min (6.78 Ndm^3/h) was obtained. A 10-cell module arrangement was sealed to a support box, and a stack containing 10 modules was tested to produce hydrogen at a maximum rate of 0.6 Nm^3/h.[8]

3.2.2 Tubular Stack Design

Westinghouse Electric Corporation has conducted steam electrolysis tests using tubular cells.[9] The cell materials consist of YSZ electrolyte, 10 mol% yttria, Sr-doped $LaMnO_3$ air electrode, and a nickel–zirconia cermet hydrogen electrode. The individual tubes are electrically connected using a Mg-doped $LaCrO_3$ interconnection layer. A combination of electrochemical vapor deposition and slurry coat processes was used to fabricate various layers of the cell. The primary difference in the material of construction is the interconnect material. Unlike the series-connected tube design, a strip of interconnect is used. The Mg-doped $LaCrO_3$ perovskite is known for its stability over the range of oxygen partial pressures the interconnect must face. Several dopants, such as Ca and Sr on the La site and Co on the Cr site of the perovskite, are known to show very high electrical conductivity. However, many of those compositions experience a phenomenon known as chemical dilation in reducing atmosphere, caused by the loss of oxygen with a concomitant loss in electrical conductivity, and in some cases, loss of strength[10] increased oxygen ion conduction to cause lattice expansion.[11,12] While Mg-doped $LaCrO_3$ exhibits a low electrical conductivity, it shows excellent stability and low loss of oxygen, minimizing ionic short circuit in the interconnect as well as change in lattice dimensions. Thus, in the tubular design where a small cross-section of the interconnect is used to connect the cells, a low-conductivity material is favored for its stability.

3.2.3 Planar Stack Design

During the 1990s significant research effort was made to develop planar cells for SOFC operation. Thus, much of the recent work on SOEC has focused on the development of planar cells. The advantage of the planar design stems from the fact that the current path of the device has a much larger area and shorter lengths favoring low electrical resistance. In addition, much thinner cells can be fabricated in comparison to a tubular design.

The Japan Atomic Energy Research Institute conducted a steam electrolysis study[13] using both tubular and planar cells. The tubular cells fabricated using a plasma spray technique achieved low efficiency, and air electrode delamination was observed after one thermal cycle. The planar cells, fabricated by Fuji Electric using porous metal as the support, produced hydrogen continuously at 950°C. But the faradaic efficiency was low and was attributed to nonuniform current density distribution and edge seal leak, highlighting one of the primary challenges in developing high temperature for planar devices.

Other groups have also reported testing planar cells in the HTE mode. Technology Management, Inc., has reported[14] testing stacks of up to 12 circular planar cells with a reversible cell efficiency, defined as the ratio of fuel cell to electrolysis voltage at the same cell current and hydrogen/steam feed composition of 90.8% at 925°C and 50 mA/cm^2. Risø National Laboratory of Denmark reported testing their planar cells in both SOFC and SOEC modes. While the area-specific resistance, which is the slope of the current density vs. cell voltage curve, did not change appreciably between the two modes, the degradation rate of the performance in the SOEC mode was found to be much higher than that in the SOFC mode.

Ceramatec, in partnership with the Idaho National Laboratory, has been evaluating cell and stack performance in the HTE mode of operation. Photographs of components and a manifolded 10-cell stack are shown in figure 3.4.

Scandia-stabilized zirconia electrolyte with standard SOFC electrodes is used to construct the cells. Stacks are constructed using stainless steel interconnects. The surfaces of the stainless steel are treated to provide an electrically conductive scale with low-scale growth rate.[15] The performance characteristics of a single cell (2.5 cm^2 active area) and a 25-cell stack (active area of 64 cm^2 per cell) are shown in figure 3.5 and figure 3.6. The performance stability of the stack is shown in figure 3.7.

One interesting aspect of comparing the performance of a single cell (no interconnects) and a stack of identical cell materials is the difference in performance. It is not uncommon to have a stack resistance 50 to 100% higher than the single cell resistance. Characterization of interconnect components shows very low resistance contributions, at least in the initial stages of testing, and does not account for the difference. As the reactant is utilized over the larger cell area, the local Nernst potential continues to decrease along the flow direction, and this will cause an increase in the apparent area-specific resistance. However, much of the contribution is expected to come from the joining of the cells and the interconnects. During the assembly of a stack, a conductive material is typically applied between the electrode and the interconnect. Commonly used materials include doped lanthanum cobaltite or lanthanum manganite on the air electrode and nickel–cermet on the hydrogen electrode. Sintering of these layers occurs over time, leading to reduced reactant access to the electrodes. Delamination and cracking of the joining layers may also contribute to high in-plane resistance. Investigation of appropriate materials composition and joining methods is also an area of considerable interest.

Another recent advance in cell fabrication technology for planar SOFC is transitioning into evaluation of the SOEC mode of operation. Thin-film YSZ supported on a hydrogen electrode, nickel–YSZ cermet, has shown good performance characteristics as an SOFC. Hydrogen electrode-supported thin YSZ cells have been successfully tested in the electrolysis mode. However, the long-term tests in the electrolysis mode have been reported to have a higher degradation rate than in the fuel cell mode.[16]

3.3 MODES OF OPERATION

Unlike the SOFC mode where the reaction is exothermic, the electrolysis mode of operation is endothermic. In both modes of operation heat is released from the ohmic

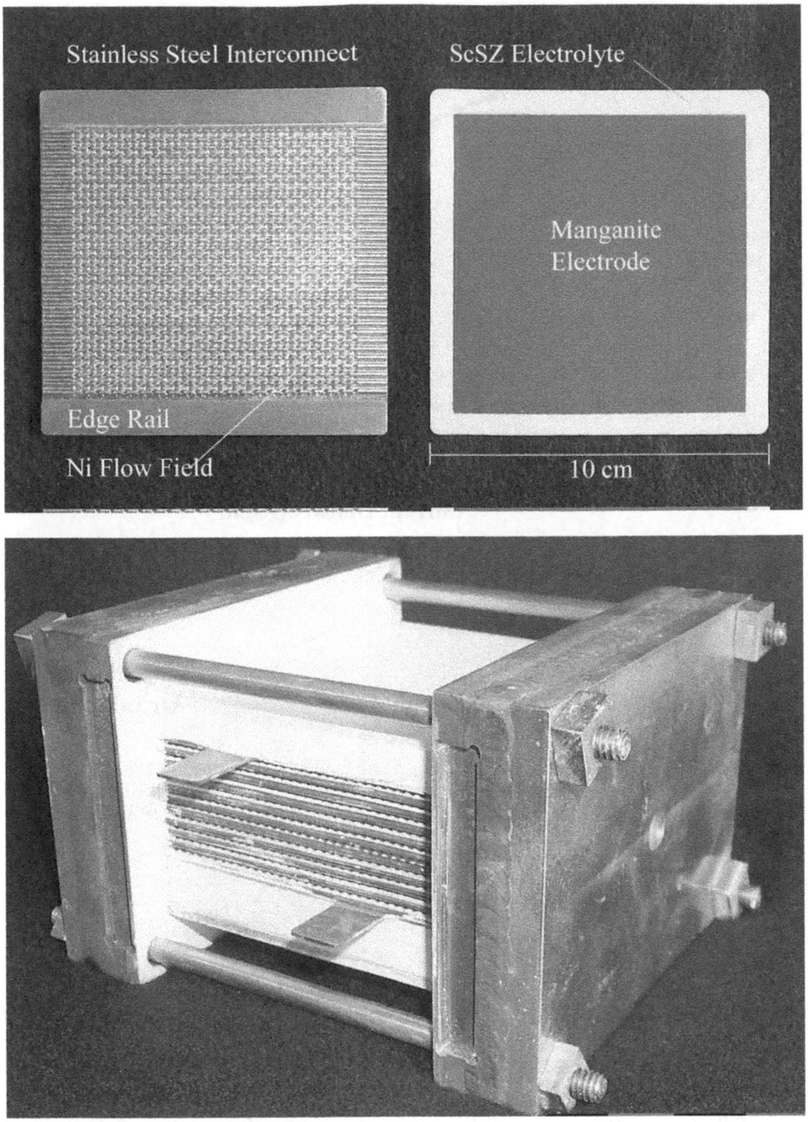

FIGURE 3.4 Stack components and a manifolded stack.

loss due to the resistance to current flow. In the SOFC mode, as the stack voltage is decreased, the current increases, causing the stack to release heat. In fact, heat removal is one of the challenging design and operational issues that limits materials selection, operating point (i.e., current density), and stack footprint. In contrast, the endothermic electrolysis reaction and the exotherm of ohmic loss move in opposite directions. At a certain cell operating voltage, the two balance, resulting in no net heat release. This voltage is referred to as the thermal neutral voltage, E_{tn}, defined as

$$E_{tn} = \frac{\Delta H}{nF}$$

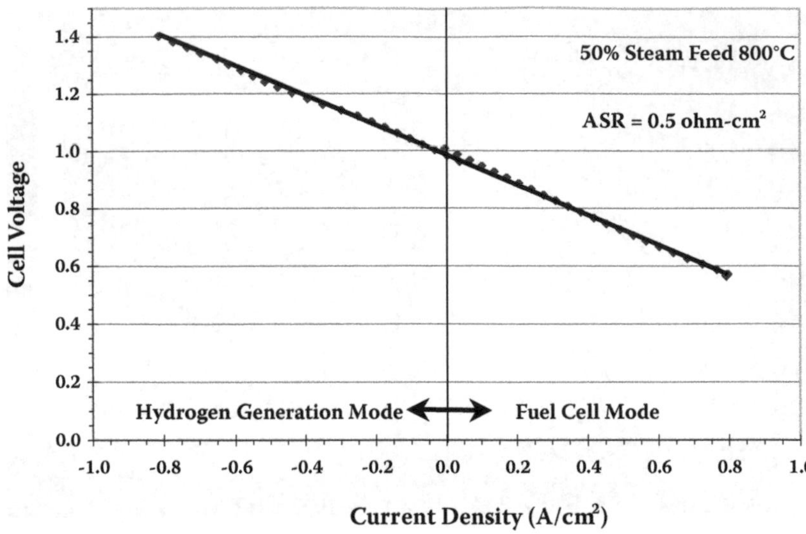

FIGURE 3.5 Single-cell performance curve using Sc-doped zirconia electrolyte.

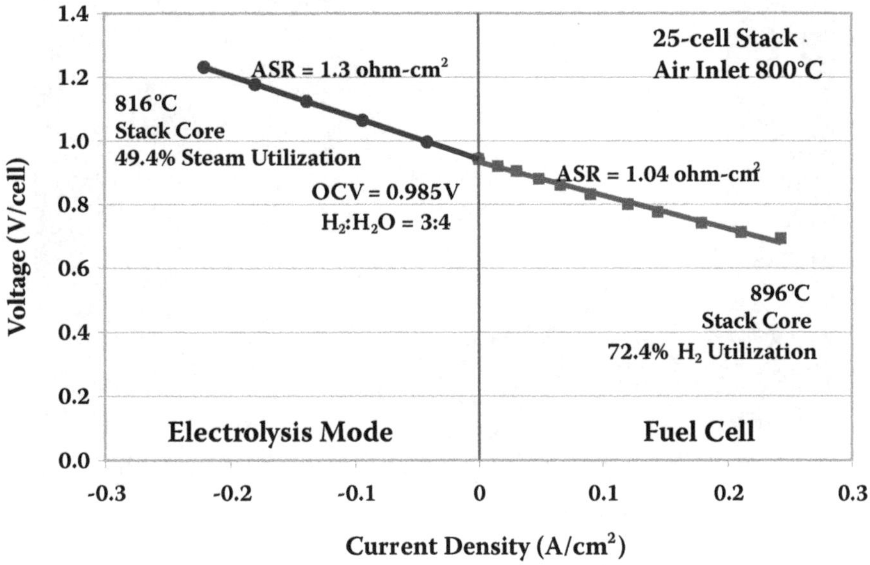

FIGURE 3.6 Stack performance using Sc-doped zirconia electrolyte and metal interconnects.

When an SOEC stack is operated at the thermal neutral voltage, the stack operation is isothermal, whereas it is exothermic above and endothermic below that voltage. In general, operating the stack near E_{tn}, which is approximately 1.3 V, has certain benefits, in particular the reduced need for cooling air for heat removal, or the need to supply the heat for the reaction. The stack components generally have

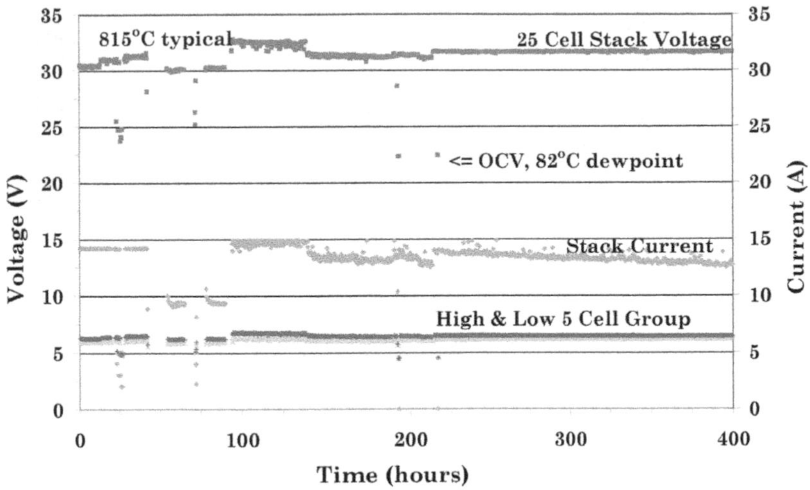

FIGURE 3.7 Performance stability of an electrolysis stack.

an upper limit to the operating temperature. In a nonisothermal condition, only a small region of the stack plane may be operating at the upper limit, while the rest of the area will operate at a lower temperature. This results in a significant reduction in average current density, and thus the hydrogen production. The temperature inhomogeneity, however, is not as severe in an SOEC stack as it is in an SOFC stack. Numerical modeling results of the cell temperature distributions at various operating potentials are shown in figure 3.8.[17]

3.4 ALTERNATIVE MATERIALS FOR HIGH-TEMPERATURE ELECTROLYSIS

As indicated earlier, the materials and design for electrolysis cells closely track the development of SOFC. In principle, a set of materials that function well in the fuel cell mode can also be used in the electrolysis mode. While some attention has been drawn to the irreversibility of the electrode to function well in both modes,[18] in general the high-temperature operation allows good reversibility. The activation polarization near the open-circuit voltage normally observed in low-temperature devices is not seen, and the slope of the current voltage trace does not change when transitioning from one mode to the other, as can be seen from the performance of the zirconia electrolyte stack shown in figure 3.6 (25-cell stack) near the point of current reversal. However, as the current density increases, the heat release in the fuel cell mode is much higher than in the electrolysis mode, as indicated by the model. For example, figure 3.6 shows a temperature increase of 96°C in the fuel cell mode at a current density of 0.25 A/cm^2, whereas only a 16°C temperature rise is seen in the electrolysis mode at the same current density and airflow rate. The increase in temperature in the fuel cell mode lowers the effective resistance of the stack. It also results in an operating temperature above the design's continuous operating limit.

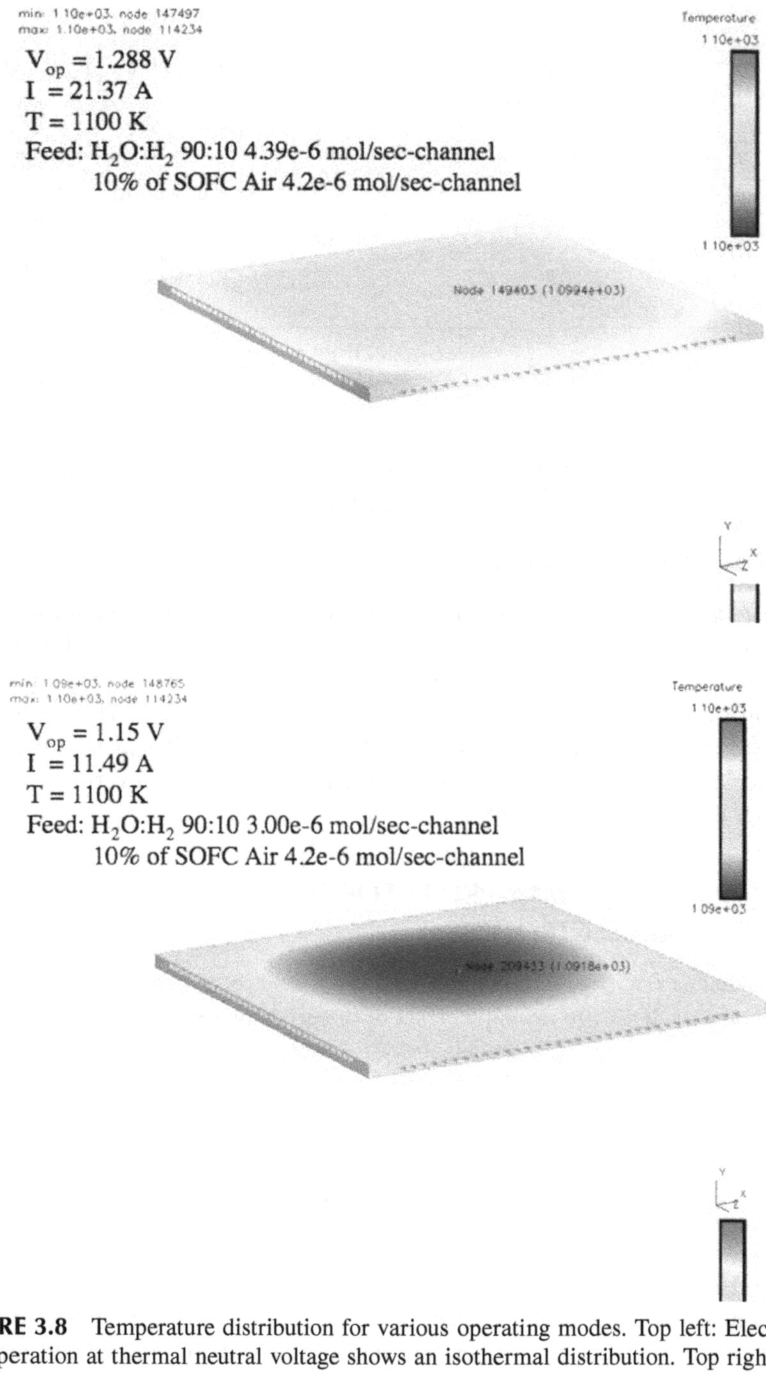

FIGURE 3.8 Temperature distribution for various operating modes. Top left: Electrolysis cell operation at thermal neutral voltage shows an isothermal distribution. Top right: Electrolysis cell operation above thermal neutral voltage shows an increase of 10°C. Bottom left: Electrolysis cell operation below thermal neutral voltage shows a decrease of 8°C. Bottom right: Fuel cell operation with high airflow shows an increase of ~40°C, even with 10 times the airflow rate of an electrolysis cell.

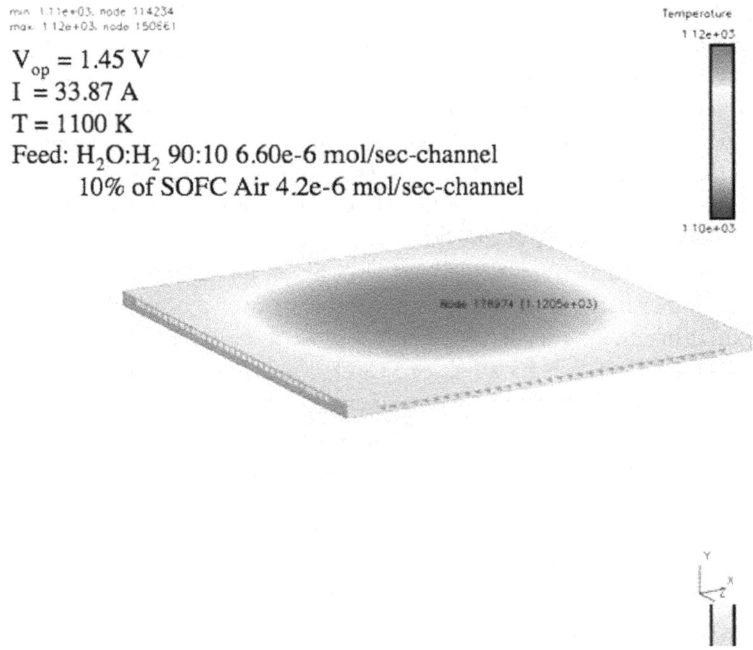
Vop = 1.45 V
I = 33.87 A
T = 1100 K
Feed: H2O:H2 90:10 6.60e-6 mol/sec-channel
10% of SOFC Air 4.2e-6 mol/sec-channel
Temperature

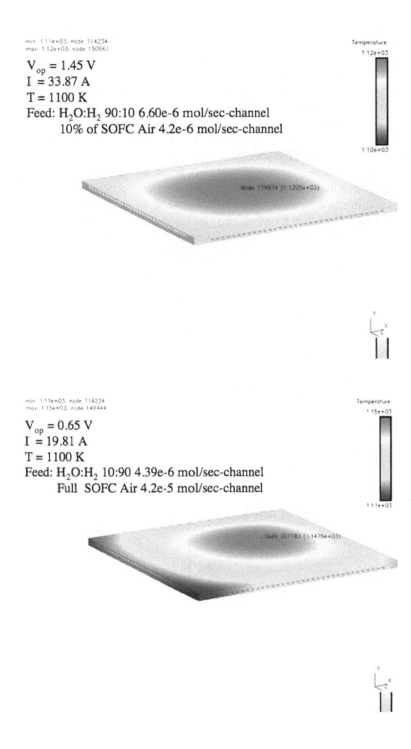

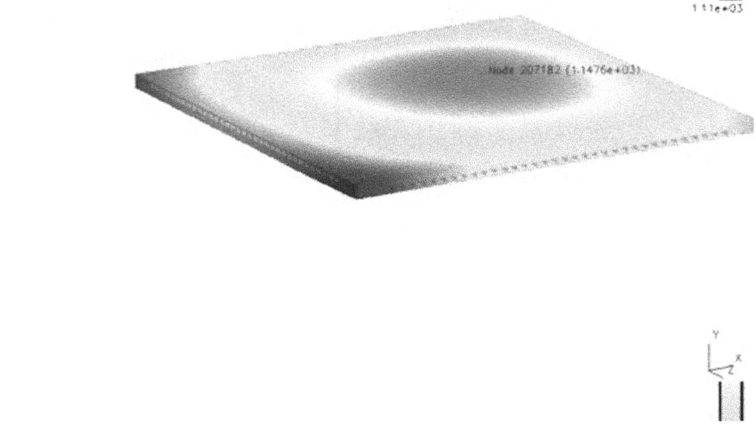
Vop = 0.65 V
I = 19.81 A
T = 1100 K
Feed: H2O:H2 10:90 4.39e-6 mol/sec-channel
Full SOFC Air 4.2e-5 mol/sec-channel
Temperature

Recent trends in lowering the operating temperature of SOFCs have resulted in evaluating a variety of new materials sets. They include doped cerium oxide and doped lanthanum gallate. Both materials are excellent oxygen ion conductors. The cerium oxide, however, undergoes a partial reduction under low oxygen partial pressures to result in a mixed ion–electron conductor. This results in internal shorting of the cell, leading to lowering of fuel cell efficiency.[19] A similar shorting problem will occur in the electrolysis mode. At temperatures around 500°C or lower, the mixed conduction of ceria becomes negligible. However, such low temperatures negate the benefit of high-temperature electrolysis.

Lanthanum gallate, typically doped with Sr on the La site and Mg on the Ga site, on the other hand, is an oxygen ion conductor over a broad range of temperatures, exhibiting an ion transference number close to unity. A single cell tested with $La_{0.8}Sr_{0.2}Ga_{0.8}Mg_{0.2}O_{3-\partial}$ using a $La_{0.8}Sr_{0.2}CoO_{3-\partial}$ air electrode and nickel–ceria cermet hydrogen electrode showed good reversibility and a very low area-specific resistance. Three different steam concentrations were used to identify the effect of steam starvation in the electrolysis mode. As can be seen in figure 3.9, the electrolyte–electrode materials set showed good reversibility across the open-circuit voltage. The nonlinearity in the performance curve is caused by steam starvation, as expected, at low steam concentrations of less than 50%.

An alternative electrolyte material, lanthanum-doped barium indium oxide, $(Ba,La)In_2O_{5+\partial}$, has been proposed.[20] This new material has been shown to have an ionic transference number of 1.0 with an oxygen ion conductivity better than that of YSZ.

Proton-conducting ceramic membranes have been studied as SOFC electrolytes for intermediate temperature, around 800°C, and can also be used as electrolytes in steam electrolysis. The families of $SrCeO_3$ and $BaCeO_3$ with dopants such as Y, Yb, and Nd on the Ce site show good selectivity for proton transport.[21–27] The advantage of using proton conductors for electrolysis is that pure hydrogen, without steam dilution,

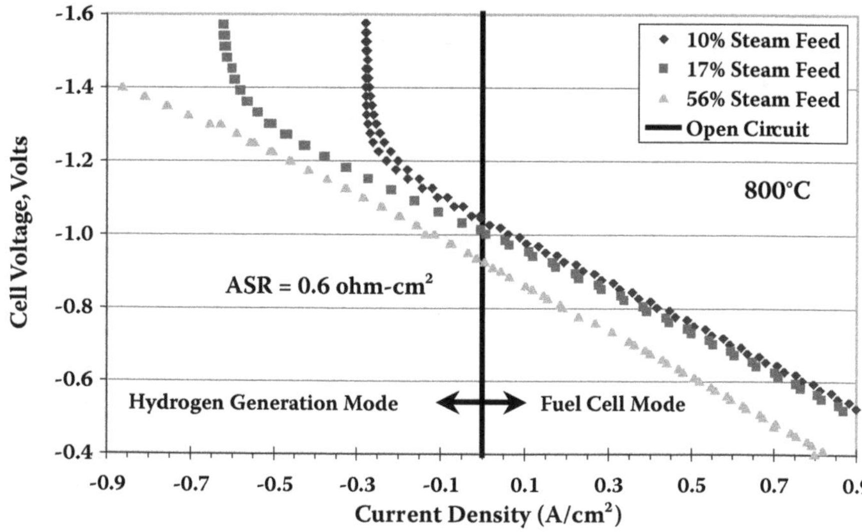

FIGURE 3.9 Performance of a gallate electrolyte cell.

can be obtained as shown by Iwahara et al.[28] However, the proton conductivity of these materials is considerably lower than the ionic conductivity of traditional oxygen ion electrolytes. At higher temperatures, while the proton conductivity values increase, the transference number for proton conduction decreases due to electronic conduction. The current efficiency is thus lowered at higher operating temperatures.[29]

3.5 ADVANCED CONCEPTS FOR HIGH-TEMPERATURE ELECTROLYSIS

3.5.1 Natural Gas-Assisted Mode

Higher operating temperature allows for a reduction in the electricity needed for electrolysis. However, materials constraints such as oxidation of metal interconnect or other metallic manifold components and continued sintering of porous electrodes may result in performance degradation at high temperatures. Pham et al.[30,31] have proposed a method for reducing the voltage necessary for steam electrolysis, thereby reducing the electric power consumption. The process, known as natural gas-assisted steam electrolysis (NGASE), uses natural gas as the anode reactant in place of commonly used air or steam as the sweep gas for removing the oxygen evolved in the anode compartment. Thus, the oxygen transported through the electrolyte membrane partially or fully oxidizes the natural gas, which in effect provides a significant portion of the driving force for the oxygen transport through the membrane.

The Nernst potential, E, of an electrochemical cell is defined as

$$E = \frac{RT}{nF} \ln \frac{p_{O_2}{}^{I}}{p_{O_2}{}^{II}}$$

where $p_{O_2}{}^{I}$ and $p_{O_2}{}^{II}$ are the oxygen partial pressures of the reactants in the two chambers separated by the oxygen ion-conducting membrane, F is the Faraday constant, n is the moles of electrons involved in the reaction, R is the universal gas constant, and T is the temperature in Kelvin. When the cathode gas is a mixture of hydrogen and steam with anode gas being air, the Nernst potential is about 0.8 to 0.9 V, depending on the ratio of hydrogen and steam. The steam electrolysis then requires a voltage that is higher than the open-circuit voltage (Nernst potential at no current). When air is replaced by natural gas methane on the anode side, the Nernst potential reduces by nearly 1 V. The voltage required to electrolyze is thus lowered by an equal amount. As the hydrogen production rate is proportional to the current, the lowering of the operating voltage results in reduced power consumption.

In the NGASE operation both the anode gas (methane–steam) and cathode gas (hydrogen–steam) are reducing (low p_{O_2}), and thus both electrode materials must be capable of low p_{O_2} stability. A Ni-based cermet electrode for both anode and cathode will be appropriate for this mode of operation. The authors have observed erosion of zirconia electrolyte at temperatures above 700°C under these conditions. This phenomenon may limit the usefulness of the NGASE process.

3.5.2 Hybrid SOFC–SEOC Stacks

As mentioned earlier, the SOFC mode of operation is exothermic while the SOEC mode can be endothermic, thermal neutral, or exothermic depending on the operating voltage. The hydrogen production efficiency, defined as the ratio of heating value of generated hydrogen to electric power input, is 100% at the thermal neutral voltage, higher in the endothermic mode as the operating voltage moves closer to the open-circuit voltage, and lower when the voltage is higher than thermal neutral. The efficiency can be as high as 140% near the open-circuit voltage. It should be noted that thermal inputs are required to satisfy conservation of energy when operated below the thermal neutral voltage. While high efficiency is attractive, it typically comes at high capital cost, as the production rate per unit cell area is low. Operating near the thermal neutral voltage is generally considered favorable from both the operational and hydrogen production cost[32] perspectives. As the SOEC can be operated with minimal requirement for heat supply or removal, it can potentially be scaled up to large-footprint devices, unlike SOFC, where the heat removal requirement constrains the overall footprint. Thus, in a reversible fuel cell, one that operates in SOFC and SOEC modes, the cell area is constrained by the cooling requirements in the SOFC mode.

In order to overcome the heat removal constraints, a hybrid stack concept has been proposed.[33] By integrating both SOFC and SOEC cells in a single stack, the exothermic SOFC and endothermic SOEC operations can be used to reduce the cooling air requirement, and thus allow for larger-footprint devices. A similar concept is under investigation by other researchers as well.[34]

3.5.3 Integration of Primary Energy Sources with High-Temperature Electrolysis Process

The attraction of the high-temperature steam electrolysis process comes from the fact that a portion of the required energy for the process is supplied as thermal energy, thereby reducing the electrical need. When operated at thermal neutral voltage, all input energy is in the form of electric power, but energy lost by resistance to heat is used to satisfy the endotherm. However, a judicial choice of the primary energy source must be made to take into account the cost, efficiency, and environmental impact of the overall process. The concept of using electricity to produce hydrogen, which in turn will be used to produce electricity, makes sense only if the electric power for electrolysis is inexpensive or from excess capacity, and thus the hydrogen becomes an energy carrier. Additionally, the compression of hydrogen for transport typically consumes 10% of the energy content. Considering the overall environmental effect, combining high-temperature electrolysis with a renewable energy source is a good option—in particular when the electricity generation is intermittent (for example, with windmill or solar generators) or the demand is low (as in the case where a nuclear generator paired with an electrolyzer fills the role of spinning reserve).

When the high-temperature electrolysis process for hydrogen generation is supported by nuclear process heat and electricity, it has the potential to produce hydrogen at a very high efficiency.[35] It is estimated that a high-temperature advanced nuclear reactor coupled with a high-temperature electrolyzer could achieve a thermal-to-hydrogen conversion efficiency of 45 to 55%.[36]

Alternatively, renewable sources such as wind, geothermal, and solar energy can also be used as inputs. It is estimated, for example, that only 1% of geothermal energy has been harnessed to produce electricity from geothermal steam.[37] It is further estimated that more than 17 TWh/y of hydrogen can be produced in Iceland alone. Similarly, solar cells can be integrated to provide the electricity for the electrolysis process. High cost due to relatively low efficiency of photovoltaic conversion of solar to electric energy has been the hindrance for such integration. A wavelength separator, which separates shorter and longer wavelengths of solar radiation and converts them into thermal and electrical energies, respectively, has been suggested.[38] A parabolic concentrator and a spectrally selective filter are used for the separation. A combined system efficiency of 22% has been estimated.[39] This could more than double with advanced multijunction cells now becoming available.

3.6 MATERIALS CHALLENGES

The SOFC materials can be in general applicable to SOEC stacks. The primary materials issues in an SOFC are related to high-temperature operation. At the operating temperature, both physical and chemical changes to the cell materials can lead to performance degradation. For example, the nickel in the fuel electrode can coarsen over time, causing the loss of interparticle connectivity. Both electrodes could also densify during operation, resulting in high gas diffusion resistance in the electrode, causing overall stack resistance to increase with time. Chemical reaction between layers, in particular the air electrode, typically a lanthanum manganite perovskite, and the zirconia electrolyte, could also result in insulating compounds such as lanthanum zirconate ($La_2Zr_2O_7$).

The most critical component is the interconnect that joins the individual cells to form a stack. While much of the criteria, such as thermal expansion match, electrical conductivity, and gas tightness, are identical to those of SOFC, there are distinct differences in the SOEC mode that must be taken into account in selecting the appropriate materials set. The cathode stream has a very high steam content, especially near the inlet. Typically only a small fraction of the inlet stream needs to be hydrogen to maintain a fully reduced metallic nickel electrode, and thus 90 to 95% of the inlet gas is steam. On the anode side, it is conceivable that steam could be used as the sweep gas. Typically, a stainless steel is selected as an interconnect alloy for thermal expansion match and low cost. In order to provide an oxide scale that is electrically conductive, Fe-Cr alloys are selected. When oxidized in air, a dense, continuous chromia scale forms on the surface. The chromia scale is conductive and to a certain degree reduces the oxidation rate of the alloy. However, exposure to high humidity is found to produce a mixture of chromia and iron oxide. Additionally, when the interconnect faces a dual-atmosphere condition, oxidizing gas on one side and reducing on the opposite side, even without moisture present on the oxidizing side, the oxide scale forms a similar nonprotective mixed oxide. It is suggested[40] that hydrogen may diffuse through the alloy to form water molecules on the oxidizing side underneath the scale, disrupting the scale formation. In contrast to the fuel cell mode, the electrolyzer evolves high-purity oxygen on the anode side, leading to a potentially severe environment for scale growth. High sweep gas flow could be used on the anode to reduce the effect of oxygen on the interconnect. However, process economics may dictate the use of steam to recover the value-added high-purity oxygen. Thus, the combination of steam and oxygen poses a severe corrosion condition. While it is still an ongoing area of research, some progress has been made by heat treating the metal surface to form a thin chromia scale prior to assembly. An additional thin layer of $LaCrO_3$ is applied on top. Figure 3.10 compares the scales formed during long-term testing in a dual atmosphere. Figure 3.10a shows the poorly adhered scale of an untreated stainless 400 series alloy after 200 h to contain a large nodule of an Fe-rich region surrounded by a Cr-rich region. Figure 3.10b shows the pretreated alloy after 1,000 h of test to contain about a 2-μm-thick Cr-rich scale. Both samples had a constant current of 0.2 A/cm^2.

The potential for efficiently producing ultra high purity hydrogen using the electrolysis process dictates that the gas streams on each side of the electrolyte are isolated. This is all the more critical when by-product oxygen is collected. The evolution oxygen on the anode side shows a detrimental effect on the seal area. While the same alkaline earth silicate glass appears to be unreactive in the SOFC mode, severe oxidation occurs in the edge seal area in the SOEC mode. The corrosion scale near the seal area could result in poor sealing over time, causing direct mixing of the reactants. The potential to electrically short adjacent interconnect plates is another serious problem. Application of an inert dielectric layer largely reduces the seal reactivity.

Evaporation and condensation of chromium vapor from the interconnect onto the air electrode are considered critical mechanisms of performance degradation in an SOFC. While no details of similar effect have been reported for the electrolysis stacks, such a mechanism is likely to be operative, leading to performance degradation. This is another area that requires attention to address and mitigate the problem in order to achieve long-term stable operation of SOEC stacks. Much of the work in this area is being conducted for SOFC applications.

3.7 SUMMARY

The primary attraction toward the use of a high-temperature ceramic membrane process for water electrolysis is the potential to produce ultra high purity hydrogen, with no need for a subsequent cleanup reaction, at a very high efficiency. The high-temperature process is also very well suited for integration with renewable energy sources. While much of the materials issues are common to solid-oxide fuel cell technology, which is receiving considerable research focus, there are some significant challenges that are unique to the electrolysis process. Evolution of high-purity oxygen poses a considerable challenge in the selection of anode and interconnect materials. The process also imposes much stricter requirements for high-temperature seal materials to derive the benefit of a high-purity hydrogen generation process. The recent thrust and advances in high-temperature materials and the focus on hydrogen technologies provide the basis for rapid progress in this area.

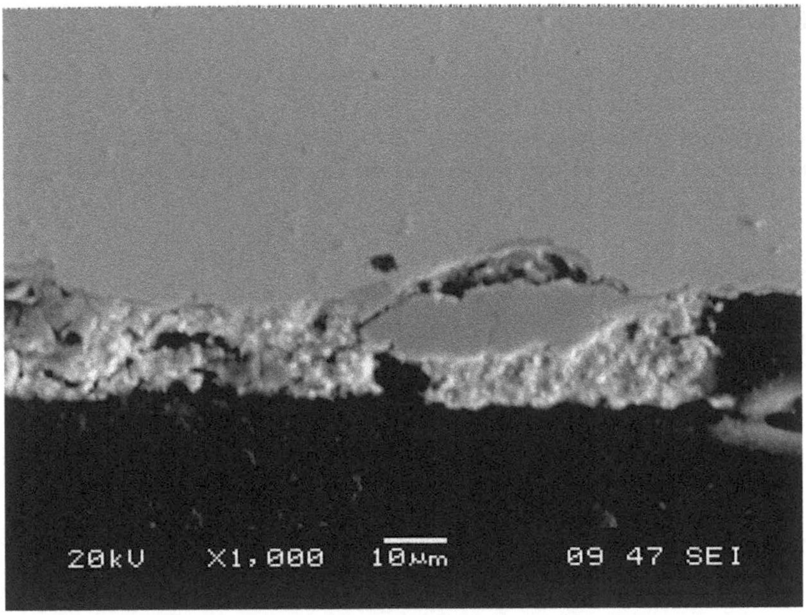

FIGURE 3.10A Corrosion of stainless steel in cell operating conditions. (a) Untreated, air side of dual atmosphere under constant current for ~200 h.

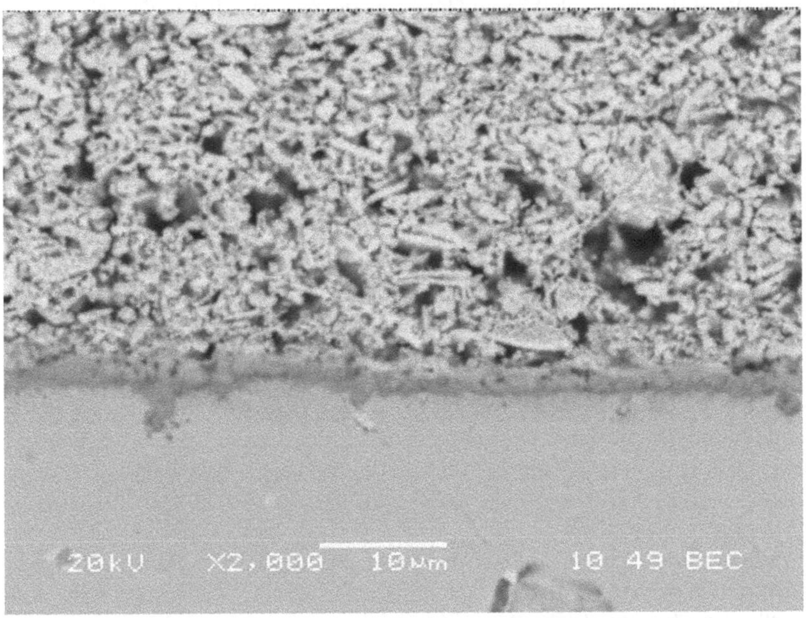

FIGURE 3.10B Corrosion of stainless steel in cell operating conditions. (b) Treated coupon air side of dual atmosphere under constant current for 1,000 h. The porous structure above the scale is a lanthanum cobaltite layer added for electrical connection.

ACKNOWLEDGMENTS

The electrolysis tests were performed under a subcontract from Idaho National Laboratory under the U.S. Department of Energy's Office of Nuclear Energy Science and Technology — Idaho Operations Office, Contract DE-AC07-05ID14517. The metal treatment work was performed under the Department of Energy's Cooperative Agreement DE-FC2602NT41569.

REFERENCES

1. Dönitz, W., Schmidberger, R., and Steinheil, E., Hydrogen production by high temperature electrolysis of water vapor, *Int. J. Hydrogen Energy*, 5, 55, 1980.
2. Schmidbergerm, R., U.S. Patent 4,174,260, November 1979.
3. Schaefer, W. and Schmidberger, R., U.S. Patent 4,789,561, December 1988.
4. Dönitz, W., Dietrich, G. Erdle, E., and Streicher, R., Electrochemical high temperature technology for hydrogen production or direct electricity generation, *Int. J. Hydrogen Energy*, 13, 283–287, 1988.
5. Dönitz, W. and Schmidberger, R., Concepts and design for scaling up high temperature water vapour electrolysis, *Int. J. Hydrogen Energy*, 7, 321–330, 1982.
6. Dönitz, W. and Erdle, E., High-temperature electrolysis of water vapor: status of development and perspectives for application, *Int. J. Hydrogen Energy*, 10, 291, 1985.
7. Schmidberger, R. and Donitz, W., U.S. Patent 4,197,362, April 1980.
8. Dönitz, W. Erdle, E., Schamm, R., and Treicher, R., Recent Advances in the Development of High Temperature Electrolysis Technology in Germany, paper presented at the 7th World Hydrogen Energy Conference, Moscow, September 1988.
9. Maskalick, N.J., High temperature electrolysis cell performance characterization, *Int. J. Hydrogen Energy*, 11, 563, 1986.
10. Milliken, C., Elangovan, S., and Khandkar, A., Mechanical and electrical stability of doped $LaCrO_3$ in SOFC applications, in *Proceedings of the Third International Symposium on Solid Oxide Fuel Cells*, S.C. Singhal and H.I. Iwahara, ed., 335, Pennington, NJ: The Electrochemical Society, 1993.
11. Yasuda, I. and Hishinuma, M., Lattice expansion of acceptor-doped lanthanum chromites under high-temperature reducing atmospheres, *Electrochemistry*, 68, 526, 2000.
12. Yakabe, H., Hishinuma, M., and Yasuda, I., Static and transient model analysis on expansion behavior of $LaCrO_3$ under an oxygen potential gradient, *J. Electrochem. Soc.*, 147, 4071, 2000.
13. Hino, R., Aita, H. Sekita, K., Haga, K., and Iwata, T., Study on Hydrogen Production by High Temperature Electrolysis of Steam, Report JAERI-Research 97-064, September 1997.
14. Milliken, C.E. and Ruhl, R.C., Low-Cost High-Efficiency, Reversible Fuel Cell System, paper presented at *Proceedings of the 2002 U.S. DOE Hydrogen Program Review*, NREL/CP-610-32405.
15. Elangovan, S., Balagopal, S., Timper, M., Bay, I., Larsen, D., and Hartvigsen, J., Evaluation of ferritic stainless steel for use as metal interconnects for solid oxide fuel cells, *J. Mater. Eng. Perform.*, 13, 265, 2004.
16. Jensen, S.H., Høgh, J., and Mogensen, M., High Temperature Electrolysis of Steam and Carbon Dioxide, In *Energy Technologies for Post Kyoto Targets In the Medium Term,* paper presented at Riso International Energy Conference Proceedings, Risø, (DK), May 19–21, 2003.

17. Hartvigsen, J., Elangovan, S., O'Brien, J., and Stoots, C., Operation and Analysis of SOFCs in Steam Electrolyis Mode, paper presented at the Sixth European SOFC Forum, Lucerne, Switzerland, June 2004.
18. Erdle, E., Dönitz, W., Scham, R., and Koch, A., Reversible and Polarization Behavior of High Temperature Solid Oxide Electrochemical Cells, paper presented at the Hydrogen Energy Conference Proceedings, Honolulu, HI, 1990.
19. Milliken, C., Elangovan, S., and Khandkar, A., Characterization and performance of ceria based SOFCs, in *Solid Oxide Fuel Cells IV*, M. Dokiya, O. Yamomoto, H. Tagawa, and S.C. Singhal, Eds., Pennington, NJ: Electrochemical Society, 1995, p. 1049.
20. Shioji, M., Research project for production—and utilization—technologies of hydrogen by the JSPS, in *Proceedings of the 11th Canadian Hydrogen Conference: Building the Hydrogen Economy*, G.F. McLean, Ed., Victoria University, British Columbia, Canada, June 2001.
21. Iwahara, H., Esaka, T., Uchida, H., and Maeda, N., Proton conduction in sintered oxides and it's application to steam electrolysis for hydrogen production, *Solid State Ionics*, 3/4, 359, 1981.
22. Iwahara, H., Uchida, H., and Maeda, N., High Temperature fuel and steam electrolysis cells using proton conductive solid electrolytes, *J. Power Sources*, 7, 293 - 301, 1982.
23. Iwahara, H., Uchida, H., and Yamasaki, I., High-temperature steam electrolysis using $SrCeO_3$-based proton conductive solid electrolyte, *Int. J. Hydrogen Energy*, 12, 73, 1987.
24. Iwahara, H., High temperature proton conducting oxides and their applications to solid electrolyte fuel cells and steam electrolyzer for hydrogen production, *Solid State Ionics*, 28/30, 573, 1988.
25. Bonanos, N., Knight, K.S., and Ellis, B., Perovskite solid electrolytes: Structure, transport properties and fuel cell applications, *Solid State Ionics*, 79, 61, 1995.
26. Stevenson, D.A., Jiang, N., Buchanan, R.M., and Henn, F.E.G., Characterization of Gd, Yb and Nd doped barium cerates as proton conductors, *Solid State Ionics*, 62, 279–285, 1993.
27. Guan, J., Dorris, S.E., Balachandran, U., and Liu, M., Transport properties of $BaCe_{0.95}Y_{0.05}O_{3-\alpha}$ mixed conductors for hydrogen separation, *Solid State Ionics*, 100, 45–52, 1997.
28. Iwahara, H., Asakura, Y., Katahira, K., and Tanaka, M., Prospects of hydrogen technology using proton-conducting ceramics, *Solid State Ionics*, 168, 299, 2004.
29. Iwahara, H., Uchida, H., and Yamasaki, I., High-temperature steam electrolysis using $SrCeO_3$-based proton conductive solid electrolyte, *Int. J. Hydrogen Energy*, 12, 73, 1987.
30. Pham, A.-Q., See, E., Lenz, D., Martin, P., and Glass, R., High efficiency steam electrolyzer, in *Proceedings of the 2002 U.S. DOE Hydrogen Program Review*, NREL/CP-610-32405, Golden, CO, 2002.
31. Martinez-Frias, J., Pham, A.-Q., and Aceves, S., A natural gas-assisted steam electrolyzer for high-efficiency production of hydrogen, *Int. J. Hydrogen Energy*, 28, 483–490, 2003.
32. Werkoff, F., Marechal, A., and Pra, F., Techno Economic Study on the Production of Hydrogen by High Temperature Steam Electrolysis, paper presented at the European Hydrogen Energy Conference, Grenoble, France, September 2003.
33. Co-Generation of High Purity Hydrogen and Electric Power, NASA SBIR contract NAS3-03025, http://sbir.gsfc.nasa.gov/SBIR/abstracts/02-2.html.
34. Tao, G., Armstrong, T., Virkar, A., Benson, G., and Anderson, H., A Reversible Planar Solid Oxide Fuel-Fed Electrolysis Cell and Solid Oxide Fuel Cell for Hydrogen and Electricity Production Operating on Natural Gas/Biogas, paper presented at the 2005 DOE Hydrogen Program Annual Review, Arlington, VA, May, 2005.

35. *Hydrogen as an Energy Carrier and Its Production by Nuclear Power*, IAEA-TEC-DOC-1085, International Atomic Energy Agency (IAEA), May 1999.
36. O'Brien, J.E., Stoots, C.. Herring, J.S., Lessing, P.A., Hartvigsen, J.J., and Elangovan, S., Performance Measurements of Solid-Oxide Electrolysis Cells for Hydrogen Production from Nuclear Energy, paper presented at the Proceedings of ICONE12, the 12th International Conference on Nuclear Engineering, Arlington, VA, April, 2004.
37. Sigurvinsson, J., Mansilla, C., Arnason, B., Bontemps, A. Maréchal, A., Sigfusson, T.I. , and Werkoff, F., Heat Transfer Problems for the Production of Hydrogen from Geothermal Energy, paper presented at Heat SET 2005, Heat Transfer in Components and Systems for Sustainable Energy Technologies, Grenoble, France, April 2005.
38. Lasich, J.B., Production of Hydrogen from Solar Radiation at High Efficiency, U.S. Patent 5,658,448, August 1997.

4 Materials Development for Sulfur–Iodine Thermochemical Hydrogen Production

Bunsen Wong and Paul Trester

CONTENTS

4.1 INTRODUCTION

The sulfur–iodine (S-I) cycle is a thermochemical water-splitting process that utilizes thermal energy from a high-temperature heat source to produce hydrogen (H_2). It is comprised of three coupled chemical reactions, as shown in figure 4.1. First, the central low-temperature Bunsen reaction (Section I) is employed to produce two

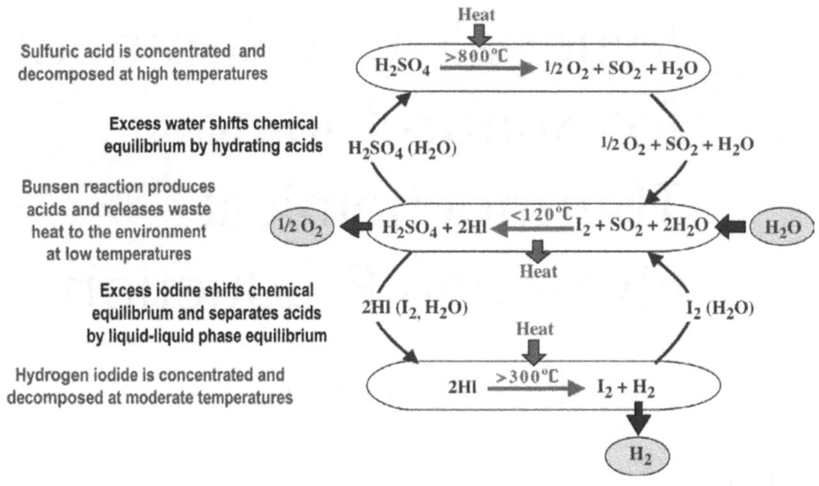

FIGURE 4.1 The coupled chemical reactions of the S-I cycle.

liquid phases from sulfur dioxide (SO_2), iodine (I_2), and water (H_2O). Under proper conditions, these two phases become immiscible and can be readily separated. The lighter upper phase is sulfuric acid (H_2SO_4), and the denser lower phase is an aqueous complex of HI, H_2O, and I_2 (HI_x). After separation, the two liquid phases are sent to the two other sections for decomposition. Section II (H_2SO_4 decomposition) first concentrates the sulfuric acid that has been received from Section I and then decomposes it into SO_2, O_2, and H_2O at high temperature. The decomposed products are returned to Section I to continue the S-I cycle. In Section III (HI decomposition) HI is distilled from HI_x and is then decomposed into H_2 and I_2 at intermediate temperature. H_2 is separated for external use and iodine is cycled back to Section I to support the Bunsen reaction. The key advantages of the S-I cycle are that it has no effluent and the reactants are in easily transportable liquid or gaseous form. All the chemicals used are recycled, and the only required process inputs are heat and water.

Heat sources that are capable of delivering the high temperature required by H_2SO_4 decomposition reaction include the modular helium reactor (MHR),[1] high-temperature solar tower, and coal- and natural gas–burning plant. Figure 4.2 shows a schematic of a conceptual S-I hydrogen–electricity co-generation plant design that is supported by the heat generated by a high-temperature nuclear reactor. The heat that is required to drive the two decomposition reactions and the electric turbine is delivered through an intermediate heat exchanger that employs helium gas as the heat transport medium. Other intermediate-loop mediums, such as molten fluoride salt, have also been proposed for this application.

The S-I cycle is capable of achieving an energy efficiency of 50%, making it one of the most efficient cycles among all water-splitting processes.[2] In addition, the S-I cycle is similar to other chemical production processes in that it is highly suitable to scaling up to large-scale production of H_2. Hence, it has good potential to deliver large quantities of low-cost hydrogen.

The baseline design for the current S-I work is the system configuration described in a report titled *High Efficiency Generation of Hydrogen Fuels Using Nuclear*

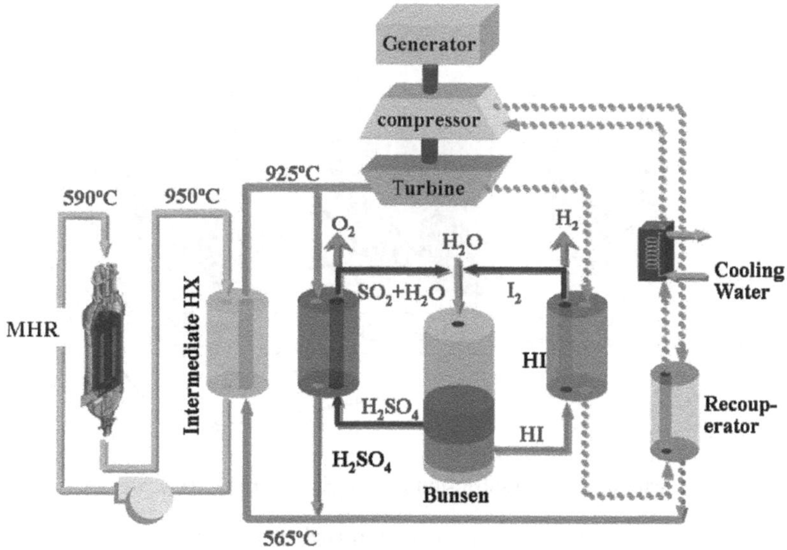

FIGURE 4.2 A schematic for a nuclear S-I hydrogen and electricity co-generation plant.

Power,[3] which describes the equipment and component requirements in detail. Due to the chemicals involved and the high reaction temperatures, the S-I cycle presents a very corrosive working environment. In order to realize a safe, stable, and functional hydrogen production plant, materials used to fabricate the boilers, heat exchangers, and other components within each section must be carefully selected. The specific requirements will be determined by the process flow sheet for the individual section and the associated processing steps.

4.2 S-I CYCLE DEMONSTRATION

The S-I cycle was invented at General Atomics in the mid-1970s and was studied extensively in the U.S. during the late 1970s and early 1980s.[4–6] The chemical reactions within the different sections of the cycle were successfully demonstrated using glass apparatus. In addition, a high-temperature metallic H_2SO_4 decomposition system was built and tested using the solar power tower at the Georgia Institute of Technology in 1984.[7] The decomposition was successfully demonstrated with the aid of a solar heat source. For the past 20 years, researchers globally, especially those in Japan, have continued research and development on the cycle. Since 1988, a complete laboratory-scale S-I test loop has been in operation in Japan by the Japan Atomic Energy Research Institute (JAERI). The system was constructed using glass equipment and is capable of delivering 30 l/h of hydrogen.[8,9]

To demonstrate the feasibility for large-scale hydrogen production, the DOE Nuclear Hydrogen Initiative (NHI) is funding the construction of a bench-scale S-I loop fabricated with proper materials of construction (figure 4.3). In addition to process demonstration, the materials and designs used in this integrated laboratory-scale (ILS) loop are chosen so that they will be applicable to future scale-up. This

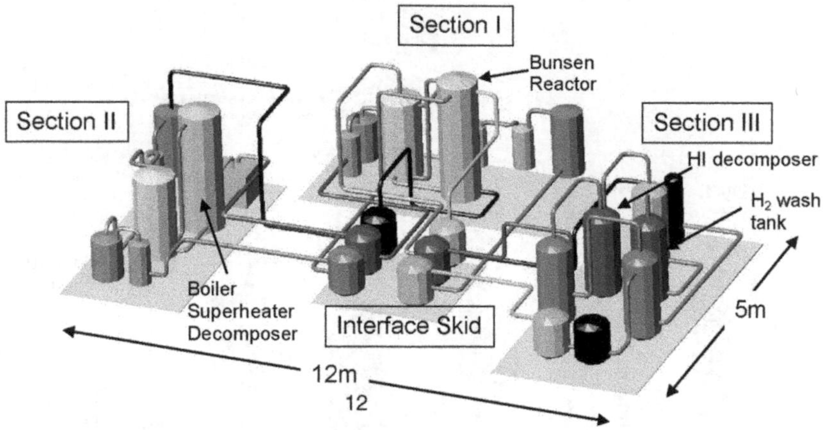

FIGURE 4.3 A schematic of the sulfur-iodine integrated laboratory scale demonstration.

S-I loop is being constructed jointly by Commissariat à l'Énergie Atomique (CEA; Section I), Sandia National Laboratory (SNL; Section II) and General Atomics (GA; Section III). The three sections will be integrated for a bench-scale demonstration capable of delivering 200 to 1,000 l/h of H_2 in 2008.[10]

4.3 S-I PROCESS FLOW SHEET

An extensive amount of flow sheet work has been performed to define the reaction conditions and process flow streams within the S-I cycle, which in turn determine the materials performance requirements. Advances in materials technology can redefine these flow sheets, as better-performing materials will allow the reactions to be conducted more efficiently and extend component lifetime. This can lead to an increase in the overall cycle efficiency and reduce the cost of H_2 produced. It is therefore essential for one to be familiar with these flow sheets and understand the material performance requirements in order to develop and identify better construction materials.

4.3.1 Section I: Bunsen Reaction

Section I receives the recycled SO_2 gas from Section II, and liquid I_2 from Section III. They are then placed into contact with excess H_2O to promote a spontaneous reaction (Bunsen reaction) at 120°C. By adding excess I_2, the reaction products will form two immiscible liquid phases that can be separated via gravity.[6,11] The upper lighter phase is H_2SO_4 and the lower denser phase is an acid complex of HI, H_2O, and I_2 (HI_x). This reaction is exothermic, and if the reaction heat is not channeled away efficiently from the reactor, sulfur will form as a result of side reactions. Hence, the materials used to construct the reactor will need to be an efficient heat exchanger in addition to being corrosion resistant to the mixture of chemicals and gases that are present. There are four basic processing steps in this section (figure 4.4), with the Bunsen reaction being the key reaction. The associated process conditions are listed in table 4.1.

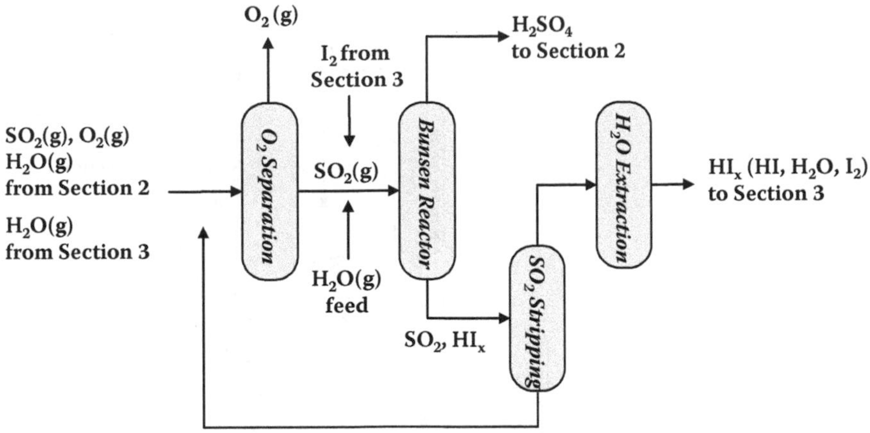

FIGURE 4.4 Flowsheet for Section I.

TABLE 4.1
Reaction Conditions for the Processing Steps in Section I

Reaction	Temperature	Pressure	Comments
O_2 separation	10°C	12 bars	Condensation and wash of incoming stream O_2 separation membrane
Bunsen reactor	120°C	10 bars	HIx–13 HI–9.5 H_2O–77.5 I2 (wt%) H2SO4 (57 wt%) Phase separation by density H_X (heat out) required
SO_2 stripping	120°C	1 bar	H_2O wash SO_2 separation membrane
H_2O extraction	120°C	1 bar	Separation membrane Electro-electrodialysis

Note: H_X = heat exchanger.

Other than the main Bunsen reaction, the other three steps in this section involve separation of gases and H_2O from the flowing chemical stream. O_2, which forms as a result of water splitting, is removed from the recycled stream to avoid formation of complexes in the rest of the section. SO_2 is washed/separated from the Bunsen reactor output to prevent any side reaction downstream. The last step in Section I involves the extraction of H_2O from the HI_x product stream before it is sent to Section III. This will reduce the energy requirement, as H_2O evaporation processes impose a high heat demand based on the current flow sheet. Hence, any reduction in H_2O content in the Bunsen reaction products will be beneficial to the overall cycle efficiency.

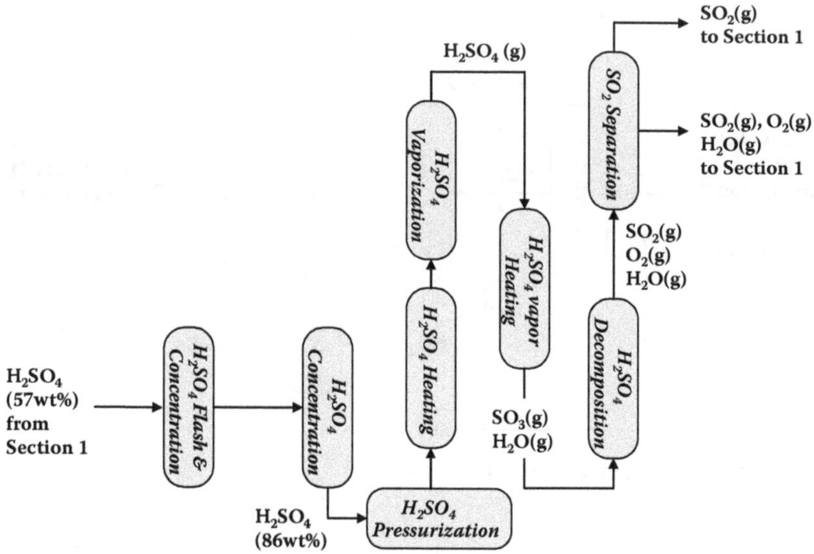

FIGURE 4.5 Flowsheet for Section II.

4.3.2 Section II: Sulfuric Acid Decomposition

This section receives H_2SO_4 from Section I and decomposes it into SO_2, O_2, and H_2O at high temperature. These gases are then returned to Section I (figure 4.5) and SO_2 is used to support the Bunsen reaction. The decomposition is carried out in a number of steps under the conditions listed in table 4.2. First, the incoming acid is concentrated from 57 to 86 wt%. This is followed by pressurizing the acid stream to 70 bars to match the pressure of the He gas in the intermediate heat exchange loop. This will minimize the stress on the heat exchange interface and reduces creep effects at the high SO_3 decomposition temperature. The acid is heated up to around 475°C for vaporization. Starting at about 500°C, H_2SO_4 gas will begin to decompose into H_2O and SO_3. Further decomposition of SO_3 into SO_2 and O_2 is accomplished by a catalytic reaction at a temperature around 850°C. Some flow sheet design has even proposed conducting the decomposition at 925°C to attain higher equilibrium conversion. This makes efficient heat conductivity an important material property.

Since $SO_3 \rightarrow SO_2 + O_2$ is an equilibrium reaction, one can enhance the reaction rate by removing SO_2 from the reactor. The only practical means to accomplish this at the current decomposition temperature will be through the use of a SO_2-selective permeable membrane. The lifetime permeability and structural integrity of such a membrane at these operating temperatures are important issues that will need to be addressed.

TABLE 4.2
Reaction Conditions for the Processing Steps in Section II

Reactor	Temperature	Pressure	Comments
H_2SO_4 flash and concentration	120°C	1.85s bar → 0.1 bar	H_2SO_4 (57 wt%)
H_2SO_4 concentration	120°C → 170°C	0.1 bar	H_2SO_4 (86 wt%) H_X (heat in) required
H_2SO_4 pressurization	170°C → 175°C	0.1 bar → 70 bars	
H_2SO_4 heating	175°C → 475°C	70 bars	H_X (heat in) required
H_2SO_4 vaporization	475°C → 500°C	70 bars	H_2SO_4 (86 wt%) H_X (heat in) required
H_2SO_4 heating	500°C → 750°C	70 bars	SO_3 formation H_X (heat in) required
H_2SO_4 decomposition	750°C → 900°C	70 bars	Catalytic reactor H_X (heat in) required
SO_2 separation	750°C → 900°C	70 bars	SO_2 membrane separator

Note: H_X = heat exchanger.

4.3.3 Section III: HI Decomposition

This section receives HI_x from Section I for decomposition and produces H_2. Hydrogen iodide (HI) within the HI_x feed stream is distilled and decomposed into H_2 and I_2, and there are two alternatives to carry out this process: *extractive distillation* or *reactive distillation.* The reaction conditions and chemicals used in these two processes are different, and their flow sheet will be addressed separately.

4.3.3.1 Extractive Distillation

Extractive distillation utilizes concentrated phosphoric acid (H_3PO_4) to extract HI and H_2O from HI_x, as they, unlike I_2, are soluble in H_3PO_4. In addition, H_3PO_4 breaks the azeotrope between HI and H_2O, thus permitting the distillation of HI from the acid complex followed by decomposition.[4,11] A schematic of the process flow sheet and the corresponding reaction conditions are shown in figure 4.6 and table 4.3. There are four separate steps: (1) *iodine separation*, (2) *HI distillation*, (3) *phosphoric acid concentration*, and (4) *gaseous HI decomposition.* In the iodine separation step, concentrated phosphoric acid (>96 wt%) is added to the HI_x feed from Section I and results in the formation of a two-phase liquid mixture of I_2 and HI + H_2O + H_3PO_4. The denser I_2 stream is separated by gravity and returned to Section I to support the Bunsen reaction. The upper lighter HI–H_3PO_4 acid complex is sent to the distillation column to carry out HI distillation. HI gas is distilled from boiling HI + H_2O + H_3PO_4 and is passed on to the decomposition column for HI decomposition [$HI(g) \rightarrow H_2(g) + I_2(g)$] in the presence of a catalyst. The decomposition can be carried out catalytically at either high temperature (~350 to 450°C) in the gas phase or low temperature (150 to 300°C) in the liquid phase under pressure. At 450°C, the equilibrium conversion rate of HI into $H_2 + I_2$ is approximately 22%. Hence, removal of H_2 from the decomposition chamber can enhance the one-pass conversion rate

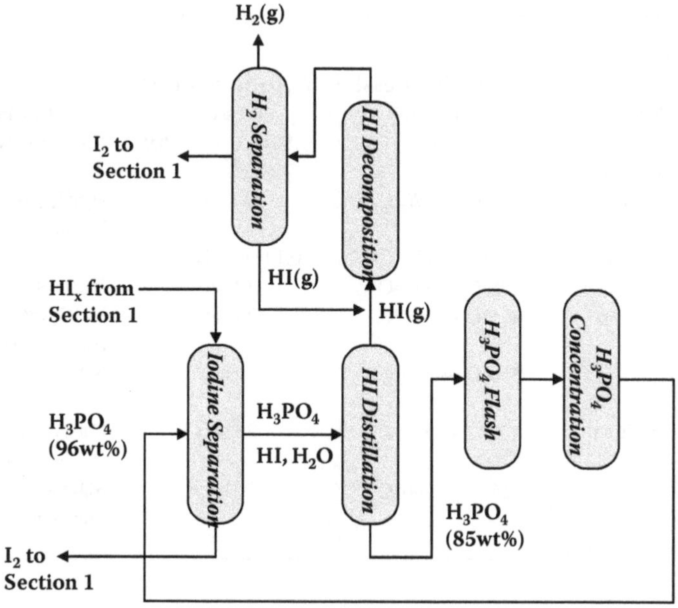

FIGURE 4.6 Flowsheet for Section III.

TABLE 4.3
Reaction Conditions for the Processing Steps in Section III Using Extractive Distillation

Reactor	Temperature	Pressure	Comments
I_2 separation	120°C	1 bar	HIx–13 HI–9.5 H_2O–77.5 I2 (wt%) H_3PO_4 (96 wt%)
HI distillation	120°C → 170°C	1 bar	H_X (heat in) required
HI decomposition	300°C → 450°C	1 bar	HI gas catalytic decomposition H_X (heat in) required
H_2 separation	<300°C	1 bar	H_2O wash to separate trace HI H_2 membrane separator
H_3PO_4 flash	160°C	1 → 0.1 bar	H_3PO_4 acid (87 wt%)
H_3PO_4 concentration	165°C → 220°C	0.1 bar	87 → 96 H_3PO_4 acid (wt%) H_X (heat in) required

Note: H_X = heat exchanger.

and reduce the amount of HI that needs to be recycled through the decomposer. The phosphoric acid ($H_3PO_4 + H_2O$) that remains in the distillation column is concentrated from 87 to 96 wt% through a series of boilers via vacuum recompression (H_3PO_4 concentration). The concentrated acid is added to the incoming HI_x feed to anew the extraction process.

4.3.3.2 Reactive Distillation

Reactive distillation is in theory a simpler process than extractive distillation, but it has yet to be demonstrated experimentally.[12] There are two key differences between reactive and extractive distillation. First, unlike the extractive process, the HI_x azeotrope* is not broken, so the composition in both the liquid and vapor phases is the same. Second, the reactive process must be conducted under pressure. Figure 4.7 shows a schematic of the reactive distillation flow sheet, and the processing conditions are listed in table 4.4. In this process, azeotropic HI_x is distilled inside a pressurized reactive column and the HI gas within the HI_x vapor stream is decomposed catalytically, resulting in a gas mixture of HI, I_2, H_2, and H_2O. To accomplish this, the HI_x feed from Section I is first heated to 262°C from 120°C and is then fed into the reactive column. At the bottom of the column, the HI_x is brought to a boil at around 310°C, and this boiling HI_x vapor results in an equilibrium vapor pressure of 750 psi inside the distillation column.

The distilled HI_x (HI, I_2, and H_2O) vapor flows through a bed of catalysts at the top half of the reactive column, and HI within the vapor stream is decomposed into H_2 and I_2 gases at around 300°C. A condenser at the top of the reactive column condenses any unreacted HI, I_2, and H_2O and the liquid is reflux back down the column.

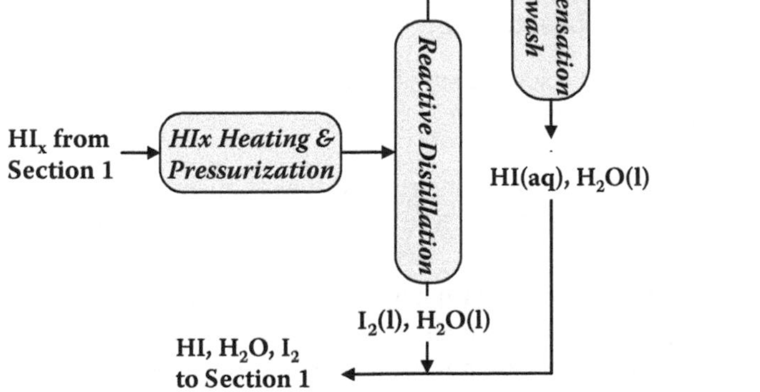

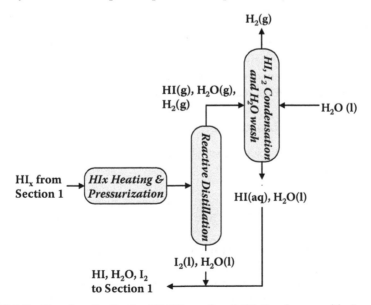

FIGURE 4.7 Flowsheet for Section III (HI reactive distillation decomposition).

* HI and H_2O form an azeotrope at 57 wt% HI.

TABLE 4.4
Reaction Conditions for the Main Reactions in Section III Using Reactive Distillation

Reactor	Temperature	Pressure	Comments
HIx heating	120°C → 260°C	1 bar → 50 bars	HIx–13 HI–9.5 H_2O–77.5 I2 (wt%) H_X (heat in) required
Reactive distillation	310°C	50 bars	HIx gas phase catalytic decomposition H_X (heat in) required Column bottom: 1.1 HI–0.7 H_2O–98.2 I2 (wt%)
HI, I_2 condensation	150°C	50 bars	

*HI and H_2O form an azeotrope at 57 wt% HI.

Consequently, the HI_x at the bottom of the reactive column is I_2 rich and is fed back to Section I as I_2 supply for the Bunsen reaction. H_2 is bled off the column top as a compressed gas for storage or use. The environment of HI_x and high temperature and pressure required in the heat exchanger in the reactive distillation process make it potentially one of the most corrosive environments within the S-I cycle.

4.4 MATERIALS DEVELOPMENT FOR THE S-I CYCLE

There are three ongoing material research areas that are important for the eventual success of the S-I cycle. First and foremost is the development of construction materials that can handle the corrosive environment for the lifetime of the process equipment, especially heat exchangers, boilers, and reactors. The other two areas involve gas-permeable membranes and catalyst development. Identifying suitable membranes and reaction catalysts can improve the efficiency of the overall cycle and make H_2 production more economical. Effort is currently ongoing in all three disciplines to find the optimal material solutions.

4.4.1 Materials of Construction

The general guidelines for selecting suitable construction materials for the S-I cycle can be summarized as follows:

- Materials must be resistant to the corrosive working environment.
- Materials used to build the heat exchanger must have good thermal conductivity.
- Components manufactured from qualified materials must have suitable mechanical and creep properties, especially those operating at high temperatures.
- Metallic alloys materials must possess good hot and cold formability, weldability, and availability to make building of a hydrogen production plant practical.
- Materials components design should allow for nondestructive testing while they are fabricated and when they are in service.

Using the above guidelines and the process conditions outlined in table 4.1 to table 4.4, the following sections will review work pertaining to construction materials development for the S-I cycle.

4.4.1.1 Materials of Construction for Section I

The maximum temperature for the two corrosive acids in this section, HI_x and H_2SO_4, is 120°C. These two acids are also present in the other two sections but at higher temperatures. Liquid HI_x is used in Section III at temperatures up to 310°C, and H_2SO_4 acid can be heated to 300°C and above during the concentration process in Section II. Based on the conventional assumption that corrosion can be exponentially accelerated by an increase in temperature, construction materials developed for use in the other sections will be applicable to the lower-temperature environment in Section I, or they can be the baseline for future materials development. Hence, material development for Section I is limited at the present time.

To identify suitable materials for an acid complex such as HI_x, one can begin by surveying materials applicable to the individual acid/chemical. Table 4.5 to table 4.7 list the corrosion properties of various materials in I_2, HI acid, and H_2SO_4. I_2 is a strong oxidizer, especially in liquid form at high temperature. The corrosion rates of a number of corrosion-resistant materials in I_2 at 300 and 450°C are listed in table 4.5.[13] Even though the data show that gold and platinum are stable in an I_2 environment, they have been found to dissolve in HI_x.[14] Refractory metals such as Ta and Nb alloys are probably the best candidates within the I_2-rich environment in Section I.

HI is a strong reducing acid with a negative pH. Even though it is a common reagent in organic chemistry, corrosion data of materials in HI acid at elevated temperatures are limited. Table 4.6 shows a summary of the available data.[15] Noble and refractory metals have shown low corrosion rates, but the temperatures at which the data were taken are lower than those in the Bunsen reaction environment. The corrosion mechanism of H_2SO_4 depends on temperature and concentration. Within the

TABLE 4.5
Corrosion Rate of Alloy in Pure Iodine at High Temperature

Alloy	Corrosion Rate(mm/yr) 300°C	450°C
Platinum	0.00	0.22
Tungsten	0.00	0.32
Gold	0.00	0.95
Molybdenum	0.12	1.30
Tantalum	0.16	34.66
Alloy B	2.24	18.12
Alloy 600	4.21	21.27

TABLE 4.6
Corrosion Rate of Alloy in HI Acid

Alloy	Temperature(°C)	Corrosion Rate(mm/yr)	Comments
Titanium	23	0.16	Concentration 57%
Niobium	<100	0.00	For all concentrations
Gold	25	<0.04	Dilute
Palladium	25	65.73	
Zr702	127	<0.04	Concentration 57 wt%

Bunsen reactor environment, it is a reducing acid. Materials capable of containing H_2SO_4 will be considered in more detail in the next section.

Based on the available data, construction materials for the Bunsen section will need to withstand a combination of reducing and oxidizing chemicals. Ta and Nb refractory alloys can be suitable construction materials, as they are known for their performance under both reducing and oxidizing conditions, but comprehensive test data of these metals in an S-I environment are still lacking. Although the lower reaction temperatures of the Bunsen reaction will make corrosion less severe than in the other two sections, hydrogen embrittlement can become an issue at this temperature range.

In addition to metallic materials, ceramics such as SiC, Si_3N_4, Al_2O_3, and mullite are also materials that will most likely perform well under the harsh S-I environment, and their applicability should be explored. Since the temperature does not exceed 120°C in Section I, fluoro-polymer coatings such as Teflon or glass-lined steel can also be viable options. The choices will depend more on the application.

The only corrosion experiment specifically designed for Section I was conducted by Trester and Staley.[14] In it, they inserted welded rods of Zr702 and Nb–1 Zr into a glass liquid phase separator within the Bunsen section. Borosilicate glass is resistant to these acids at this temperature. The immiscible gap at the HI_x and H_2SO_4 interface moved along the surface of the metal rods as the reaction progressed. Rapid localized corrosion was observed in Zr702, whereas Nb–1 Zr remained stable after a 4-h exposure at 105°C. Trester and Staley[14] have also conducted an extensive corrosion study of materials in various composition of HI_x, which will be discussed in Section 4.4.1.3.1. More recent results from internal testing at General Atomics also showed that Zr705 corrodes rapidly in the HI_x–H_2SO_4 environment. Figure 4.8 shows a

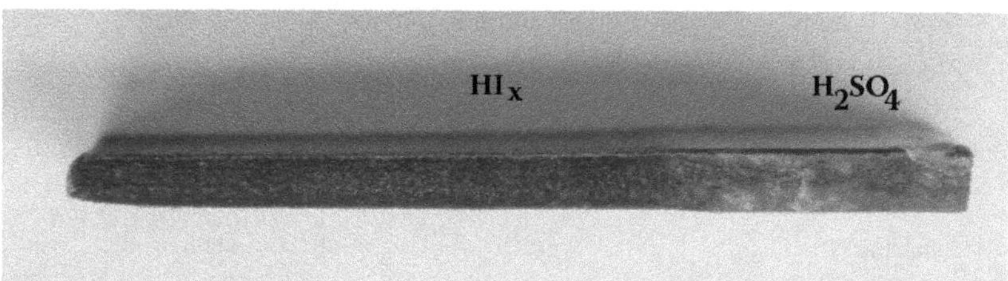

FIGURE 4.8 Side view of a Zr7O5 coupon tested in the static Bunsen separator acid for 450 h.

Zr705 coupon that has been immersed in a HI_x–H_2SO_4 static acid mixture. The test was conducted at 120°C for 450 h. Rapid corrosion of Zr705 in the area where it is in contact with HI_x has been observed.

Onuki et al.[16] have screened a number of materials in an acid mixture of H_2SO_4 (50 wt%) and HI (0.1 wt%) at temperatures up to 120°C (table 4.7). This simulates the composition of the upper liquid phase. They found Ta, Zr, Pb, and quartz glass to be corrosion resistant in this acid complex, whereas common construction material such as stainless steel and Hastelloy did not possess acceptable corrosion rates. PFA (Teflon) also showed satisfactory corrosion performance, but I_2 absorption by perfluoroalkoxy or PFA has been observed, which raises questions about its long-term viability.

From the currently available data, Ta, Nb, and their alloys, glass-lined steel, and probably SiC-based materials are good candidates. They will need to undergo more extensive testing in the Bunsen reaction environment. Ta and SiC are good candidates for the reactor because of their good thermal conductivity, but the final thermal properties requirements will depend on the heat exchanger design. The Bunsen section components for the bench-scale S-I cycle demonstration experiment will be manufactured from glass-lined steel since glass has been demonstrated to be inert to all the chemicals present. The drawback is the uncertainty of the heat-exchanging capability of such a construction to channel the heat away from the reactor.

4.4.1.2 Materials of Construction for Section II

The sulfuric acid decomposition reaction is divided into three different steps: *concentration*, *vaporization*, and *decomposition*. Taking this into consideration, construction material requirements can be divided into two different categories. The first group consists of materials that can withstand the extreme corrosiveness of H_2SO_4 acid up to 450°C. The second group includes materials that have suitable high-temperature (850°C) mechanical properties and good thermal conductivity and are also resistant to a gaseous oxidation and sulfidation environment.

TABLE 4.7
Corrosion Test Results in (50 H_2SO_4 + 0.1 HI wt%) Solution at 120°C

	100-h Test		1000-h Test	
	I	II	I	II
Quartz glass	NC	0.20	NC	0.00
Silicon carbide	NC	2.95	NC	1.97
Silicon nitride	NC	3.35	NC	0.39
PFA		0.00	NC	0.00
PPS	NC	+	NC	+
Ta	NC	0.20	NC	0.00
Zr	NC	0.98	NC	0.00
Pb	NC	1.77	R	1.97
Fe–15 Si	R	1.77		
Hastelloy C-276	SD	850.70		

Note:
I: Surface appearance. NC = no change; SD = severely damaged; R = roughened; BR = brown colored.
II: Corrosion rate in mm/yr. + = weight gain.

In general, a liquid environment is much more corrosive than a gaseous environment due to a few factors. Ion transport, which is the driving force of all corrosion reactions, is much higher in the liquid phase, thus enhancing the corrosion redox reaction. In a gaseous environment, the corrosion product will most likely stay on the reaction surface and may act as a barrier to decrease any further reaction. On the other hand, flowing liquid chemicals can dissolve or remove such corrosion product and lead to accelerated corrosion.

The step of H_2SO_4 concentration presents a very challenging corrosion problem. This is because the corrosion mechanism of liquid H_2SO_4 is linked to the temperature and the concentration of the acid. H_2SO_4 is reducing in nature when the acid concentration is below 85 wt% at room temperature or less than 65 wt% at higher temperatures. At high concentration and high temperature, it becomes an oxidizing agent. Hence, construction materials for the acid concentration application will face both the oxidizing and reducing conditions. Table 4.8 shows a summary of the materials that exhibit corrosion to the higher concentrations of H_2SO_4.[13] Ta and Au have low corrosion rates in higher concentration H_2SO_4 liquid phase environments. Precious metals such as Au and Pt have shown complete resistance at all concentrations of H_2SO_4. On the other hand, their inherent high cost and relatively low yield strength make them only viable as corrosion resistance coatings or cladding for components. Techniques to apply precious metals coating to a base structure are readily available, but care must be taken to deal with inherent defects such as pinholes that can compromise the integrity of the component.

Effort to develop materials for the H_2SO_4 concentration process has emphasized developing noble metals and engineering alloys and SiC-based ceramics for H_2SO_4 concentration application.[14,17–21] For low-temperature applications within the Section II process, Hastelloy B, Incoloy 825, Alloy 20, and glass-lined, Teflon-lined, and coated steel can be used, as their behavior in H_2SO_4 is well documented.

Savitsky et al.[21] reported a Ni-Cr alloy (80 Ni–15 Cr–5 Al) that has improved corrosion performance in 94 wt% boiling H_2SO_4 acid. The corrosion rate is about 4 mm/yr, compared to 11 mm/yr for a Cr-Ni steel alloy (Fe–18 Ni–18 Cr–2 Ti). However, they demonstrated that by using Cr^+, Ni^+, and Al^+ ion implantation to alter the surface properties of the Cr-Ni steel alloy, they were able to reduce the corrosion rate by a factor of 3. This is an interesting technique that needs to be explored further.

TABLE 4.8
Corrosion Rate of Alloy in Sulfuric Acid (wt%)

Alloy	Temperature(°C)	Corrosion Rate(mm/yr)	Comments
Inconel 625	80	90.19	Concentration 80%
Niobium	Boiling	50.02	60%
Gold	250	<1.97	For all concentrations
316 stainless	93	4001.42	Concentration 80%
Tantalum	200	<0.04	Concentration 98%

The casting alloy Duriron (Fe–14.2 Si–0.8 C wt%), with its high Si content, has been shown to be extremely resistant against concentrated H_2SO_4, but the brittle nature limits machinability, assembly, and maintenance of components. In light of this, iron-based alloys with lower Si content (4 to 6 wt%), such as Saramet 23, Saramet 35, and ZeCor, are being tested for this application. These alloys have good ductility and forming characteristics and can be joined by traditional means such as arc welding. Investigators at Idaho National Laboratory have established a screening effort studying the applicability of these alloys and other materials to the H_2SO_4 concentration process. Figure 4.9 shows the corrosion rate of various engineering alloys tested in 96 wt% H_2SO_4 at 200°C. The initial corrosion rate is relatively high but is reduced with test time. This can be attributed to the formation of a Si-rich passivation layer on the alloy that slows the corrosion process and consequently retards the corrosion rate (figure 4.10). When the test is conducted at 375°C, the corrosion rate has a similar trend, but the Si-rich passivation layer grows at a faster rate due to the faster kinetics. The corrosion resistance of the tested alloys is thought to depend to a large extent on adherence of the Si-rich layer to the alloy substrate. This is especially true in a boiling H_2SO_4 acid environment. New experiments will need to be conducted to probe the effect of long-term operation and process variations on the corrosion resistance of these candidate alloys.

Another approach to this problem is to employ ceramic materials. SiC-based materials are preferred candidates, as they have both excellent corrosion resistance and high thermal conductivity. Researchers at JAERI have been exploring the use of ceramics materials in boiling H_2SO_4.[17] They have shown that Si-based ceramics such as SiC, Si-SiC, and Si_3N_4 have extremely good corrosion resistance in concentrated H_2SO_4 acid at high temperature, and are even better than high-Si steel (table 4.9).

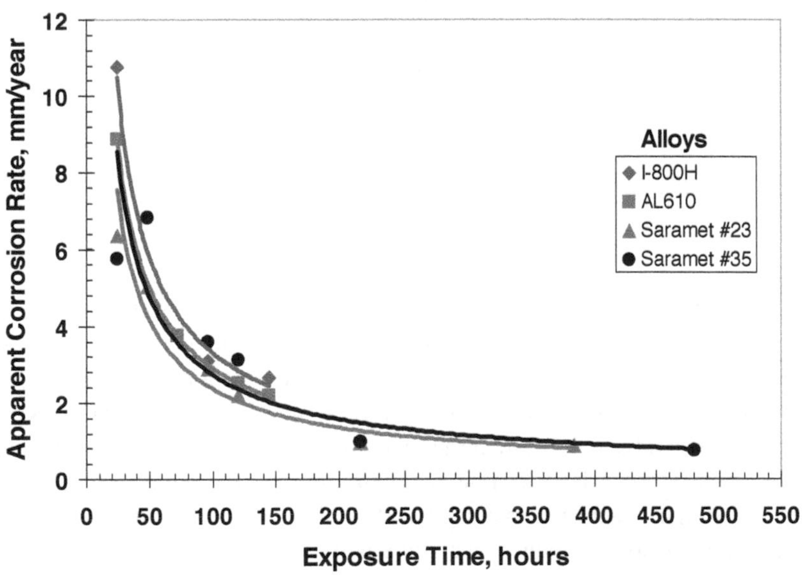

FIGURE 4.9 Corrosion rate of selected alloys as a function of exposure time.

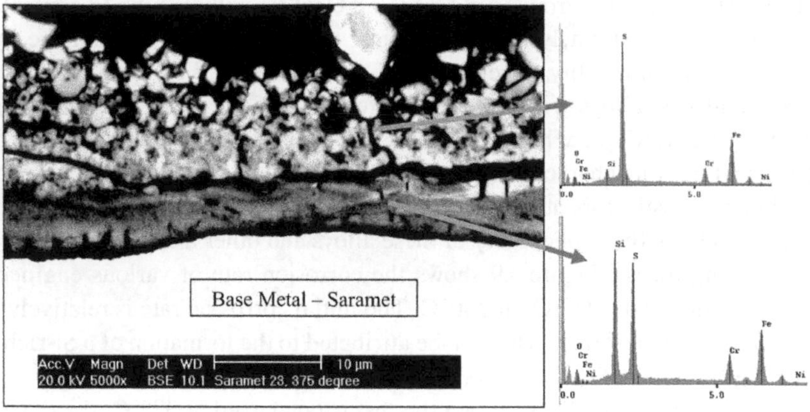

FIGURE 4.10 Cross-section through the surface film that formed on a Saramet #23 sample exposed to sulfuric acid at 375°C.

TABLE 4.9
Corrosion Rate of Si-Based Ceramics and High-Si Steel in Concentrated H_2SO_4 at High Temperature

	H_2SO_4 Acid Concentration and Temperature			
	95 wt%(460°C)		85 wt%(380°C)	75 wt%(320°C)
	100 h	1,000 h	100 h	100 h
SiC	–0.1	–0.002	0	0
Si-SiC	0	–0.006	NA	0
Si_3N_4	0	–0.007	0	0
Fe–20 Si	1.1	0.13	NA	NA
Fe-Si (annealed)	–0.12	0.065	0	0
Ni–Cr–Si steel	–0.28	0.96	NA	5 (17 h)

Note: Original corrosion rate (g/m²h) from the reference is used. Minus sign indicates weight gain.

Construction materials capable of handling H_2SO_4 vapor were studied extensively during the early stages of the S-I cycle development, as it was thought to be the most critical materials issue of the cycle. Because of the high operating temperature involved, materials candidates were chosen from those that derive their strength from solid-solution strengthening instead of precipitation hardening, as overage conditions can lead to a decrease in strength. Different researchers have come to a similar conclusion that steel with a high Ni-Cr content is most suitable for the decomposition reaction, especially when the operation temperature is above 850°C. At this temperature, it was found that corrosion due to H_2SO_4 vapor is similar to that in air,[22] and thus many qualified high-temperature engineering alloys can be suitable. Alloys that have been tested in this environment include Incoloy 800H, AISI310, Inconel 600, stainless 304, a 70 Fe–10 Ni–18 Cr alloy, etc.[13,14,22–26] The high Cr content in all these alloys helps to passivate the metal surface. Among all the different candidates, Incoloy 800H has shown the best performance, with a corrosion rate of less than 100 μm/yr after a 9,000-h test at 900°C. AISI310 was judged to be applicable to lower-temperature application.[23]

Construction materials for vaporization and decomposition of H_2SO_4 have largely been defined by the solar decomposition demonstration that was carried out at Georgia Institute of Technology in 1985.[19] A schematic of the solar decomposition experimental setup is shown in figure 4.11. In this work, concentrated ambient H_2SO_4 (98 wt%) is fed into a dry wall boiler constructed from Hastelloy C-276 operating at around 600°C. Inside the boiler are heated Denstone ceramic balls (56 SiO_2–38 Al_2O_3 wt%). The cold H_2SO_4 acid is vaporized to about 400°C upon contact with these balls. The H_2SO_4 vapor is heated to above 600°C as it passes through a superheater also made with Hastelloy C-276. At this stage, the vapor begins to decompose into H_2O and SO_3. This gas mixture is sent to a decomposer operating at temperatures above 800°C in order for the $SO_3 \rightarrow SO_2 + O_2$ catalytic decomposition to take place. This decomposer was actually a heat exchanger constructed from Incoloy 800H tubes filled with Fe_2O_3 catalyst pallets. Thermal radiation and air heated by the solar tower circulate outside the tubes and manifolds to provide heat for the decomposition reaction. Some of the tubes had an aluminized coating on their inner wall to test the corrosion resistance of such coating. Downstream the unreacted acid was then condensed in an Incoloy 825 coiled tube and collected. Table 4.10 is a summary of the corrosion performance of the various construction materials tested in the experiment. The oxidation of the high-temperature tubing by SO_2, SO_3, and O_2 was found to be the dominant corrosion mechanism, and the degree of corrosion is similar to exposure to air. The performance of the selected alloys was satisfactory, but oxidation and evidence of corrosion were observed. Moderate corrosion was experienced by all the components, and it will need to be reduced in order to ensure the long-term viability of components.

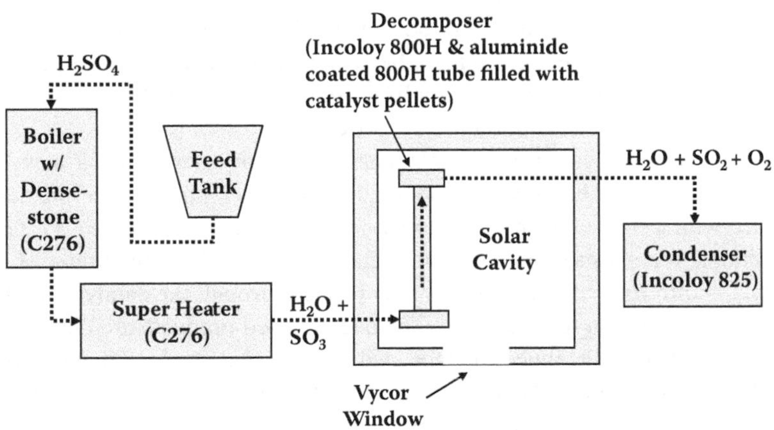

FIGURE 4.11 Schematic of the experimental set up for the solar H_2SO_4 decomposition experiment.

TABLE 4.10
Principal Materials of Construction for the H_2SO_4 Solar Decomposition Experiment

Component	Materials	Service Temperature and Media(°C)	Estimated Depth of Corrosion(mil)
Boiler	Hastelloy C-276	>330 (l, v)	10
Superheater	Hastelloy C-276	600 (v)	10
Decomposer tube	Incoloy 800H	850 (v)	10
Condenser	Incoloy 825	100–400 (l, v)	40

Note: l = liquid; v = vapor.

The long-term corrosion performance and creep characteristics of the alloys in the present environment will need to be addressed in the H_2SO_4 decomposition environment. One must also guard against carbide sensitization, stress corrosion, and intergranular crack formation when these alloys are heated for an excessive period at high temperature. This is in addition to possible sulfidation of the alloy surface at moderate temperatures due to incomplete chemical reaction where the SO_3/SO_2 ratio favors sulfide formation. Only limited long-term testing has been carried out, and the initial signs have been positive, as no stress corrosion cracking has been observed in Incoloy 800H, Inconel 600, and Hastelloy XR tested at 850°C.[24] Testing is currently underway to study crack formation and growth and creep properties in Alloy 800H, Hastelloy C-276, and other high-temperature alloys, including Inconel 617 in the H_2SO_4 decomposition environment. This hopefully will provide insight not only on the high-temperature mechanical behavior of these alloys, but also on the effect of a sulfidizing environment.

In addition to engineering alloys, effort is also ongoing to develop a SiC-based heat exchanger for this application. SiC is extremely stable in this high-temperature oxidation environment. The challenge in using these ceramic materials is the processing and joining of materials to accommodate the brittle nature of ceramics. Effort is ongoing to develop a microchannel plate heat exchanger, and success hinges on the ability to join these plates together. Figure 4.12 shows a prototype of such a microchannel plate that is manufactured from C–SiC composite infiltrated with liquid Si at high temperature. Such a manufacturing technique is capable of producing a complex profile. It is expected that a heat exchanger constructed from a stack of such plates will be a viable option in the future.

Recently, researchers at Sandia National Laboratory developed a new H_2SO_4 decomposer design with which the vaporization and decomposition of the acid can be carried with a single bayonet SiC boiler–reactor. Figure 4.13a shows a schematic of this design. It consists of an outer SiC feed tube with an SiC outlet tube. H_2SO_4 is vaporized within the outer tube and is then pushed through the catalyst bed for the decomposition reaction to take place. The decomposed products are channeled out of the reactor through the inner SiC tube, and its heat is recuperated in the lower section of the heat exchanger. This design is very efficient, but the manufacturing and joining of the various SiC parts remains an obstacle that needs to be overcome.

Since the final design and operation conditions of the nuclear S-I hydrogen loop are still being finalized, materials of construction development for H_2SO_4

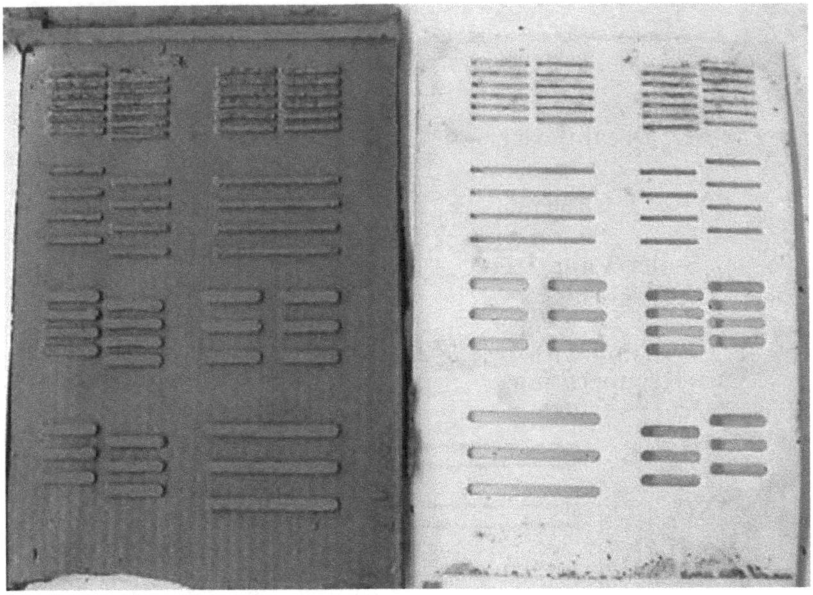

FIGURE 4.12 Prototype of a C-SiC microchannel plate.

decomposition will need to proceed with a broad scope. This ensures materials applicable to the final specification will be in place when needed. Table 4.11 lists the various construction material candidates applicable to the sulfuric acid decomposition process. The candidates can be classified into four different categories: superalloys, ceramics, noble metals, and intermetallics. Selection will depend on the chemical environment and manufacturing technique. Even though the surveyed materials have shown good corrosion resistance in H_2SO_4, the effect of chemical contaminants such as traces of HI and I_2 and corrosion species on their corrosion performance is not known. These factors will need to be considered in the development testing process.

4.4.1.3 Materials of Construction for Section III

As discussed above, there are two pathways to carry out HI decomposition, and the respective process steps are listed in table 4.3 and table 4.4. Construction materials development has focused on identifying materials that can withstand the different acids and chemicals at the processing conditions. HI_x acid and vapor are present in both distillation processes, whereas H_3PO_4 is only used in extractive distillation. Hence, the following discussion of candidate construction materials for Section III will be based on chemical contents instead of processing environment, as was the case for the other two sections. First, data for general corrosion will be reviewed, followed by the effects of stress corrosion and chemical contamination.

4.4.1.3.1 Materials for HI_x

Corrosion data of materials in HI_x are extremely limited. The most comprehensive set is from Trester and Staley.[14] Table 4.12 lists a summary of the immersion coupon test results from their work. The test temperature is similar to that for extractive

(a)

(b)

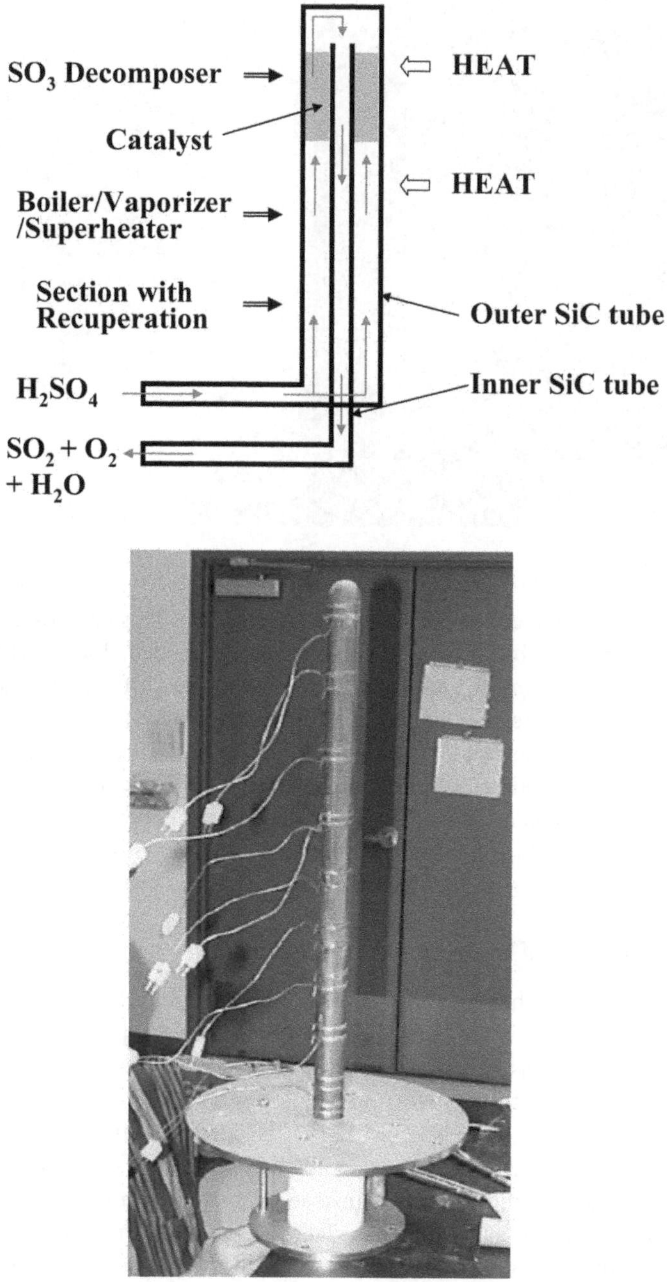

FIGURE 4.13 A schematic and a prototype of a SiC bayonet H_2SO_4 vaporizer/decomposer.

TABLE 4.11
Summary of Materials Options for the Sulfuric Acid Decomposition Process

Process Regime	Conditions	Candidate Materials	Compatibility	Comments
H_2SO_4 concentration	25–180°C, 50 wt% 25–150°C, 50–75 wt% 180–450°C, 75–95 wt% H_2SO_4, iodine species, impurities	Glass-lined steel, plastics, ceramics Hastelloy B-2 and C-276 Incoloy 800H, AL610, high Si steel, Au or Pt plating/cladding	Hastelloy B-2 < 0.1 mm/yr for concentrations up to 60% High Si steel corrosion ~0.1 mm/yr for concentrated acid, higher for low concentrations	Concentration may require multiple materials Options identified for all concentrations and temperatures Evaluate coatings, plating, and cladding B-2 promising @ low concentration High, Si steel fabrication issues
H_2SO_4 vaporization	350–550°C $H_2O + SO_3$, iodine, other contaminants	Structural: Incoloy 800H AL610, high-Si steel SiC, Si_3N_4 Hastelloy G and C-276	800H, 800HT High-Si steel (SiO2) < 5 mpy SiC ~ no corrosion in 1000-h test @ 75 to 95% acid (JPN) C-276 ~ 1 mm/yr @ 476 h	Coated materials (Pt) cost issue Ceramics promising, but have fabrication and joining issues Dry wall boiler design with ceramics may be option Data needed with iodine contamination
H_2SO_4 decomposition	550–950°C H_2O, H_2SO_4, SO_3, SO_2, O_2	Structural: Incoloy 800HT, Incoloy 800H (with aluminide coatings), AL610 Ceramics, Pt or Au coatings on superalloy structural materials	Incoloy, Inconel Bare—2–4 mg/cm2 in 1000 h @ 900°C Aluminide coatings—approximately 1 mg/cm2 in 1,000 h @ 900°C Intergranular corrosion observed for 800H Noble metal coatings may provide corrosion protection	Incoloy 800HT may address intergranular corrosion C–SiC composites should be examined Pt coating may serve function of catalyst and reduce corrosion Corrosion benefits of noble metal coatings must be demonstrated

TABLE 4.12
GA Results (1977–1981) Summary from Trester and Staley[4]

Test	HIwt% (molar%)	I2	H2O	Temperature	Pressure	Time	Materials Tested: Excellent	Fair	Poor
1	20% (4%)	20% (2%)	60%(93%)	25°C	atm	8,760 h	Mo, Nb–1% Zr, Ta, Ta–10% W, Ti (as cast), Ti–0.5% Pd, Zr, Zircaloy2TFE, FEP, Kalrez 1050, Kel-F 3700, Fluorel 2174, Viton A, Parker V-834-70	Chlorimet 2 and 3, Hastelloy B2 and C276PVC, polycarbonate, Vespel sp. 1, CPE, FETFE	Inconel 600, Monel, Haynes, Hastelloy G, 304 stainless Nylon, mylar, silicone
2	30% (15%)	50% (13%)	20%(72%)	300–500°C	13.1–17.2 MPa	5–10 h	Mo, Ta	Ti	Inconel 600
3	11% (11%)	82% (40%)	7%(49%)	100°C	atm	3,170 h	Mo, Nb–1% Zr, Ta, Zr, TFE, FEP, PFA, Tefzel (Teflon)SiC, alumina, boronsilicate glass	Ti–0.2% Pd (annodized), Hastelloy B2, Durichlor 51, Zircaloy2, Zr702Kynar 450Zirconia	Duriron D, Chlorimet 2, Hastelloy B2 and C276, Ti–0.5% Pd, gold, platinumCPVC, polypropylene
4	11% (11%)	82% (40%)	7%(49%)	120°C	atm	500 h	Mo, Nb, Nb–1% Zr, TaAlumina, vitreous carbon	TFE, FEP, PFA	LeadViton VTX 5362, Viton B, Carbonrundum
5a	24% (12%)	55% (14%)	21% (74%)	135°C	atm	178 h	Mo, Nb, Nb–1% Zr, Ta, Zircaloy2, Ta–10% W, Zr	Ti–0.5% Pd	Ti–0.2% Fe–0.25% O (anodized Ti)

[a] Circulating HI_x.

distillation and is about 100 to 200°C lower than that required by reactive distillation. Based on these data, Ta, Nb, Mo (refractory metals), Zr (reactive metals), SiC (ceramics), and carbon-based materials have the best prospect of being compatible with HI_x at higher temperatures.

Immersion coupon tests have been conducted at General Atomics to study the corrosion resistance of a variety of construction material candidates in HI_x at temperatures used for extractive distillation.[26] Figure 4.14 shows the progression of an immersion coupon test of a Ta–2.5 W coupon in HI_x at 310°C for 2,000 h. No evidence of corrosion, including the weld region, can be observed. This can be compared with a Zr705 coupon that shows extensive dissolution after only 120 h in the same environment (figure 4.15). Immersion coupon testing has shown that Ta and Nb alloys have the best general corrosion characteristics among the different metals when tested against HI_x at high temperatures (table 4.13). Tests are ongoing to understand the effect of a HI_X environment for occurrence of nucleation and the growth of cracks in Ta and Nb alloys, which showed good corrosion properties.

SiC-based materials have also shown very good corrosion resistance in HI_x. Both sintered and chemical vapor deposition (CVD) SiC have very low corrosion rates when tested in HI_x at elevated temperatures. In addition, Si-infiltrated C-based materials (Si-SiC) also have good potential. This method of manufacturing may become an attractive option in the future, as it promises an extremely low-cost alternative to manufacture SiC-based corrosion-resistant materials[25] and reduce the potential joining problems. Effort is continuing to resolve the manufacturing techniques and improve the inherent mechanical properties of these SiC-based materials.

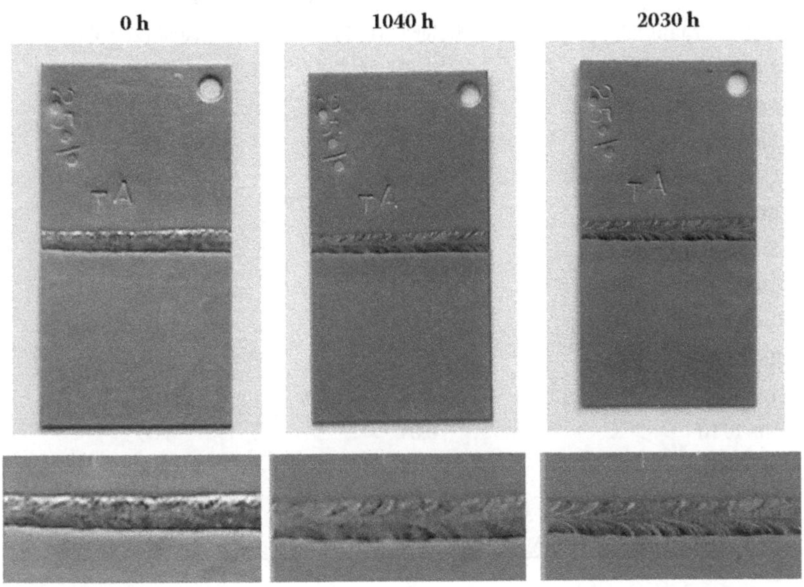

FIGURE 4.14 A Ta-2.5W coupon with an e-beam weld that has been immersion tested in HI_x at 310°C.

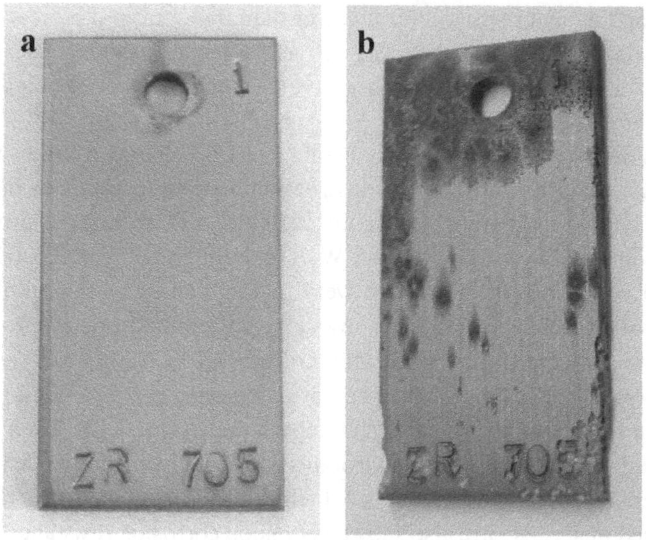

FIGURE 4.15 Zr705 coupon before and after a 120-h test in HIx at 310°C.

TABLE 4.13
Corrosion Rate of Various Materials in HI_x at High Temperatures

	Corrosion Rate (mm/yr)	
Material	**Boiler (310°C)**	**Feed (262°C)**
Nb–7.5 Ta	–3.90	0.39
Splint Si-SiC	–3.31	0.00
SiC (sintered)	–2.60	0.00
Ceramatec SiC (sintered)	–1.06	0.00
Mo-47Re	–0.67	0.00
SiC (CVD)	–0.55	–0.55
Ta	–0.51	0.08
Ta–2.5 W–2	0.00	0.00
Ta–2.5 W–1	0.04	0.00
Ta–10 W	0.04	–0.24
Nb–10 Hf	0.04	0.00
Ta–40 Nb	0.28	–0.08
Nb	0.43	0.00

Note: Minus sign indicates weight loss.

Other ceramic materials, such as Al_2O_3 or mullite, have also been shown to be stable in the presence of I_2 and HI_x, but their application in a traditional heat exchanger design is limited, as they have very low thermal conductivity even relative to SiC. On the other hand, recent modeling results from a microchannel heat exchanger indicated that there may be an advantage in using low thermal conductivity material in these designs. Even though there are still many obstacles to using them in the near future, ceramic-based components will most likely play an important role as the S-I cycle develops.

4.4.1.3.2 *Materials for Phosphoric Acid*

Phosphoric acid is a chemical reagent commonly used in the chemical industry. However, most of the corrosion data of materials in H_3PO_4 are for acid concentrations up to 85 wt%, whereas the H_3PO_4 concentration in Section III ranges from 85 to 96 wt%. High Mo stainless steel such as Alloys 28 and G-30 is commonly employed in the chemical industry to contain H_3PO_4. In addition, Ni-Mo alloys are also widely used. Table 4.14 lists the corrosion rate of a number of metals in 85 wt% H_3PO_4 [M14]. At high acid concentration and at the boiling temperature of 158°C, it has been shown that Ta, Nb, and their alloys have good corrosion resistance in the acid and are good candidates for materials of construction. Table 4.15 shows the corrosion rate of Ta and Nb alloys in 80% concentration H_3PO_4 at 150 and 200°C, respectively.[23]

Since both HI and H_3PO_4 are reducing in nature, it is possible to use materials that are common to both in the iodine separation reaction (see table 4.3). The material of construction used to fabricate the I_2 separation reactor must be able to resist the combination of HI_x and H_3PO_4. Preliminary test results show that the corrosion behavior of the various materials tested in the HI_x–H_3PO_4 acid mixture is similar to that in HI_x at high temperatures, with Ta and Nb alloys and SiC-based materials the most promising construction materials candidates.

4.4.1.3.3 *Materials for HI + H_3PO_4 and Iodine (Iodine Separation)*

Construction materials used for the iodine separation step in Section III will encounter a flowing mixture of HI_x and H_3PO_4, a light HI + H_3PO_4 upper phase, and a

TABLE 4.14
Corrosion Rate of Alloys in H_3PO_4

Alloy	Concentration	Temperature (°C)	Corrosion Rate (mm/yr)
316 stainless	85	115	5.91
Hastelloy C	85	Boiling	44.90
Durimet 20	75–85	115	9.06
Haynes 556	85	Boiling	33.08
Inconel 617	85	Boiling	25.99
Monel 400	85	124	10.24
Tantalum	85	100	0.00
Niobium	85	100	5.12
Silver	85	140	1.97

TABLE 4.15
Corrosion Rate of Ta and Nb Alloys in 80% Concentration H_3PO_4 at 150 and 200°C

Alloy	Corrosion Rate (mm/yr)	
	150°C	200°C
Niobium	59.06	NA
Nb–20 Ta	11.81	NA
Nb–40 Ta	8.66	492.13
Nb–60 Ta	2.76	82.68
Nb–80 Ta	0.39	29.92
Tantalum	0.05	5.31

denser iodine-rich lower phase that forms in the separator (see section 4.3.3.1). Materials candidates for use in HI_x and H_3PO_4 have been discussed previously. However, mixing of various acids may lead to a synergistic corrosion effect. Figure 4.16 shows a Nb–10 Hf coupon that has been tested in a static HI_x and H_3PO_4 mixture that represents the iodine separator environment. A scale has formed on the coupon where it is in contact with the H_3PO_4-rich upper phase. This can be compared with a Ta–10 W coupon tested in the same environment (figure 4.17). Table 4.16 shows a summary of the corrosion rate of various materials that have been tested in HI + H_3PO_4, and the results show that Ta and its alloys and SiC-based materials are suitable for this environment.

Based on data obtained so far, it appears that liquid processes in Section III will rely heavily on Ta and Nb alloys as construction materials. Since these metals are expensive in nature and have low yield strength, one must explore different manufacturing means to reduce the overall component cost and enhance the mechanical properties of components. For example, cladding can be used to bond a Ta layer to a base material to take advantage of its corrosion characteristics while keeping the

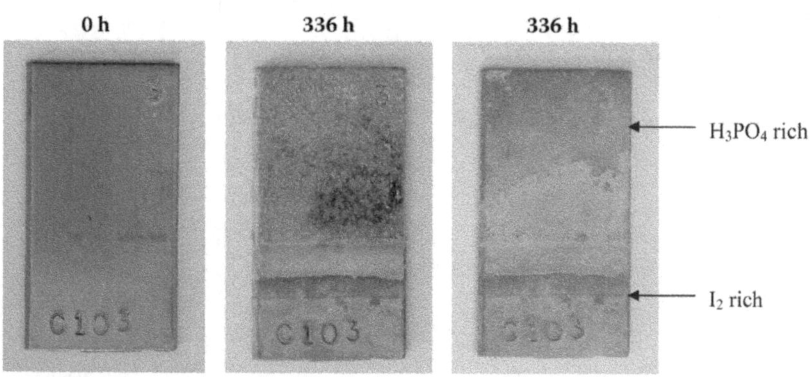

FIGURE 4.16 A Nb-10Hf coupon tested in a static HIx-H3PO4 mixture at 140°C for 336 h new, post test, and post test with scale removed.

TABLE 4.16
Corrosion Rate of Various Materials Tested in a HI_x–H_3PO_4 Mixture at 140°C for 120 to 1,100 h

Alloy	Corrosion Rate (mm/yr)
Ta–10 W	0.02
Ta–2.5 W	0.03
SiC	0.08
Ta	0.11
Mo	0.45
Nb–1 Zr	24.88
Nb	38.96
Nb–10 Hf	40.55
Zr705	91.44
Hastelloy B2	137.35
C-276	140.08
C-22	147.28
Nb–7.5 Ta	187.10
Monel	225.77

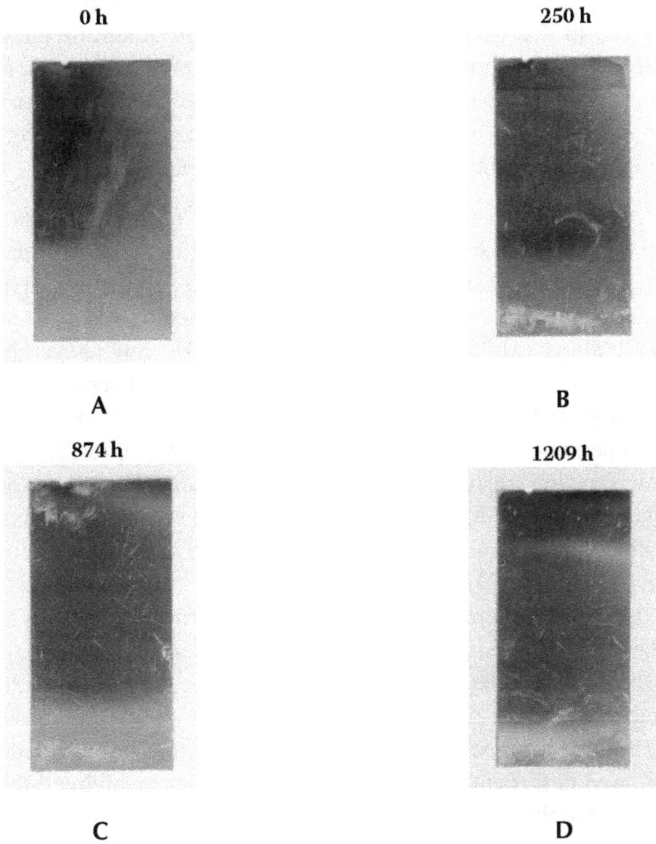

FIGURE 4.17 A Ta-10W coupon tested in HIx-H3PO4 at 140°C for 1,209 hours. (A) 0 h, (B) 250 h, (C) 874 h & (D) 1,209 h..

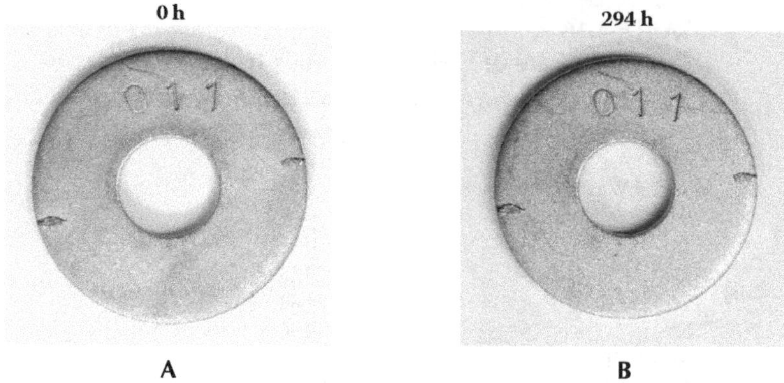

FIGURE 4.18 A stainless washer that has been plated with a Ta layer and tested in HI_x-H_3PO_4 acid mixture at 140°C for 294 h. .

cost down. Techniques for cladding or lining are well established, and suitable means will need to be selected for the application. Figure 4.18 shows a Ta-plated stainless washer that has been tested in a HI_x and H_3PO_4 acid mixture. Its corrosion performance is similar to that of Ta alloys in the same environment (figure 4.18). Other surface-coating or modification techniques, such as PVD (physical vapor deposition) processes and ion implantation, may help to fabricate components in addition to cladding, lining, or plating. The application of such materials processing technology to manufacture process components will need to be explored.

4.4.1.3.4 Materials for HI + I_2 + H_2 (Gaseous HI Decomposition)

In extractive distillation, the HI → H_2 + I_2 decomposition reaction is carried out in the gas phase between 300 and 450°C (see table 4.3). Given that both HI and I_2 will be in gaseous form, the corrosiveness of the environment is reduced. Since refractory Ta and Nb alloys exhibit internal oxidation at 350°C and above, their application in this environment will be limited. Onuki et al.[16] had screened a number of engineering alloys and corrosion-resistant materials in a $HI/I_2/6H_2O$ gaseous environment at temperatures between 200 and 400°C [M6]. Metallic Ta, Zr, Ti, and SiC and SiN show very low corrosion rates at all temperatures. Based on the screening results, they conducted 1,000 h of long-term testing on selected materials, and their results are shown in table 4.17. Scale formation and discoloring were observed in all specimens. Onuki et al.[16] suggested that high-temperature oxidation of the metal surfaces leads to a protective scale that limits the corrosion of metals in this environment. The H_2O fraction in their test environment is much higher than that stated in the flow sheet. Hence, the applicability of the engineering alloys will need to be verified. Trester and Staley[14] also conducted preliminary testing of Hastelloy B and C-276 in a HI + I_2 environment in the absence of moisture. They concluded that both alloys can be used in this environment.[14] Unlike reactions involving liquid phases, it appears that commonly available corrosion-resistant engineering alloys can be used for the HI gaseous decomposition process, but this will require more long-term testing to confirm. Table 4.18 presents a compatibility chart of construction materials that are applicable to the different processes in Section III.

TABLE 4.17
Corrosion Test Results in a Vapor Medium of $HI–I_2–H_2O$ (1/1/6) Vapor at Various Temperatures for 1,000 h Duration

	200°C			300°C			400°C		
	I	II	III	I	II	III	I	II	III
Fe–1 Cr–0.5 Mo	BK	R	3.94	DB	R	3.94	BK	R	7.88
SUS444	BK	P	3.94	DB	R	7.88	GB	R	7.88
SUS315L	BK	N	3.94	DB	R	7.88	BR	R	7.88
Inconel 600	BK	N	0	DB	N	0.05	BK	R	3.94
Hastelloy C-276	BK	N	0	DB	N	0	BK	R	1.87
Ti	N	N	0	N	N	0	GR	N	0

Note:
I: Surface appearance. N = no scale; BK = black; DB = dark brown; GB = greenish brown; BR = brown; GR = gray.
II: Surface appearance after descaling. N = no change; P = pitted; R = roughened.
III: Corrosion rate in mm/yr.

4.4.1.3.5 *Effect of Stress Corrosion and Chemical Contaminants*

In terms of general corrosion, one or more preliminary construction material candidates for each of the S-I cycle processing steps have been identified. However, materials are subjected to processing and environment effects that could lead to susceptibility to the stress corrosion phenomenon. Fabrication methods such as welding, machining, and hot and cold forming can modify the corrosion resistance of materials that may have shown good general corrosion characteristics. Investigative work has been ongoing to study such environmentally assisted crack initiation and growth phenomena in C-ring, U-bend, double cantilever bean (DCB), and compact tension (CT) specimens in both liquid and gaseous test environments at General Atomics. Figure 4.19 shows a Zr705 C-ring specimen that was immersed in HI_x at 310°C for 120 h; a crack was observed after testing. This can be compared to a Hastelloy C22 U-bend specimen that has been tested in the HI gaseous decomposition environment at 450°C for 1,570 h, in which no crack was observed (figure 4.20). Since Section III will be operated in a pressurized environment, stress corrosion properties of candidate materials will need to be considered.

Another key issue that needs to be addressed during materials development is the effect of contaminants within the process streams. There are a number of liquid–liquid separation steps within the S-I cycle that result in a trace amount of chemical contamination in the separated liquid stream. The presence of such contaminants may affect the corrosion performance of materials. Table 4.19 summarizes the contaminants that can exist in the different process fluids. The only previous experiment that addressed this involves the effect of a minor amount of HI in the H_2SO_4 of Section I (see table 4.7). More recent investigations on this subject at General Atomics have illustrated the potential pitfalls of contamination. Figure 4.21 shows the effect of contaminants on the corrosion performance of C706 (Cu-Ni alloy) in concentrated phosphoric acid. Even though C706 exhibits a satisfactory corrosion rate in 96 wt%

TABLE 4.18
Materials Options for Section III: Hydrogen Iodide Decomposition

Process Regime	Conditions	Candidate Materials	Compatibility	Comments
Reactive distillation—reactor HI_x feed ($HI + I_2 + H_2O$)	250°C, ~40 bars Impurities, H_2S, S, etc.	Ta and alloys, Nb and alloys, Mo SiC	Pure Mo and Ta < 0.1 mm/yr Hydrogen embrittlement effects	Processing effects unknown Evaluate coatings, plantings Fabrication and cost issues
Reactive distillation—reactor bottom HIx with high I_2 concentration (<85%)	310°C, ~40 bars Sulfur species contaminants	Ta and alloys, Nb and alloys SiC, Si-SiC, mullite	W, Mo, and Ta < 0.01 mm/yr in pure iodine at 300°C Gold and Pt work well in pure iodine but perform poorly in HI_x	SiC composites should be examined Materials need to be compatible with the HI_x feed for reactor application
Extractive distillation—iodine separation HI_x and concentrated H_3PO_4	120°C, atmospheric pressure Contaminants from other sections—H_2SO_4	Ta and alloys SiC and carbon composite	Ti and Hastelloy B2 compatible with HI_x below 150°C	Carbon composite can be considered for lower temperatures
Extractive distillation—phosphoric acid concentration	250°C, atmospheric pressure Contaminants from other sections—HI_x, H_2SO_4	Ta and alloys SiC and carbon composite		Systems-level design considerations in the integration of different materials
Extractive distillation—HI gaseous decomposition	300–450°C, atmospheric pressure Contaminants from other sections—H_2SO_4	Hastelloy B and C-276, Ti SiC and carbon composite		

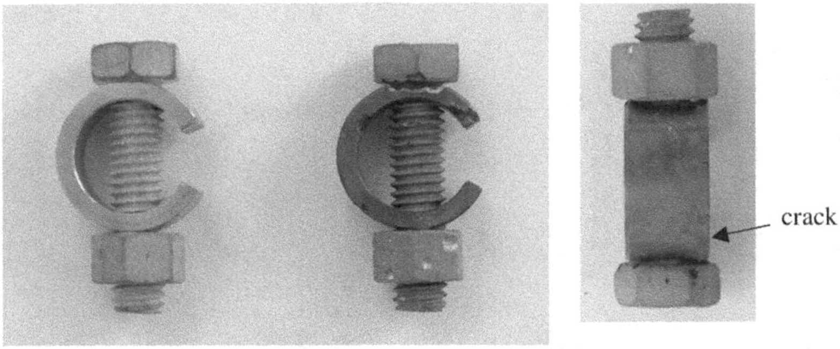

FIGURE 4.19 A Zr705 C-ring specimen under tensile loading to 98% of yield stress that has been tested in HI_x acid.

H_3PO_4 acid with a trace amount of HI, it corrodes rapidly when a trace amount of I_2 is added into the acid complex. This shows that much more work remains to be done to understand contamination effects.

4.4.2 Separation Membranes

Separation membranes can be extremely important to the economic success of the S-I cycle, as they can improve the overall efficiency of the cycle through removal of water, SO_2, and H_2 from the chemical stream. This can reduce the amount of excess heat needed to boil off the water and improve the conversion efficiency of decomposition reactions. There are three potential membrane applications within the S-I cycle:

1. H_2O separation from the HI_x liquid phase from the Bunsen reaction to reduce the heat input required in the subsequent distillation processes
2. SO_2 separation from the H_2SO_4 decomposition gas product stream to improve the conversion efficiency and perhaps shift the reaction to a lower temperature
3. H_2 separation from the HI gaseous decomposition reaction to reduce the amount of recycle

4.4.2.1 H_2O Separation

Nafion-117, a perfluorinated polymer membrane, has been studied for use in a direct separation of H_2O from HI_x. Its ability to remove H_2O from a flow stream of HI_x at 125°C has been successfully demonstrated (figure 4.22). The permeability or flux through the Nafion-117 membrane is temperature dependent and inversely correlated to the water concentration.[27,28]

Nafion membranes have also been used to carry out other concentration schemes. Hwang et al.[29] employed an electro-electrodialysis approach to raise the concentration of HI within the HI_x solution at 110°C. Nafion acts as a cation exchange membrane in an electrolysis cell in which HI is formed at the cathode. This raises the HI concentration within the HI_x catholyte (figure 4.23). In addition to this, the increase in HI content helps to break up the azeotropic between HI, I_2, and H_2O and facilitates the distillation of HI from HI_x. Researchers at JAERI have taken this concept

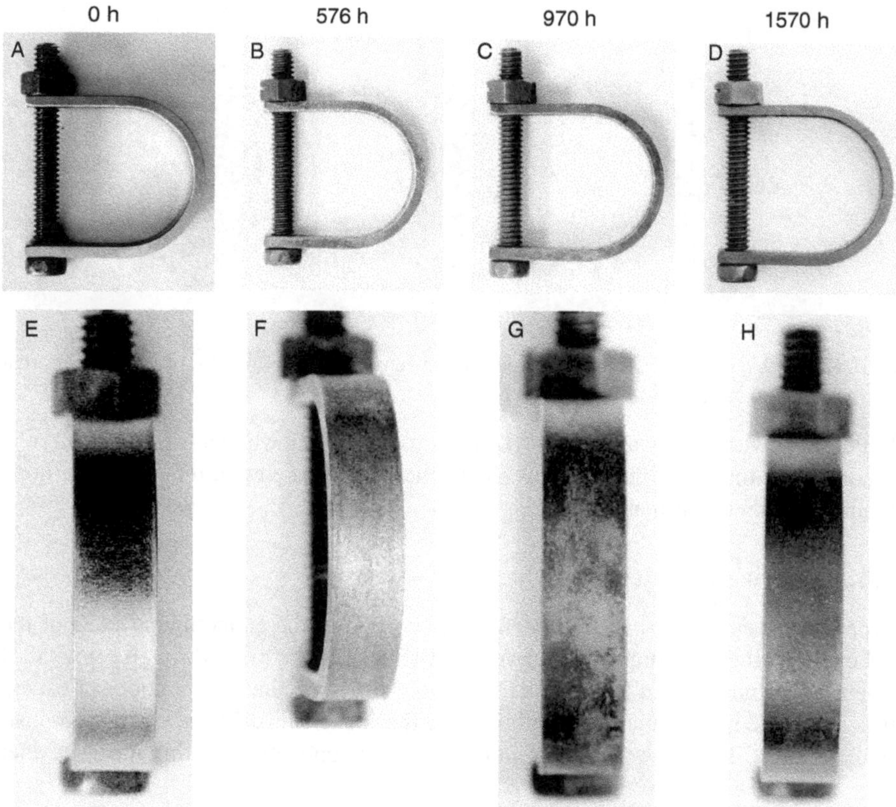

FIGURE 4.20 C22-U-bend specimen coupon tested in the gaseous HI gaseous decomposition environment.

TABLE 4.19
Contaminants in Process Streams That May Affect the Corrosion Environment

Section	Process	Contaminants
II	H_2SO_4 concentration	HI_x from Section I Corrosion products from other sections
III	Iodine separation	H_2SO_4 from Section I Corrosion products from other sections
III	H_3PO_4 concentration	H_2SO_4 and HI_x from Section I and iodine separation Corrosion products from other sections
III	HI distillation (reactive and distractive)	H_2SO_4 from Section I Corrosion products from other sections

671 h in 96wt% H_3PO_4 acid	380 h in H_3PO_4-3HI-5H_2O (wt%)	490 h in H_3PO_4-14.4HI-0.7I_2-16.1H_2O (wt%)

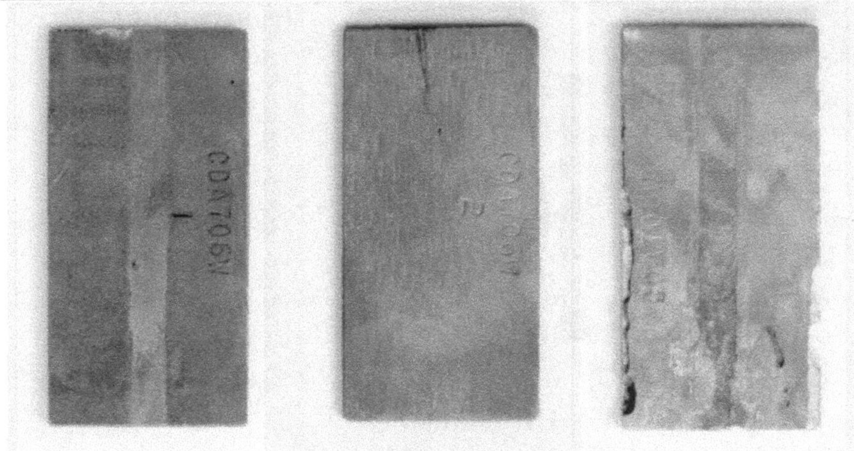

FIGURE 4.21 C706 alloy coupons tested in conc. H_3PO_4 acid, conc. H_3PO_4 acid with a trace amount of HI and conc. H_3PO_4 acid with a trace amount of HI and I_2.

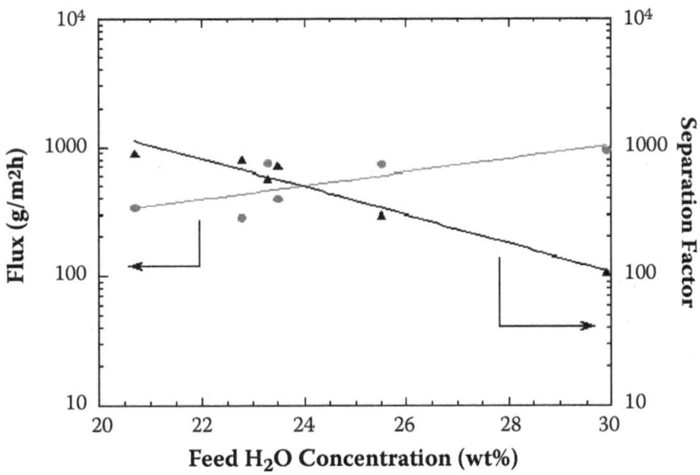

FIGURE 4.22 Naflon-117® performance using an HI/water/I_2 feed at 125°C.

a step further and explored the possibility of using the Nafion membrane to run an electrochemical membrane Bunsen reactor.[30,31]

Long-term testing has shown stable transport characteristics in Nafion when operated at the Bunsen reaction temperature, but more testing is needed.[28] The transport kinetics, permeability, and capability to manufacture a membrane of substantial area are key issues that need more research and development. There are ample opportunities to test other membranes for this application.

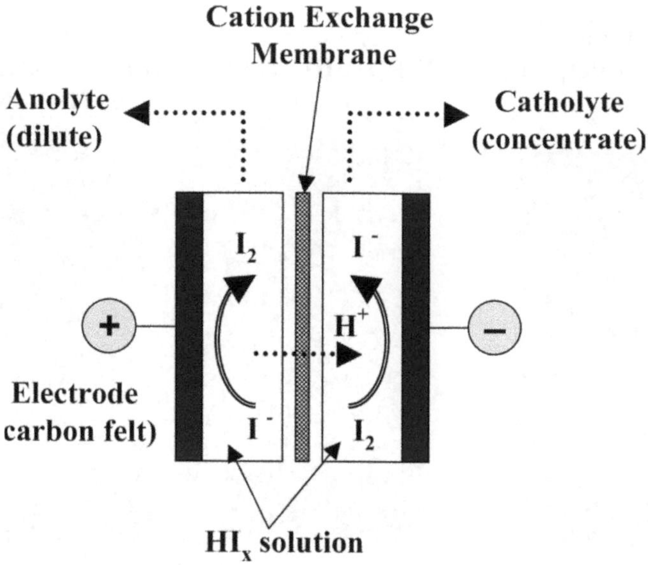

FIGURE 4.23 Schematic of the electro-electrodialysis process to concentrate the HI_x acid feed from the Bunsen reaction.

4.4.2.2 SO_2 Separation

The second possible application of membrane in the S-I cycle is to separate SO_2 from SO_3 and O_2 gases during the $SO_3 \rightarrow O_2 + SO_2$ decomposition reaction. The biggest challenge is that the membrane will have to operate at temperatures between 800 and 950°C, and no existing commercial membrane has been shown to meet these requirements. Work is underway at Oakridge National Laboratory to identify and test suitable porous ceramics that are capable of this application.

4.4.2.3 H_2 Separation

In the gaseous HI decomposition reaction, H_2 separation membranes can play a critical role. The decomposition reaction $HI \rightarrow H_2 + I_2$ is an equilibrium reaction that has a conversion efficiency of around 22% at 450°C. One can enhance the decomposition rate by removing hydrogen from the reactor. Therefore, functional membranes that can separate hydrogen from the reactor will lead to a reduction in the amount of HI gas that needs to be recycled through the reactor. In addition, an effective separation membrane can maintain the purity of the H_2 gas produced. A number of hydrogen separation membranes have been developed for many applications. Unfortunately, most of them employ metals such as Pd, Pd-Ag, and Zr. Since these materials have been shown to be susceptible to corrosion in an iodine-rich environment, they are not suitable for this application. Hwang et al.[32] have prepared silica membranes by CVD that have shown high permeability of H_2 from a H_2–H_2O–HI gaseous mixture at elevated temperatures (figure 4.24). A separation factor of more than 600 between H_2 and HI has been observed.[32,33] The importance of such a separation membrane has been demonstrated by Nomura et al.[34] They calculated that the use of

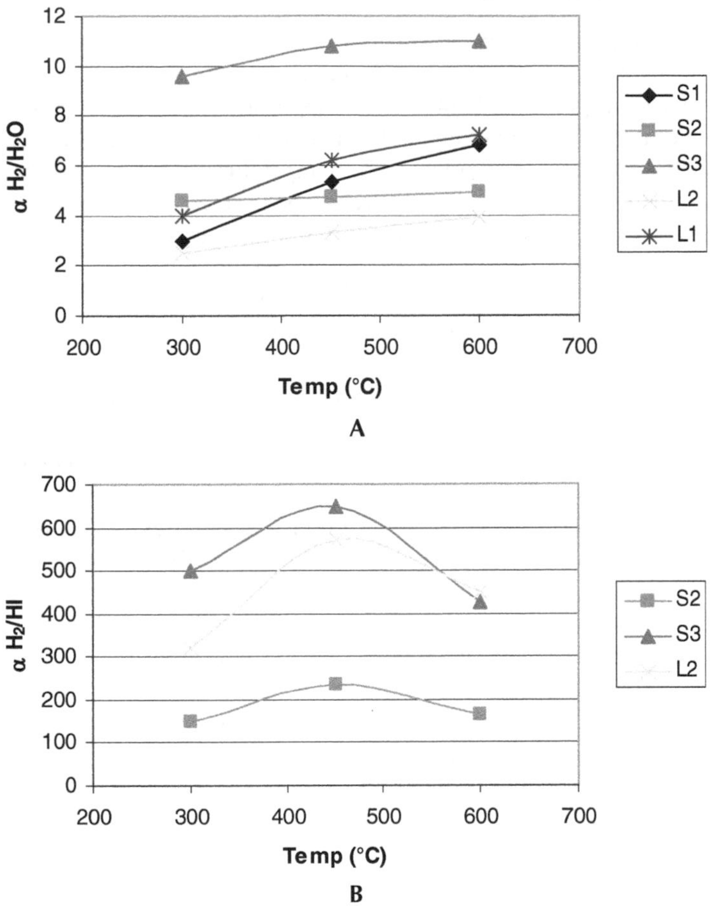

FIGURE 4.24 Separation factor of H_2 from H_2-HI-H_2O for various silica membranes.

such membranes can improve the overall cycle thermal efficiency by 1%. Ideally, the membrane should have a high thermal conductance so that a membrane reactor can be used.[35] There is a need for the development of such a membrane, possibly using porous SiC or SiC-based materials. The manufacturing process of the silica and other membrane will need to be addressed to improve the reliability and cost of such a tubular membrane reactor.

4.4.3 Catalysts

The sulfuric acid and the HI decomposition reactions in Sections II and III are both catalytic processes. A variety of oxides, activated charcoal, and platinum have been employed as the catalyst for these reactions. Ongoing research in this area is trying to identify the optimal catalyst support to minimize the overall cost and integration of the catalyst into the process systems.

4.4.3.1 Sulfuric Acid Decomposition

As mentioned in above, the catalytic decomposition reaction $SO_3(g) \rightarrow SO_2(g) + \frac{1}{2} O_2(g)$ is carried out at 850°C and above in order to obtain an appreciable conversion rate. The reaction is greatly enhanced by the application of a catalyst. However, catalysts have been observed to fail due to formation of volatile acid, which causes support poisoning and catalyst attrition at this temperature.[36,37] If an optimum catalyst can be identified, the reaction temperature can be lowered.

The first solar demonstration H_2SO_4 decomposition experiment employed Fe_2O_3 pellets as the catalyst. They were packed inside the tubes of a heat exchanger. Adequate decomposition took place, but breakdown and minor wear of the pellets were observed. Figure 4.25 shows a plot of SO_3 conversion as a function of temperature for various catalysts.[38] The carrier gas is N_2, which contains 4 mol% of SO_3 flowing at atmospheric pressure. Based on experimental results, it was found that the order of activities is $Cr_2O_3 > Fe_2O_3 > CuO > CeO_2 > NiO > Al_2O_3$. At temperatures higher than 700°C, the activity of Cr_2O_3 is similar to that of platinum at atmospheric pressure. One of the drawbacks in using oxides as a catalyst in this reaction is that they have a tendency to form metal sulfate at lower temperatures, which reduces their activity for the reaction. In addition, the vapor pressure of these oxides is not negligible at this high of a working temperature, which may affect their long-term performance. Work is ongoing to identify an oxide-based catalyst that does not subscribe to sulfate formation. For now, platinum, despite its high cost, is the preferred catalyst for sulfuric decomposition.[36]

The Pt catalyst needs to be supported to reduce cost, and this is commonly done by plating Pt onto porous metal oxide such as Al_2O_3, TiO_2, and ZrO_2 at about 0.1 wt%. Such catalyst is commercially available. Pt-based catalysts need to have high catalytic activity and stability. They are controlled in turn by factors such as interaction with

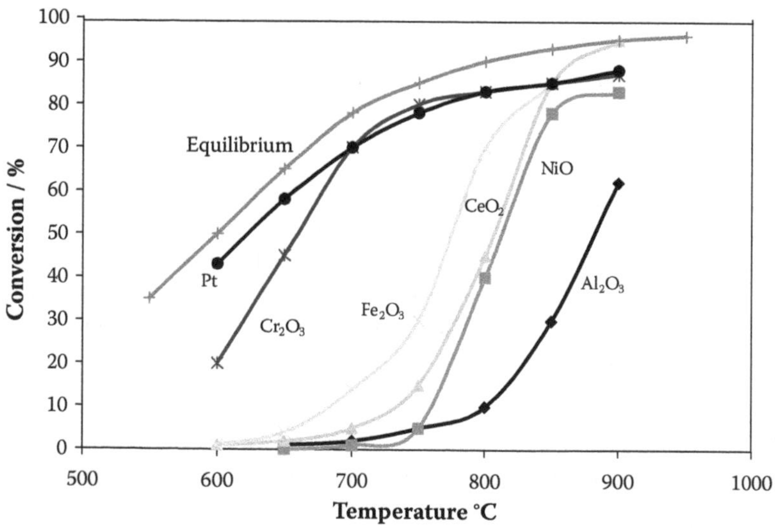

FIGURE 4.25 Relationship between conversion to $SO_2 + H_2O$ and temperature for catalytic metal oxides and Pt in a N_2 flow containing 4 mol% SO_3 at a space velocity of 4,300 h.

the reaction environment, support poisoning, amount of exposed surface area, and formation of acid salts. Of the commercially available catalysts that have been tested, the conversion rate declines as testing continues (figure 4.26). Working is ongoing to find a suitable support and loading fraction to ensure long-term performance.

A novel idea to incorporate Pt into the materials of construction for the H_2SO_4 decomposition reaction heat exchanger has been put forth by R. Ballinger.[39] By adding 1 to 5 wt% Pt to heat exchanger construction material, such as Alloy 800H and Inconel 617, the metals become self-catalytic and the decomposition of H_2SO_4 can be carried without the addition of catalyst pellets. Figure 4.27 shows the micrograph of Alloy 800H alloys with a 2 wt% Pt addition. At this low-level addition, it has been reported that no change is observable in the alloy, but an improvement in the material potential as a catalyst is observed. Development effort is ongoing to characterize the effectiveness of such alloys for aiding the H_2SO_4 decomposition reaction.

4.4.3.2 HI Decomposition

The decomposition of HI takes place in the gaseous phase between 300 and 450°C, depending on the distillation process. Due to the corrosive nature of HI, I_2, and H_2, only a limited number of catalysts have been evaluated in the process. Pt and Pd-based catalysts are not suitable, as they will exhibit severe corrosion if any measurable amount of moisture is present in the chemical stream. Activated charcoal is the only satisfactory catalyst found to date. However, it has been observed that iodine can be trapped inside the charcoal pores if the decomposition temperature is too low. Hence, a higher reaction temperature is required in order to maintain the reflux of iodine. Unfortunately, this can lead to accelerated corrosion. There is a need to identify a catalyst that can promote the decomposition at a lower temperature, which will be beneficial to the process. There is currently no work ongoing in this area.

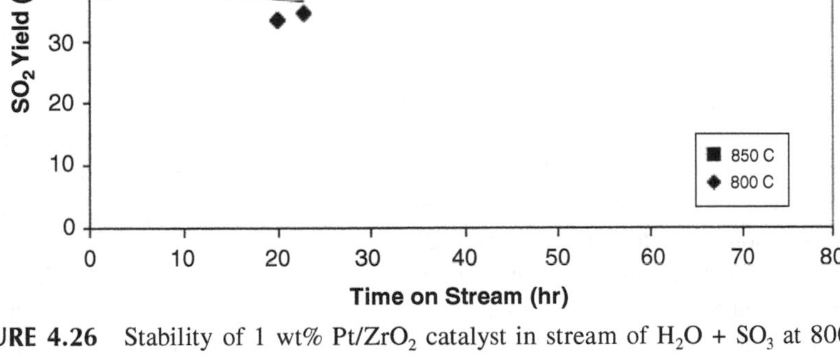

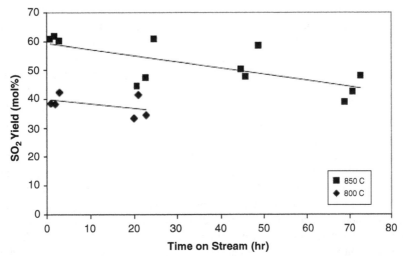

FIGURE 4.26 Stability of 1 wt% Pt/ZrO_2 catalyst in stream of $H_2O + SO_3$ at 800° and 850°C.

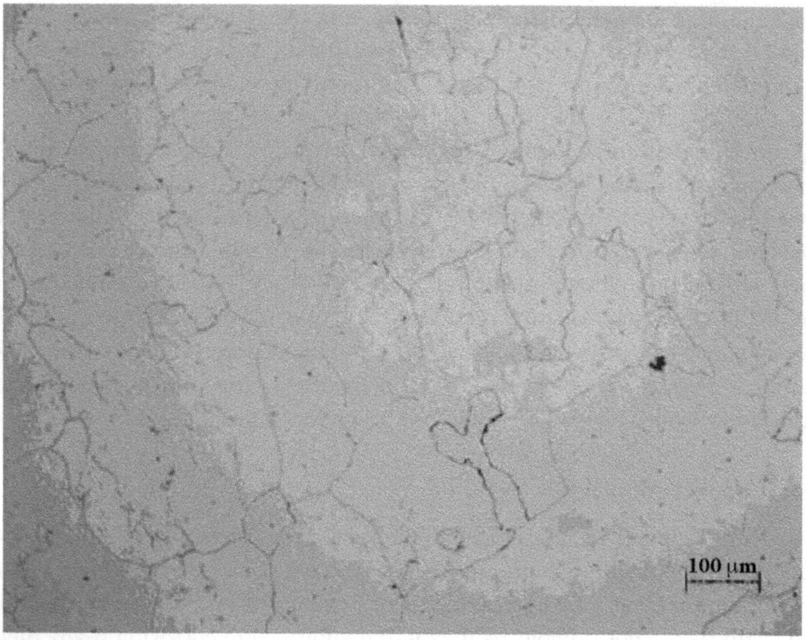

FIGURE 4.27 Micrograph of Alloy 800 + 2 wt% Pt at 100X.

4.5 SUMMARY

Commercial success of sulfur–iodine hydrogen production depends largely on the capacity to identify materials of construction that can handle the corrosive environments and on the ability to manufacture process components with these materials economically. This chapter has reviewed a cross-section of materials data generated by testing in the various process settings within the cycle. Ta alloys and SiC have been shown to have good corrosion characteristics in all the liquid environments, but individual candidates have also been suggested for use in particular conditions. Both of these materials have unique mechanical properties, and much effort is needed in order to use them for component fabrication.

In addition, the effect of cross-contamination of both process streams and corrosion products on their corrosion performance will need to be addressed. For high-temperature sulfur acid decomposition, a number of metallic alloys have been identified, but their long-term creep characteristics remain an issue. Reactors constructed from SiC will be undergoing testing in the near term. Even though they are extremely corrosion resistant, obstacles in component fabrication with such ceramic materials remain. Various Hastelloys have demonstrated very good general and stress corrosion behavior in the intermediate-temperature HI decomposition environment, but the effect of moisture in the vapor stream will need to be studied. Two other important S-I cycle-related material areas are separation membranes and catalysts. Different water and hydrogen separation membranes have been investigated. Their long-term viability and cost will need to be examined. Pt and oxide-based

SO_3 decomposition catalysts and their substrates are being studied extensively in an effort to improve their performance. Intelligent designs to incorporate Pt into the decomposer construction materials have shown promise. Similar energy is needed to study HI decomposition catalysts.

REFERENCES

1. Miyamato, Y. et al., R&D program on hydrogen production system with high temperature cooled reactor, in *International Hydrogen Energy Forum*, Vol. 2, Munich, Germany, 2000, pp. 271–278.
2. Brown, L.C., Funk, J.F., and Showalter, S.K., *Initial Screening of Thermochemical Water Splitting Cycles for High Efficiency Generation of Hydrogen Fuels Using Nuclear Power*, GA-A23373 Report, General Atomics, San Diego, 2000.
3. Brown, L.C., Besenbruch, G.E., Lentsch, R.D., Schultz, K.R., Funk, J.F., Pickard, P.S., Marshal, A.C., and Showalter, S.K., *High Efficiency Generation of Hydrogen Fuels Using Nuclear Power*, GA-A24285 Report, Rev. 1, General Atomics, San Diego.
4. Norman, J.H., Besenbruch, G.E., and O'Keefe, D.R., *Thermochemical Water-Splitting for Hydrogen Generation*, GRI-80/0105, GA-A15267 Report, General Atomics, San Diego, 1981.
5. Norman, J.H., Mysels, K.J., O'Keefe, D.R., Stowell, S.A., and Williamson, D.G., *Proceedings of the 2nd World Hydrogen Energy Conference*, Vol. 2, 1978, p. 513.
6. Norman, J.H., Mysels, K.J., Sharp, R., and Williamson, D.G., *International Journal of Hydrogen Energy*, 7, 545–556, 1982.
7. Sakuraia, M., Nakajimaa, H., Amirb, R., Onukia, K., and Shimizua, S., Experimental study on side-reaction occurrence condition in the iodine-sulfur thermochemical hydrogen production process, *International Journal of Hydrogen Energy*, 25, 613–619, 2000.
8. Ohashi, H., Inaba, Y., Nishihara, T., Inagaki, Y., Takeda, T., Hayashi, K., Katanishi, S., Takada, S., and Mano, T., Performance Test Results of Mock-Up Test Facility of HTTR Hydrogen Production System, paper presented at the Proceedings of the 11th International Conference of Nuclear Engineering, 2003.
9. Nakajima, H., Sakurai, M., Ikenoya, K., Hwang, G.J., Onuki, K., and Shimizu, S., Closed Cycle Continuous Hydrogen Production Test by Thermochemical IS Process, paper presented at the Proceedings of the 7th International Conference of Nuclear Engineering (ICONE-7), Tokyo, 1999.
10. *High Efficiency Hydrogen Production from Nuclear Energy: Laboratory Demonstration of S-I Water-Splitting.*
11. Sakurai, M., Nakajima, H., Onuki, K., and Shimizu, S., Investigation of two liquid phase separation characteristics on the iodine-sulfur thermochemical hydrogen production process, *International Journal of Hydrogen Energy*, 25, 605–611, 2000.
12. Roth, M. and Knoche, K.F., Thermochemical water splitting through direct HI-decomposition from $H_2O/HI/I_2$ solutions, *International Journal of Hydrogen Energy*, 14, 545–549, 1989.
13. Lai, G.Y., *High-Temperature Corrosion of Engineering Alloys*, ASM International, Materials Park, OH, 1990.
14. Trester, P.W. and Staley, H.G., *Assessment and Investigation of Containment Materials for the Sulfur-Iodine Thermochemical Water-Splitting Process for Hydrogen Production*, GRI Report 80/0081, 1981.
15. Craig, B.D., *Handbook of Corrosion Data*, ASM International, Metals Park, OH, 1989.
16. Onuki, K., Ioka, I., Futakawa, M., Nakajima, H., Shimizu, S., and Tayama, I., Screening tests on materials of construction for the thermochemical IS process, *Corrosion Engineering*, 46, 141–149, 1997.

17. Kubo, S. et al., Corrosion Test on Structural Materials for Iodine-Sulfur Thermochemical Water-Splitting Cycle, paper presented at the Proceedings of the 2nd Topical Conference on Fuel Cell Technology, AIChE 2003 Spring National Meeting, 2003.
18. Lillo, T., Communications on "Materials for the Sulfuric Concentration and Decomposition Section," Idaho National Laboratory.
19. *Decomposition of Sulfuric Acid Using Solar Thermal Energy*, GA-A17573 Report, General Atomics, San Diego, 1985.
20. Porisini, F., Selection and evaluation of materials for the construction of a pre-pilot plant for thermal decomposition of sulfuric acid, *International Journal of Hydrogen Energy*, 14, 267–274, 1989.
21. Savitsky E.M., Arskaya, E.P., Lazarev, E.M., and Korotkov, N.A., Investigation of corrosion resistance of materials in the presence of sulfuric acid and its decomposition products applied in the thermochemical cycle of hydrogen production, *International Journal of Hydrogen Energy*, 7, 393–396, 1982.
22. Gern, F.H. and Kochendoerfer, R., Liquid silicon infiltration: description of infiltration dynamics and silicon carbide formation, *Composites*, 28A, 355–364, 1997.
23. Robin, A. and Rosa, J.L., Corrosion behavior of niobium, tantalum and their alloys in hot hydrochloric and phosphoric acid solutions, *International Journal of Hydrogen Energy*, 18, 13–21, 1997.
24. Kurata, Y., Tachibana, K., and Suzuki, T., High temperature tensile properties of metallic materials exposed to a sulfuric acid decomposition gas environment, *Journal of Japan Institute of Metals*, 65, 262–265, 2001.
25. Robin, A., Corrosion behavior of niobium, tantalum and their alloys in boiling sulfuric acid solutions, 15, 317–323, 1997.
26. Wong, B., Brown, L.C., and Besenbruch, G.E., *Corrosion Screening of Construction Materials for Hydrogen Iodide Decomposition Heat Exchanger Fabrication*, GA-A25128 Report, General Atomics, San Diego, 2005.
27. Stewart, F.F., Orme, C.J., and Jones, M.G., Membrane Processes for the Sulfur-Iodine Thermochemical Cycle, paper presented at the AIChE Spring Meeting 2005, Atlanta, GA, 2005.
28. Orme, C.J., Jones, M.G., and Stewart, F.F., Pervaporation of water from aqueous HI using Nafion-117 membranes for the sulfur-iodine thermochemical water splitting process, *Journal of Membrane Science*, 252, 245–252, 2005.
29. Hwang, G.-J., Onuki, K., Nomura, M., Kasahara, S., and Kim, J.-W., Improvement of the thermochemical water-splitting IS (iodine-sulfur) process by electro-electrodialysis, *Journal of Membrane Science*, 220, 129–136, 2003.
30. Nomura, M., Fujiwara, S., Ikenoya, K., Kasahara, S., Nakajima, H., Kubo, S., Hwang, G.-J., Choi, H.-S., and Onuki, K., Application of an electrochemical membrane reactor to the thermochemical water splitting IS process for hydrogen production, *Journal of Membrane Science*, 240, 221–226, 2004.
31. Kasahara, S., Ikenoya, K., Kubo, S., Onuki, K., Nomura, M., Nakao, S., Okuda, H., and Fujiwara, S., Development of an electrochemical cell for efficient hydrogen production through the IS process, *Environment and Energy Engineering*, 50, 1991–1998, 2004.
32. Hwang, G.J., Onuki, K., and Shimizu, S., Separation of hydrogen from a H_2-H_2O-HI gaseous mixture using a silica membrane, *AIChE Journal*, 46, 92, 2000.
33. Yoshino, Y., Suzuki, T., Nair, B.N., Taguchi, H., and Itoh, N., Development of tubular substrates, silica based membranes and membrane modules for hydrogen separation at high temperature, *Journal of Membrane Science*, 267, 8–17, 2005.
34. Nomura, M., Kasahara, S., Okuda, H., and Nakao, S., Evaluation of the IS process featuring membrane techniques by total thermal efficiency, *International Journal of Hydrogen Energy*, 30, 1465–1473, 2004.

35. Hwang, G.J. and Onuki, K., Simulation study on the catalytic decomposition of hydrogen iodide in a membrane reactor with a silica membrane for the thermochemical water splitting IS process, *Journal of Membrane Science*, 194, 207, 2001.
36 Ginosar, D.M., Glenn, A.W., and Petkovic, L.M., Stability of Sulfuric Acid Decomposition Catalysts for Thermochemical Water Splitting Cycles, paper presented at the AIChE Spring Meeting 2005, Atlanta, 2005.
37. Tagawa, H. and Endo, T., Catalytic decomposition of sulfuric acid using metal oxides as the oxygen generating reaction in thermochemical water splitting process, *International Journal of Hydrogen Energy*, 14, 11–17, 1989.
38. Ishikawa, H., Ishii, E., Uehara, I., and Nakan, M., Catalyzed thermal decomposition of H_2SO_4 and production of HBr by the reaction of SO_2 with Br_2 and H_2O, *International Journal for Hydrogen Energy*, 46, 237–246, 1982.
39. Ballinger, R., The Development of Self Catalytic Materials for Thermochemical Water Splitting Using the Sulfur-Iodine Process, paper presented at the UNLV-HTHX quarterly meeting, Univerity of Nevada, Las Vegas, December 5, 2005.

5 Materials Requirements for Photobiological Hydrogen Production

Daniel M. Blake, Wade A. Amos,
Maria L. Ghirardi, and Michael Seibert

CONTENTS

5.1 INTRODUCTION

The world's energy infrastructure is under pressure from rapidly increasing demand. Recent worldwide events have increased public anxiety about the cost of gasoline and heating fuels, and the security of these resources has become an issue. Finally, the amount of pollution and CO_2 that society is generating is increasing, and everyone, from individual villages to entire countries, is looking for, or should be looking for, a sustainable, secure energy system. Currently, the world has over 6 billion individuals, all powered by solar energy. The food we eat and the oxygen we breathe come from photosynthesis. Can energy from the sun, which indirectly powers all animal life on the planet, including ourselves, also give us hydrogen from water to be used as a renewable energy carrier?

This review will discuss the development of photobiological hydrogen production processes, where microorganisms (algae or cyanobacteria) function as living photocatalysts. In this context it is appropriate to consider microorganisms a material, analogous to a traditional chemical catalyst. In the same way that chemical catalysts are developed, microorganisms require improvements in kinetics, selectivity, resistance to deactivation, and efficiency. This review will highlight improvements required in the biological system, the materials requirements for photobioreactor systems in which H_2 will be produced from water, and economic considerations for the process. This is a challenging problem, but the development of new processes that can contribute to hydrogen production from sources other than fossil carbon carriers is critical to the sustainability of the hydrogen economy.

Biological H_2 production is expected to be one of many processes that contribute to the ultimate supply of H_2. But it is useful to consider that in an area of less than 5,000 square miles (about 0.12% of the U.S. land area) of bioreactor footprint in the desert southwest, photobiological processes could in principle produce enough H_2 to displace all of the gasoline currently used in the U.S. (236 million vehicles). The underlying assumptions are that H_2 could be produced from water at 10% efficiency (the maximum theoretical efficiency of an algal H_2 production system is about 13%) and that fuel cell–powered vehicles could get 60 miles/kg of H_2 (M. Mann, personal communication).

5.2 DESCRIPTION OF THE PROCESS

The photobiological production of H_2 is a property of only three classes of organisms: photosynthetic bacteria, cyanobacteria, and green algae.[1] These organisms use their photosynthetic apparatus to absorb sunlight and convert it into chemical energy. Water (green algae and cyanobacteria) or an organic/inorganic acid (photosynthetic bacteria) is the electron donor. This review will focus on oxygenic organisms such as green algae and some cyanobacteria, which produce hydrogen directly from water, without an intermediary biomass accumulation stage. This process is considered to have the highest potential sunlight conversion efficiency to H_2, which, as mentioned above, could be on the order of 10 to 13%.[2,3] To accomplish this, green algae and cyanobacteria can utilize the normal components of photosynthesis to split water into O_2, protons, and electrons, deriving the requisite energy from sunlight. However, instead of using the protons and electrons to reduce CO_2 and storing the energy as starch molecules (the normal function of photosynthesis), these organisms can recombine the protons and electrons and evolve H_2 gas *under anaerobic conditions* (figure 5.1) using a reaction catalyzed by an induced hydrogenase enzyme.

The enzymes responsible for H_2 production are metallocatalysts[4,5] that belong to the classes of [NiFe]- (in cyanobacteria) or [FeFe]- (in green algae) hydrogenases. Although sustained and continuous photobiological H_2 production has been achieved with green algae (see section 5.2.2), the establishment and maintenance of culture anaerobiosis is currently a *sine qua non* for the process. This requirement results from the following biological realities:

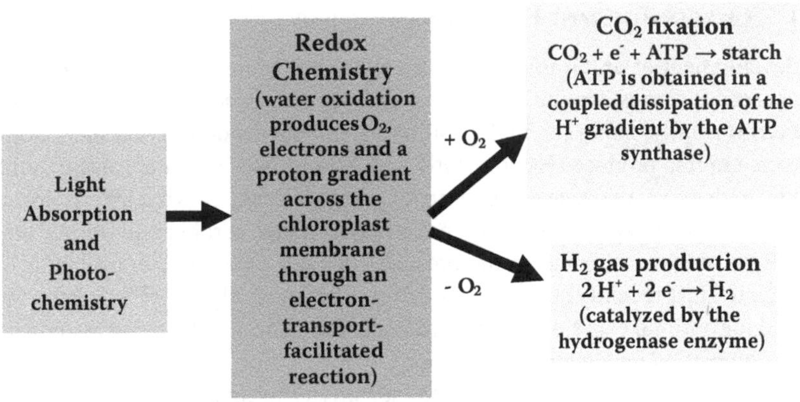

FIGURE 5.1 Representation of the major steps required for photosynthetic CO_2 fixation (upper right box) and hydrogen production (lower right box).

1. The hydrogenase genes in green algae and in some cyanobacteria are not expressed (the genes are not turned on) in the presence of O_2,[6–9] and a variable period of anaerobiosis is required to induce gene transcription.[10–12]
2. The expression and function of the genes that catalyze the assembly of the catalytic metallocluster of the algal [FeFe]-hydrogenases require anaerobiosis,[13] while in most cyanobacteria, the [NiFe]-hydrogenase genes are expressed and the protein assembled in the presence of O_2; in this case, the proteins are assembled in an inactive form, but can be activated quickly when exposed to anaerobic conditions.[14,15]
3. The activity of the catalytic metallocluster of the hydrogenases is inhibited reversibly (in cyanobacteria) or irreversibly (in green algae) by the presence of O_2.[4,16–18]

Until recently, culture anaerobiosis was achieved by a variety of means, including continuous purging with inert gases,[19,20] addition of exogenous reductants,[21] or addition of O_2 scavengers to the medium.[21,22] All of these methods are expensive, especially considering the large scale of potential commercial applications. Two approaches have been developed recently to circumvent the O_2 sensitivity challenge and may lead to the development of inexpensive H_2 production photobiological systems. The first involves (1) engineering [FeFe]-hydrogenases for increased O_2 tolerance[23] or (2) expressing bacterial [NiFe]-hydrogenases, known to have high O_2 tolerance, in cyanobacteria (P.-C. Maness, personal communication). The second approach involves the manipulation of green algal physiology to induce culture anaerobiosis, expression of the hydrogenase enzyme, and sustained H_2 photoproduction from water.[24] Although the first approach may eventually yield an enzyme that is fully functional in the presence of O_2, it is clearly a longer-term effort that will require intensive molecular engineering of enzymes and metabolic pathways. The feasibility of the second approach, on the other hand, has been demonstrated in the laboratory, and its economics have been estimated based on its maximum potential performance.[25,26] In the next two sections, we will describe other research issues that need to be addressed in order to further develop an O_2-tolerant hydrogenase (section 5.2.1) or induce culture anaerobiosis (section 5.2.2).

5.2.1 Oxygen-Tolerant Hydrogenase Systems

The development of an O_2-tolerant hydrogenase (by either mutagenesis of existing hydrogenases or transforming O_2-tolerant hydrogenase genes from bacteria into cyanobacteria) is only one of the biological requirements needed to obtain a photosynthetic system that produces H_2 efficiently and at high rates under aerobic conditions. As indicated above, the genes' encoding for an O_2-tolerant enzyme and its accompanying assembly enzymes must also be expressed under aerobic conditions, and the assembly process must occur in the presence of O_2 as well. These are major issues that we have just recently started to investigate in green algae,[13,27,28] and the factors required for optimal expression of the hydrogenase genes in algae and cyanobacteria are not completely understood at present.

However, even if the hydrogenase-O_2-sensitivity problem is solved, other major issues remain to be addressed before a biological system might become a commercial reality. The first, which is common to both green algae and cyanobacteria, is the existence of competing pathways for photosynthetic reductants. In both organisms, the major competing pathway under aerobic conditions is the CO_2 fixation pathway. In green algae, photosynthetically reduced ferredoxin donates electrons to FNR (ferredoxin–NADPH–oxidoreductase), FTR (ferredoxin–thioredoxin reductase), nitrite reductase, sulfite reductase, glutamate synthase, or hydrogenase.[29] The binding affinity of ferredoxin to each of these enzymes varies from 2.6 μM (in the case of FNR[30]) to 20 μM (in the case of nitrite reductase[31]) and has been estimated to be 10 to 35 μM in the case of the hydrogenase.[32–34] These estimates suggest that the interaction between ferredoxin and hydrogenase will have to be genetically manipulated in order to ensure that most of the photosynthetically generated reductants will be utilized for H_2 production.

In the case of cyanobacteria, two additional competing pathways must be taken into account as well. One of them involves the uptake hydrogenase found in N_2-fixing cyanobacteria,[35–37] which consumes the H_2 gas produced by either the bidirectional hydrogenase or nitrogenase. This problem can be easily addressed by genetically knocking out the uptake hydrogenase gene in the organism of choice.[38,39] The second competitive pathway is a homologue of respiratory Complex I present in the membranes of cyanobacteria and proposed to form a complex with the bidirectional hydrogenase through a diaphorase subunit.[14,40] This means that although functionally able to accept reductants from the photosynthetic electron transport chain, as indicated in figure 5.1, the cyanobacterial hydrogenase may also play a physiological role in using these reductants (in the form of NADPH) to support respiration. Indeed, this role of hydrogenases may explain the lack of H_2 accumulation in cyanobacteria under illumination, and the detection of measurable light-induced H_2 production only in mutants defective in the NADPH dehydrogenase complex, homologous to Complex I.[15] Although desirable, the redirection of photosynthetic reductants away from the CO_2 fixation pathway and toward hydrogenase has dire implications for the rates of H_2 production, at least in the case of green algae. In the absence of CO_2 fixation, the ATP generated by the operation of the photosynthetic electron transport pathway is not consumed. ATP is generated by dissipation of the proton gradient established across the two sides of the thylakoid membranes, which in turn is coupled to the function of the ATPase enzyme. Thus, it is clear that the loss of the CO_2 fixation pathway will inevitably lead to a static, maximum proton gradient between the lumenal and stromal sides of the membrane. Under these conditions, photosynthetic electron transport in green algae is downregulated,[41–43] and the rates of H_2 photoproduction become limited. This problem is currently being addressed by designing and inserting artificial proton channels across the thylakoid membrane, to be expressed only under conditions of H_2 photoproduction.[42] It is obvious that more focused research is required in this area.

The existence of an undissipated proton gradient in cyanobacteria, and its possible effect on photosynthetic electron transport rates, is less clear[16] and, to our knowledge, is not currently being addressed by any research group.

Finally, in order to ensure that close to the theoretical 13% light conversion efficiency is achieved under solar illumination, the low-light saturation properties of green algae and cyanobacteria in mass culture need to be considered as well. Due to the presence of large amounts of light-harvesting antenna pigments, the light reactions of photosynthesis in both organisms saturate at less than 1/5 that of full solar intensity.[44] As a result, more than 80% of the absorbed photons are wasted, reducing the culture productivity to very low levels.[45] It may be possible to eliminate this energy inefficiency by truncating the pigment antennae complex of algae/cyanobacteria to their minimum functional size (about 40 Chl/Photosystem II and 96 Chl/Photosystem I). At these stoichiometries, the light conversion apparatus of photosynthesis consists only of reaction centers and their proximal light-harvesting antennae.[46,47] Many researchers have reported mutants with decreased antenna size and increased photosynthetic productivity.[25,48–56] However, the H_2-production capability of these mutants has not yet been evaluated.

In summary, an optimal H_2-photoproducing biological system would have to be able to function at high rates and with about 10% solar light conversion efficiency in an aerobic atmosphere, simultaneously evolving H_2 and O_2 gases at a ratio of 2:1. Materials and gas separation issues, as well as photobioreactor designs to address these, will be discussed in section 5.3.

5.2.2 Anaerobic Hydrogenase Systems

In contrast to a photosynthetic system that utilizes an O_2-tolerant hydrogenase under development to photoproduce H_2 from water, an alternative process based on manipulation of the physiology of algae was developed by some of us a number of years ago.[24] As originally published, the process was based on the specific effects of sulfate deprivation on the O_2-evolving properties of *Chlamydomonas reinhardtii*. Under this condition, *C. reinhardtii* is unable to maintain the fast turnover of the D1 Photosystem II reaction center protein.[57] As a consequence, the photosynthetic O_2 evolution activity of the cells gradually decreases to a level lower than their respiratory activity, and this leads to culture anaerobiosis. Once the cultures become anaerobic, the hydrogenase gene is induced and the culture starts producing H_2 gas. The H_2-producing activity is temporary, however, due to the eventual nonspecific effects of sulfur deprivation on other cellular functions. Hydrogen production can be reactivated by a short incubation (about 2 days) of the cultures in sulfur-replete medium in order to reconstitute photosynthetic function and critical cellular enzymes.[58] Alternatively, continuous H_2 production can be maintained by the use of a two-reactor system, where O_2 evolution and H_2 production are physically separated in different photobioreactors.[59] Hydrogen production has been observed continuously for 6 months in such a system, but at lower specific rates and low cell suspension concentrations.

Although sulfur-deprived cultures can be manipulated to produce H_2 continuously in the light, the maximum rates of H_2 production observed are only about 25% of the corresponding maximum potential of photosynthetic electron transport,[41,60] suggesting that there are either (1) limitations in the activity of critical electron transport components or (2) downregulation of photosynthesis by, perhaps, the accumulated proton gradient (see section 5.2.1). In addition, the presence of a large light-harvesting antenna is also a disadvantage when using sulfur-deprived algae in mass cultures for producing H_2, as discussed above.

Recently we demonstrated that sulfur-deprived algal cells can be immobilized onto glass fibers in order to increase the culture density and potentially the light conversion efficiency of a mass culture. Under these conditions, the algae produced H_2 for longer periods at only slightly lower specific H_2 production rates when compared to algal suspension cultures, but at twice the rate per volume of photobioreactor.[61] However, this was achieved with constant purging with argon gas and liquid medium replacement, which may have a significant role in process cost.

A recent development[62] may contribute to further increase the rates of H_2 production by sulfur-deprived cultures. It was observed that a mutant of *C. reinhardtii* that overaccumulates starch is defective in state transitions, has decreased rates of cyclic electron transfer around Photosystem I, and photoproduces H_2 gas at rates significantly higher than those of its parental wild-type (WT) strain. These results support the fact that starch degradation during sulfur deprivation plays a triple role in H_2 production: (1) as an electron donor to the photosynthetic electron transport chain,[41,63] (2) as a regulatory factor for hydrogenase gene transcription,[27] and (3) as a substrate for aerobic respiration, responsible for keeping the cultures anaerobic.[58,63] It remains to be seen whether the H_2 production activity of the mutant can indeed be optimized for commercial applications since the WT organism, in the case of Kruse et al.'s[62] work, was a very poor H_2 producer compared to other *C. reinhardtii* strains.

The process of sulfate uptake by *C. reinhardtii* occurs in two steps: sulfate anions are transported into the cytosol through the operation of a plasma membrane transporter system,[64] and from there they are translocated into the chloroplasts through a newly discovered chloroplast enveloped–localized sulfate permease holocomplex.[65–67] One of the sulfate permease genes, SulP, was attenuated by antisense technology. The resulting transformants displayed different levels of sulfate permease activity and exhibited phenotypes of sulfur-deprived cells.[66] The H_2-production capability of these transformants could be detected even in the presence of 100 μM sulfate in the medium, and varied between 60 and 100% of the activity of the wild-type strain measured in the absence of sulfate. In principle, these transformants could be used in photobioreactors for H_2 production under controlled sulfate levels, as long as the cultures were able to maintain robust respiratory activity to sustain anaerobiosis in the light. Clearly, this will require a regulated expression of the antisense SulP gene, to allow the cultures periods of normal photosynthetic activity to replace storage material required for respiration.

Until recently, the sulfur-deprivation process was dependent on the presence of acetate in the medium, at least during the initial steps of sulfur deprivation.[63,68] As the cultures became anaerobic, the active uptake of acetate ceased, and the cultures started using degraded starch as the substrate for respiration instead of extraneous acetate.[63] Although the cost of added acetate was not a major contributor to the cost of the system,[26] it was believed that the consumption of an extraneous organic source of carbon by the cultures detracted from the claims that the system is based on direct biophotolysis.[69] Two recent publications, however, report H_2 production by sulfur-deprived cultures resuspended in the total absence of added acetate.[70,71] For high H_2-production rates, the photoautotrophic system requires careful control of the light intensity during different phases of the process in order to balance the rates of O_2 evolution and starch degradation.[71] This development demonstrates that H_2 photoproduction can be dependent totally on photosynthetic water oxidation, and that it is possible to optimize a system that does not require any added sources of organic carbon.

Finally, it is worth mentioning that H_2 photoproduction induced by sulfur deprivation is not a property exclusive to *C. reinhardtii*. For, example, the marine unicellular alga *Platymonas subcordiformes* has been sulfur deprived in the absence of acetate in the medium and has been reported to produce some H_2 gas.[72,73] The experimental conditions for optimal H_2 production by this organism are still being identified, and research is ongoing in many laboratories aimed at identifying other organisms from nature that exhibit similar properties.

An optimal anaerobic H_2-photoproducing system will consist of two stages: one for production of cellular storage material by photosynthesis, and the other for H_2 photoproduction under sulfur deprivation (either physiologically or genetically induced). The maximum solar energy conversion efficiency of this system is expected to be about 1%. This is due to the fact that the maximum rate of H_2 photoproduction cannot be higher than the corresponding rate of respiration required to maintain anaerobic cultures (maximum respiration in *C. reinhardtii* is about 20 μM O_2•mg Chl^{-1}•h^{-1}, and maximum H_2 photoproduction rate is 400 μM H_2•mg Chl^{-1}•h^{-1}). As is the case with an O_2-tolerant hydrogenase system, an anaerobic H_2-photoproducing system will also require mutants with truncated antenna in order to operate at maximum efficiency under solar illumination. However, the latter will only produce H_2 gas, so that gas separation issues do not play a role or contribute to the overall cost of the system. Immobilized systems may turn out to have cost advantages over suspension culture systems in that changes between sulfur-replete and sulfur-deprived conditions are as easy as turning a valve.

5.3 REACTOR MATERIALS

Engineering design of full-scale photobioreactors and the balance of the facility for photobiological hydrogen production has not been considered beyond very general concepts. Consequently, there has been little effort to identify construction material and establish boundaries for specifying materials performance and properties. In

2004 and 2005 the U.S. Department of Energy started a small project at the National Renewable Energy Laboratory (NREL) to provide an initial look at the question of material requirements.

5.3.1 Photobioreactors

There is a substantial amount of literature that has investigated biotechnology for algal mass cultivation as well as photobioreactor systems for growing algae.[74] A full range of open and closed photobioreactors are shown schematically in figure 5.2. A recent review[2] and a cost analysis[75] have begun to discuss solar photobioreactor concepts and specific challenges for practical application of a closed biological H_2 production system. Both used what is considered to be an optimistic assumption of 10% efficiency for conversion of sunlight to H_2. Assuming a production capacity required to meet the needs of an average service station supplying fuel for the transportation sector and a maximum insolation of 1 kW/m^2 at full sun, one can calculate a required reactor area on the order of 110,000 m^2 (about 27 acres). The area will be sensitive to solar insolation averaged over a year, which will depend on location.[76] For the purposes of this discussion, the above bioreactor area will serve as the commercial scale required for the production of useful amounts of H_2. Open ponds and raceways for commercial production of algae approach this scale, but no closed bioreactors of this size have been constructed.[77] Designs for shallow solar ponds for thermal applications may also provide some guidance for the reactor application.[78–80]

Capturing energy from the sun for the production of H_2 creates unique challenges for reactor design and materials of construction. The reactors must be closed to contain H_2 and exclude O_2. The low energy density of the sunlight dictates a large area for collecting solar energy for use by direct, light-driven photobiological water-splitting technologies. The total area of the solar collector will be about the same whether the sunlight is used at ambient intensity (one sun) or concentrated before being directed into a reactor. The difference will be in the reactor design. A one-sun reactor has the same geometric area as the light aperture. The reactor for a solar concentrating system will be more compact, but the geometric area of the sunlight collection optics will likely be somewhat greater than for the same production level from a one-sun system. This is because of the losses inherent in concentrator elements and those that transmit light into the reactor. The trade-off is in the bioreactor cost and performance vs. the cost of the concentrating optics. Concentrating reactor systems would also require planar or optical fiber light-transmitting elements to carry the light from the line focus or dish concentrating collector into the reactor.[76] Only the specular component of the sunlight is concentrated. A nonconcentrating system will use both the specular and diffuse components of the sunlight. Hence, they are less affected than are concentrating systems by cloud cover, other atmospheric effects that cause scattering, and are more tolerant to such things as dust, soil, and surface imperfections in the transparent reactor cover material.

Some work has been done on photoreactor concepts that would collect sunlight using dish solar concentrators that direct concentrated sunlight onto the aperture of a light pipe or optical fiber system.[76,81,82] An optical fiber system would then carry light into the reactor and disperse it evenly throughout the bioreactor volume. Such a

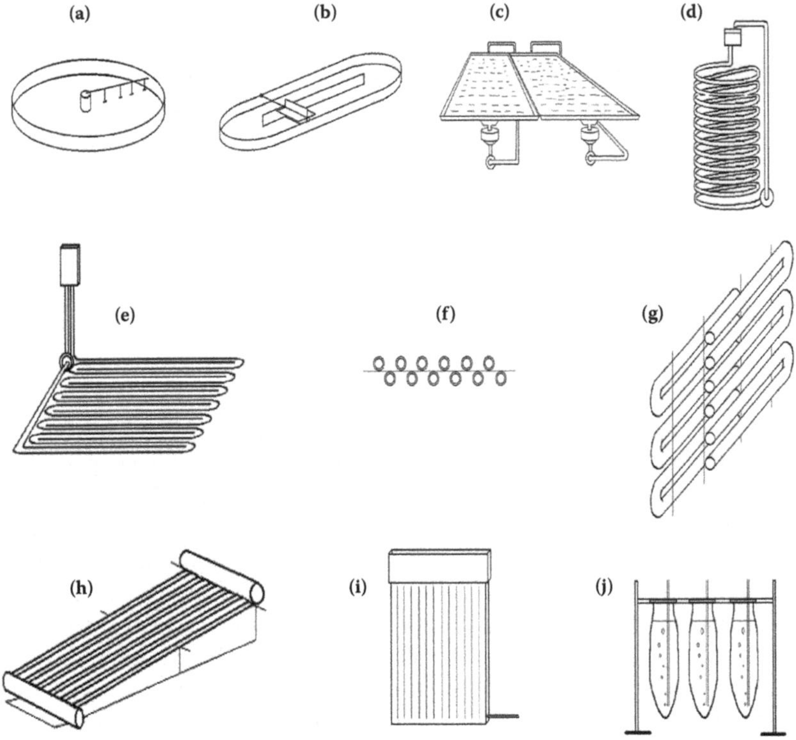

FIGURE 5.2 Schematic diagrams of the most common outdoor algal photobioreactor systems: (a) circular pond, (b) paddle wheel raceway, (c) sloping panel reactor, (d) helical tubular reactor, (e) plane tubular reactor, (f) two-plane tubular reactor, (g) espalier tubular reactor, (h) sloping tubular reactor, (i) vertical alveolar panel reactor, and (j) hanging sleeve reactor. (Figure 6 from Torzillo, G. and Vonshak, A., in *Recent Advances in Marine Biotechnology*, Fingerman, M. and Nagabhushanam, R., Eds., Science Publishers, Inc., Enfield, NH, 2003, p. 45. With permission.)

system would reduce the area through which light enters the H_2-production reactor, and this would reduce (but not completely eliminate) the impact of potential H_2 permeation through the transparent material. It would also move the durability requirements for optical properties from the photobioreactor cover material to the collection mirrors and secondary concentrator. Work on the engineering and performance has been done for different configurations of both open and closed reactors, mainly in the context of production of high-value products.[83]

5.3.2 Photobioreactor Materials

The bases for selecting key materials properties, identifying polymer types for consideration, evaluating properties of materials of construction, and finding sources of materials were discussed in a recent report to the U.S. Department of Energy.[84] A summary of the key operating requirements is given in table 5.1. Specifications

TABLE 5.1
Photobiological Hydrogen Photobioreactor Requirements

Property	Range
Spectral requirement	>400–900 nm (depends on the organism)
Light intensity	0.05–0.10 sun intensity (may be increased by antenna size truncation)
Hydrogen pressure	As high as practical
Oxygen pressure	ppm to few percent (depends on the organism)
Gas permeation rates	As low as practical
pH	6.5–8.2 (biological limits)

for important material properties have not been established. The list of properties would include such things as transmittance, outdoor lifetime, biocompatibility, H_2- and O_2-permeation rates, physical and mechanical properties, chemical resistance, and those properties required for particular reactor configurations. The materials that can be considered are subject to an almost unlimited range of modifications. A given type of polymer, polycarbonate, for example, can be produced with different organic groups on the polymer chain, different stabilizer packages, and different protective coatings. Each will have unique durability characteristics. It is necessary to select candidate materials and evaluate durability and performance with the solar H_2 requirements as targets. The initial cost of materials and the system maintenance costs associated with materials degradation over time are major economic considerations for solar systems.

The importance of understanding the performance and lifetime of materials exposed to sunlight and weather has long been recognized for architectural, advertising display, and greenhouse applications. It is critical for renewable energy and energy-efficiency technologies such as photovoltaic and solar thermal electric generation, solar hot water and space heating, and high-performance windows. The economics of the renewable energy and fuels sectors are strongly affected by the capital and operating costs associated with the materials used to collect, reflect, and transmit the energy from sunlight, or encapsulate the photoactive components.

Glass has many advantages as a glazing material; however, weight and cost of low-iron glass with high transmission for the solar spectrum, coupled with the large areas that are required, have fueled the search for polymers that can be used as glazings and mirrors in solar applications. Work on lifetime and durability of polymers for glazing and heat exchanger components for solar water heating systems has been reviewed.[85–89] Polymer materials for solar reflectors have been the subject of development since the push for renewable energy began in the 1970s. This has been covered in a number of U.S. Department of Energy reports.[90,91] Materials for construction, including polymers, for shallow solar ponds were reviewed in a thesis.[85] Materials of construction were also considered in an early evaluation of the cost for a solar photocatalytic H_2 production reactor.[92]

Extensive information on the durability of polymers based on outdoor and accelerated weathering tests is available from work carried out at NREL and other places over the last 25 years. In the NREL work, outdoor exposure is done at sites in Golden, CO, Miami, FL, and Phoenix, AZ. These sites are representative of moderate, hot humid, and hot dry climates, respectively. Accelerated testing is done with standard commercial equipment. Much of this work has been performed using the durability of optical properties as the indicator of lifetime since the application was for concentrating mirrors or encapsulating photovoltaic solar cells.[93]

The best performers, as measured by optical properties after accelerated and real-time weathering tests, are acrylics, polycarbonates, polyesters, and fluorinated polymers such as Teflon® and related materials.[93] It should be noted that with the exception of the fluorinated polymers, the durability depends on UV and oxidation protection additives or overlayers. Real-time and accelerated weathering effects on transmittance are shown in figure 5.3 for some of the better-performing materials.[93] The outdoor data are from the NREL Golden, CO, site, and the accelerated tests were done in an Atlas Ci5000 Weatherometer at NREL. Optical durability test results, updated with new data collected in 2005, are presented graphically as plots of percent hemispherical transmittance solar weighted between 300 and 1,200 nm vs. the total UV dose (100 MJ/m²). This allows the real-time and accelerated results to be plotted on the same scale (see top scale in figure 5.3). The UV dose in accelerated tests is converted to the equivalent amount of time it would take to achieve that dose outdoors. The equivalent exposure time is calculated by multiplying the time in accelerated test conditions by the acceleration factor. For example, in figure 5.3, the optical performance of the Sunguard® polycarbonate construction begins to fall off after a UV dose equivalent to about 5 to 6 years outdoors.

A key property of potential construction materials for solar H_2 photobioreactors is the rate of H_2 and O_2 permeation through the materials. Data on the perme-

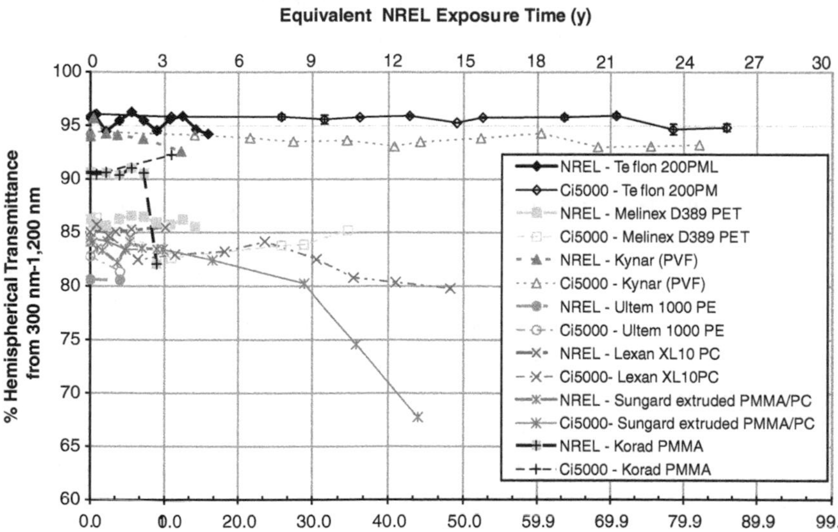

FIGURE 5.3 Accelerated and real-time weathering of polymer materials obtained at NREL. PET, polyethyleneteraphthalate; PVDF, polyvinylidenedifluoride; PE, polyethylene; PC, polycarbonate; PMMA, polymethylmethacrylate.

ability coefficient, P (cm^3·mm/m^2·day·atm), of O_2 are available for a wide range of polymers.[94] However, similar data for H_2 permeation are limited. Some data that we did find are presented in table 5.2. The temperature dependence of the permeability coefficient follows the pattern for rates of chemical processes in that the permeation rate roughly doubles for every 10°C rise in temperature. Because of the shortage of data for H_2 permeation, work was done under subcontract to NREL in 2005 to obtain data on more of the polymers of interest for solar applications. These data are shown in figure 5.4.[93] The results are preliminary but should be representative of the performance of received polymer materials. Errors are estimated to be on the order of ±10%. The higher permeability coefficients for the thicker polymers may reflect the difficulty of sealing the samples in the test fixture. Oxygen-permeation rates at NREL were measured on a Mocon Oxytran instrument. To our knowledge, there is no information available on the effect of polymer weathering on H_2 or O_2 permeation. One can assume that permeability will increase with time. It can be anticipated that the gas-permeation specification for biological H_2-reactor materials must be as low as is technically practical. A recent review of technology for reducing gas permeation of polymers in high-technology applications provides some information on the performance and cost of barrier coatings.[95] The main effort has been barrier coating to reduce O_2 and water permeation. Again, we are not aware of any work done to reduce H_2-permeation rates through different polymers.

Looking ahead to the operation of commercial photobiological H_2 production plants, one can see the potential for materials enhancements that will help with the cleaning of outside surfaces, preventing biofilm growth on inner surfaces, reducing O_2 permeation into and H_2 permeation out of the bioreactors, and increasing the lifetime of the construction materials. These challenges must be addressed within rigorous cost constraints in order for the process to compete in a commodity market.

TABLE 5.2

Hydrogen and Oxygen Permeability Coefficients: Literature Values

Polymer	Designation	Permeability Coeffient (P)cm^3·mm/m^2·day·atm Hydrogen, H_2	Permeability Coeffient (P)cm^3·mm/m^2·day·atm Oxygen, O_2
Polyethylene	HDPE	156	49
Tetrafluoroethylene	TFE	520	222
Polyester	PET	39.4	2.4
Polycarbonate	PC	Not available	67.9
Silicone		17716	19685

Source: Massey, L.K., *Permeability Properties of Plastics and Elastomers: A Guide to Packaging and Barrier Materials*, 2nd ed., Plastic Design Library/William Andrew Publishing, Norwich, NY, 2003, Appendix II.

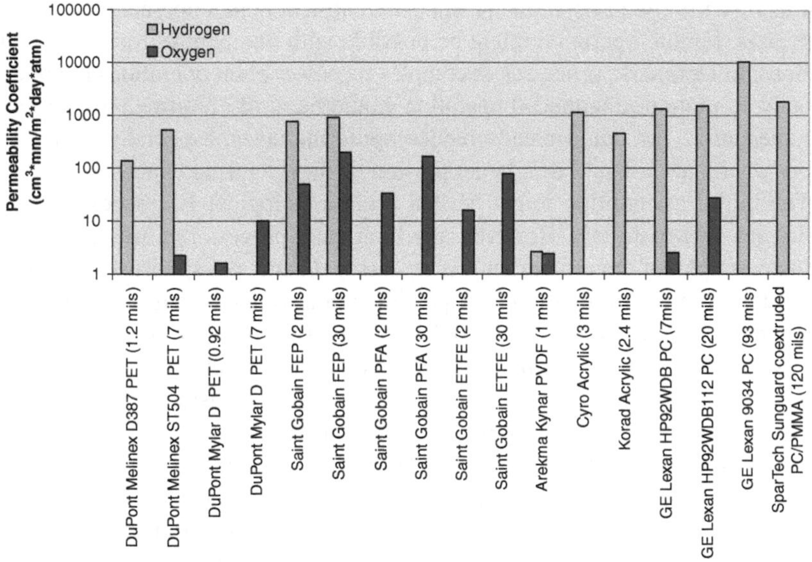

FIGURE 5.4 Permeability measurements for hydrogen and oxygen diffusion through different plastic materials.[93] FEP, fluorinated ethylene propylene; PFA, perfluoroalkoxy fluorocarbon; ETFE, ethylene-tetrafluoroethylene.

5.4 ECONOMICS AND COST DRIVERS FOR PHOTOBIOLOGICAL HYDROGEN PRODUCTION

A recent cost analysis has looked at the economics of biological H_2 production using a *C. reinhardtii* green algal system such as those described in section 5.2.[75] Although photobiological H_2 production with cyanobacteria occurs via a different pathway, many of the same design factors influence the process economics for both methods of H_2 production. The economics can be analyzed by looking at both operating and capital costs. The two cost components can be combined, and then a final H_2-production cost in dollars per kilogram of H_2 can be calculated using a discounted cash-flow analysis. In addition to items such as equipment costs or labor rates that clearly affect capital costs or operating expenses, there are some process variables that affect the overall economics, such as the specific H_2-production rate of the organism utilized.

5.4.1 Operating Costs

Many operating costs are the same as those for any conventional chemical production process, such as operating labor, raw materials, and equipment maintenance. However, some expenses, such as cleaning optical surfaces or preventing biofilm growth on the photobioreactor surfaces, are unique to photobiological processes.

Labor costs should be minimized, as with any manufacturing operation. Depending upon the robustness of the system and the amount of process control instrumentation, it might in later generations of the system be possible to achieve unattended or remote operation of a biological H_2 production facility. Using a wastewater treatment facility for comparison, many smaller treatment plants operate unattended. In other cases, remote operation might be possible with one control room for multiple locations. In Denmark, processes as complex as power plant operations are operated remotely. Remote or unattended operation would naturally require a higher level of instrumentation, but could greatly reduce operating costs, especially for a smaller facility where labor would be a larger portion of the operating expenses.

The initial assumption might be that a photobiological H_2 process would not require any raw materials. However, any biological process requires water and an adequate supply of macronutrients (i.e., nitrogen and phosphorus) and possibly some supplemental micronutrients, depending upon what nutrients and minerals are already present in the water supply to the process. One method to reduce raw material costs would be to utilize the waste stream from another process. For example, effluent streams from wastewater treatment anaerobic digesters are nitrogen rich, and blowdown streams from power plants are relatively pure.

Likewise, waste disposal is another operating cost. Generally, in any biological process some liquid is purged to prevent buildup of soluble waste components to unacceptable levels. The wastewater in the case of photobiological processes will probably require minimal treatment (i.e., primary treatment for solids removal). If a supplemental carbon source, such as acetate, is required, additional treatment would be necessary. Because of the potentially large size of a photobiological system, the water and wastewater treatment requirements might exceed the capacity of an existing publicly owned facility in the area.

With respect to solid waste disposal, it should be recognized that over time there is an accumulation of inert cell debris from dead cells. While much of the material in dead cells is released for uptake by growing cells, in biological waste treatment systems a significant percentage of the cell mass remains. This level of inert material will build up in the system if it is not periodically removed. Solid waste disposal cannot be ignored. It represents a significant expense for two reasons: (1) Photobioreactors can expect to be operated at cell concentrations of 1% cell mass or less (unless an immobilized system is employed). The solids must generally be increased to 25% solids or higher for disposal in a landfill and must be approximately 45% solids to have any net fuel value if burned. Considerable costs—both operating and capital—are associated with dewatering and drying. The second reason waste disposal cannot be ignored is because landfill costs can be hundreds of dollars per wet ton of waste.

Ideally, some beneficial use of the waste cell material could be found. Because of the high protein content, it might be used for cattle feed. The algae could also potentially be used to make biodiesel fuel. Dewatering, drying, and transportation costs should be minimized by locating the H_2 facility near the waste consumer. Transport costs for waste sludge must consider the amount of water accompanying the solid waste. Any water leaving in the waste represents a net water makeup requirement to the plant.

Electricity may be a significant expense. With large pond systems or other bioreactor configurations, a rational design is required to minimize pumping and compression costs. In conventional chemical production facilities, pumping costs are minimal and can be ignored without dramatically affecting the results of an economic analysis. However, with the potential for hundreds of thousands of gallons of water in the facility, even a small change in elevation can represent a significant pumping cost. Low-head piping designs, similar to the hydraulic design of a wastewater treatment plant, with a minimum number of valves should be considered.

Gas compression can also be significant. In some reactor designs, inert gas is bubbled through reactors or used to scour reactor surfaces to prevent cell adhesion to light-transmitting surfaces. The pressure drop through a nozzle or even a few inches of differential pressure can represent a high-electricity use item for a large process. The second area where gas compression comes into play is in the purification and storage of the H_2 product. Pressure swing adsorption (PSA) generally requires three atmospheres of pressure and works more efficiently at higher pressures. Operation of the bioreactor at elevated pressures may minimize compression costs, if it does not interfere with the organism's ability to produce H_2 or add significantly to capital and operating costs.

Maintenance costs will depend upon the amount of mechanical equipment, the complexity of the equipment, and the level of instrumentation. An additional consideration is the need to clean both the inner and outer optical surfaces of the bioreactors. Any soil on the outside surfaces will reduce light transmission, directly affecting the H_2-production capacity. The loss in transmission with time and the frequency/cost of cleaning should be included and may be seasonal. The inner surface of the bioreactor must also be kept clear. If cells attach to the inside of the light-transmitting surface, they might completely block the light from entering the reactor.

5.4.2 Capital Costs

Two major capital costs are the bioreactors and the H_2-compression/storage system, if required. As mentioned above, with a light conversion efficiency of 10% or less, a large photobioreactor area would be required to make any significant amount of H_2. Therefore, the cost of the "window" material, along with the limited lifetime of many materials when exposed to the sun, makes photobioreactor window cost a prime consideration. The goal is to have a low-cost material, such as plastic film with a lifetime of 5 to 10 years or more (see section 5.3.2). It is generally accepted that acrylics and glass are too expensive. With a stand-alone photobioreactor without storage, the window cost can be one-third of the total capital cost.[75] With processes requiring O_2 exclusion or operating at elevated pressures, gas permeability may become a concern and may require more expensive construction materials.

Hydrogen storage can be expensive, especially for a small production facility. If a photobioreactor must meet a continuous daily H_2 demand—like at a filling station—H_2 storage must be included in the cost analysis to supply H_2 at night and during cloudy periods. High-pressure compressed H_2 storage can double the cost of H_2 production.[75] A better application of an intermittent H_2 supply would be to feed a pipeline network where H_2 is pooled from other renewable or nonrenewable sources. Another option might be to locate a photobioreactor at an existing H_2 facility to take advantage of its existing storage and purification equipment and make a portion of the plant's output renewable.

The purity of H_2 produced will affect further purification costs. PSA is commonly used for H_2 purification, and it is best suited for H_2 concentrations of 50%

or higher. If by-products, such as carbon dioxide, are present, they may need to be removed. If a sweep gas is used in the bioreactor design, this will affect purification costs and options. If remote or unattended operation is planned, the amount of instrumentation will increase, with an associated increase in capital costs. Because of the presence of H_2, explosion-proof equipment might also be required, increasing capital costs.

5.4.3 General Design Considerations

The number 1 factor affecting cost is the specific H_2-production rate. With higher H_2 production rates per algal cell, the nutrient requirements are lower, the required bioreactor area drops, and waste is reduced. Examples of biological improvements to increase the H_2-production rate were discussed in sections 5.2.1 and 5.2.2. The exact photobioreactor configuration, such as the pond depth, reactor volume, and cell concentration, is affected by the specific H_2-production rate. For example, if antennae size is reduced, light penetrates deeper into the pond of algae and the lower layers are more productive. If cell concentration is lowered, light also penetrates deeper, but the overall system volume requirement increases.

Once the specific H_2-production rate has been optimized, the next most important factor is the sunlight itself. While laboratory results may be presented in kg/day or kg/hour, production is actually dependent upon both the light intensity and how long the algal cells are exposed to light. The physical size of the photobioreactor will depend upon the amount of light available at the site. The daily production will depend on the number of hours of light and the average light intensity. Finally, for stand-alone supply systems, the amount of storage will depend upon the length of the night and the number of cloudy days that occur in a row.

The best way to model a production system is to use actual solar insolation data. Hourly average data are readily available for many sites (National Solar Radiation Database, for example). In two studies, hourly insolation data were used to optimize the production unit size and the storage capacity for dedicated stand-alone systems.[75,96] If yearly data are used to estimate annual production, at least one other year should be chosen and used to check the analysis.

Along with concerns of lost production during cloudy periods, the amount of downtime associated with maintenance will also affect the overall economics. Ideally, the photobioreactor would operate continuously. If operated in batch mode, there would be a period of limited H_2 production while the cell density increased at the start of a new batch. If the system needs to be emptied for maintenance purposes, or if the cell culture is lost due to equipment failure or contamination, lost production will significantly affect the economics. The length of time required to build up cell mass in a wastewater treatment plant during start-up can be weeks, for example.

An early system designed for the sulfur-deprived process required several days to poise the cells for H_2 production.[58] This transition period represented lost production compared to continuous operation. Daily start-up of the photobioreactor at sunrise may also decrease production if it extends into the daylight hours. Some other operations, for example, H_2 liquefaction processes, would not be a logical choice for a diurnal cycle because of the losses and inefficiencies of starting up and shutting down the process every day. Gas compression, on the other hand, can be started and stopped, or be regulated to match changes in production as clouds pass overhead.

One general design concern with bioreactors is how to manage contamination. In a laboratory setting it is possible to autoclave glassware and operate small reactors with pure cultures. In the pharmaceutical field, it is common practice to steam sterilize bioreactors up to 20,000 l. However, with larger bioreactors, it would not be practical to build a photobioreactor capable of withstanding steaming at elevated pressures. Vapor sterilization might work, or chemicals can be used to reduce bioburden levels, but it would probably still be expensive. Antibiotic resistance is sometimes used in the lab to favor the growth of a desired organism over other bacteria, but this has limited application on a large scale due to high cost and concerns about the possibility of releasing antibiotic-resistant organisms into the environment.

In biological wastewater treatment plants, no attempt is made to use pure strains of microorganisms. Instead, conditions such as pH, temperature, growth rate, and substrate levels are controlled to favor and select the desired organism over other organisms. The undesirable organisms are still present, but growth pressures slow their growth rate and keep them at an acceptable level. The same approach would be desirable for the operation of a large-scale photobioreactor. For example, if there is no carbon source present, the growth of nonphotosynthetic bacteria would be limited.

One final consideration, common to all economic analyses, is the question of economy of scale. Generally the economics of most processes get better with size. The increased production rate overcomes fixed costs, and large-volume purchasing can reduce both capital and operating costs. A large power plant can be operated by the same number of people as a smaller one, meaning the labor cost per kilowatt-hour of electricity is less for a larger plant. Combining a solar H_2 process with a complementary wind process or a nonrenewable process could help take advantage of economy of scale with respect to H_2 storage and utility equipment as well as labor.

5.4.4 Case Study

For a stand-alone photobiological (sulfur-deprived, algal) H_2-production facility producing 300 kg/day of H_2, the total capital investment was estimated to be \$5 million with a H_2 selling price of approximately \$14/kg of hydrogen and a 15% return on investment. This system assumed moderate improvements in the H_2-production rate and included PSA purification with high-pressure compressed H_2 storage. The total photobioreactor area was 110,000 m^2 with a 10-cm pond depth, 0.2 g/l cell concentration, and \$10/m^2 reactor cost.[75]

By feeding a H_2 pipeline and eliminating the high-pressure compressed gas storage, the H_2 cost drops from \$13.53/kg to \$5.92/kg. Eliminating the high-pressure storage cuts the capital cost by more than \$1 million. In addition, feeding a pipeline eliminates any storage capacity limitation, so for the same 110,000 m^2 reactor area, the daily production rate increases from 300 kg/day to 446 kg/day.[75]

Eliminating the PSA only saves \$0.50/kg for comparable production systems. The effect of economy of scale was investigated by increasing the production capacity from 300 kg/day to 600 kg/day, with a corresponding doubling of the storage capacity. This resulted in a 12% reduction in H_2 selling price. One reason costs were not reduced more is because the bioreactor cost in this analysis was assumed to vary linearly with reactor surface area.[75]

Reducing the photobioreactor unit price from \$10/m^2 to \$1/m^2 dropped the H_2 selling price from \$13.53/kg to \$8.97/kg. For the pipeline delivery scenario, the H_2 selling price dropped from \$5.92/kg to \$2.83/kg with the lower reactor cost. Because industrial compressed gas storage is considered mature technology, no future cost reductions were assumed for the storage equipment. From the perspective of the U.S. Department of Energy's H_2 cost goal of \$2.50/kg, this comes very close. If an aerobic biological H_2 production system, capable of using the maximum efficiency of photosynthesis, can be developed employing low-cost materials, the H_2 price might fall to less than \$1.00/kg. Furthermore, ocean-based photobioreactors are expected to have unique cost advantages.

5.5 CONCLUSION

Photobioreactors pose some unique challenges with respect to reactor design and the intermittent nature of light-driven processes. However, innovative new materials and systems design, particularly for the photobioreactor, and taking advantage of synergies with existing processes, can go a long way to improving process economics. Most importantly, one cannot ignore factors like nutrient supply, waste disposal (or possible waste treatment in the future), pumping/compression requirements, and the possibility of co-product generation when working on a large scale. The future will depend on how clever we can be in developing the technology and how competitive the technology is compared to other alternative-energy options. The fact that organisms manufacture themselves from very cheap raw materials and can keep themselves in good repair favors biological approaches.

ACKNOWLEDGMENTS

The authors thank the Energy Biosciences (M.L.G. and M.S.) and the Hydrogen, Fuel Cell, and Infrastructure Technology (D.M.B., M.L.G., and M.S.) Programs, U.S. Department of Energy, for sponsoring this effort. M.S. also acknowledges support from the NREL DDRD Program. The authors also thank the members of their laboratories at NREL for contributions to much of the research discussed in this review.

REFERENCES

1. Boichenko, V.A., Greenbaum, E., and Seibert, M., Hydrogen production by photosynthetic microorganisms, in *Photoconversion of Solar Energy: Molecular to Global Photosynthesis*, Archer, M.D. and Barber, J., Eds., Imperial College Press, London, 2004, p. 397.
2. Prince, R.C. and Kheshgi, H.S., The photobiological production of hydrogen: potential efficiency and effectiveness as a renewable fuel, *Crit. Rev. Microbiol.*, 31, 19, 2005.
3. Ghirardi, M.L., Maness, P.C., and Seibert, M., Photobiological methods of renewable hydrogen production, in *Solar Hydrogen*, K. Rajeshwar, S. Licht, and R. McConnell, Eds., in press.

4. Adams, M.W.W., The structure and mechanism of iron-hydrogenases, *Biochim. Biophys. Acta*, 1020, 115, 1990.
5. Vignais, P.M. and Colbeau, A., Molecular biology of microbial hydrogenase, *Curr. Issues Mol. Biol.*, 6, 159, 2004.
6. Appel, J. and Schulz, R., Hydrogen metabolism in organisms with oxygenic photosynthesis: hydrogenases as important regulatory devices for a proper redox poising? *J. Photochem. Photobiol. B Biol.*, 47, 1, 1998.
7. Vignais, P.M., Billoud, B., and Meyer, J., Classification and phylogeny of hydrogenases, *FEMS Microbiol. Rev.*, 25, 455, 2001.
8. Happe, T. and Kaminski, A., Differential regulation of the [Fe]-hydrogenase during anaerobic adaptation in the green alga *Chlamydomonas reinhardtii*, *Eur. J. Biochem.*, 269, 1, 2002.
9. Forestier, M. et al., Expression of two [Fe]-hydrogenases in *Chlamydomonas reinhardtii* under anaerobic conditions, *Eur. J. Biochem.*, 270, 2750, 2003.
10. Roessler, P. and Lien, S., Purification of hydrogenase from *Chlamydomonas reinhardtii*, *Plant Physiol.*, 75, 705, 1984.
11. Ghirardi, M.L., Togasaki, R.K., and Seibert, M., Oxygen sensitivity of algal H_2-production, *Appl. Biochem. Biotechnol.*, 63–65, 141, 1997.
12. Happe, T., Mosler, B., and Naber, J.D., Induction, localization and metal content of hydrogenase in the green alga *Chlamydomonas reinhardtii*, *Eur. J. Biochem.*, 222, 769, 1994.
13. Posewitz, M.C. et al., Discovery of two novel radical S-adenosylmethionine proteins required for the assembly of an active [Fe] hydrogenase, *J. Biol. Chem.*, 279, 25711, 2004.
14. Schmitz, O. et al., HoxE: a subunit specific for the pentameric bidirectional hydrogenase complex (HoxEFUYH) of cyanobacteria, *Biochim. Biophys. Acta*, 1554, 66, 2002.
15. Cournac, L. et al., Sustained photoevolution of molecular hydrogen in a mutant of *Synechocystis* sp. strain PCC 6803 deficient in the type I NADPH-dehydrogenase complex, *J. Bacteriol.*, 186, 1737, 2004.
16. Abdel-Basset, R. and Bader, K.P., Physiological analysis of the hydrogen gas exchange in cyanobacteria, *J. Photochem. Photobiol. B Biol.*, 43, 146, 1998.
17. Frey, M., Hydrogenases: hydrogen-activating enzymes, *Chembiochemistry*, 3, 153, 2003.
18. Serebriakova, L., Zorin, N.A., and Lindblad, P., Reversible hydrogenase in *Anabaena variabilis* ATCC 29413, *Arch. Microbiol.*, 161, 140, 1994.
19. Gfeller, R.P. and Gibbs, M., Fermentative metabolism of *Chlamydomonas reinhardtii*. I. Analysis of fermentative products from starch in dark and light, *Plant Physiol.*, 75, 212, 1984.
20. Greenbaum, E., Energetic efficiency of hydrogen photoevolution by algal water splitting, *Biophys. J.*, 54, 365, 1988.
21. Randt, C. and Senger, H., Participation of the two photosystems in light dependent hydrogen evolution in *Scenedesmus obliquus*, *Photochem. Photobiol.*, 42, 553, 1985.
22. Healey, F.P., The mechanism of hydrogen evolution by *Chlamydomonas moewusii*, *Plant Physiol.*, 45, 153, 1970.
23. Ghirardi, M.L. et al., Approaches to developing biological H_2-photoproducing organisms and processes, *Biochem. Soc. Trans.*, 33, 70, 2005.
24. Melis, A. et al., Sustained photobiological hydrogen gas production upon reversible inactivation of oxygen evolution in the green alga *Chlamydomonas reinhardtii*, *Plant Physiol.*, 122, 127, 2000.
25. Melis, A., Green alga hydrogen production: progress, challenges and prospects, *Int. J. Hydrogen Energy*, 27, 1217, 2002.
26. Ghirardi, M.L. and Amos, W., Hydrogen photoproduction by sulfur-deprived green algae: status of the research and potential of the system, *Biocycle*, 45, 59, 2004.
27. Posewitz, M.C. et al., Hydrogen photoproduction is attenuated by disruption of an isoamylase gene in *Chlamydomonas reinhardtii*, *Plant Cell*, 16, 2151, 2004.

28. Posewitz, M.C. et al., Identification of genes required for hydrogenase activity in *Chlamydomonas reinhardtii*, *Biochem. Soc. Trans.*, 33, 102, 2005.
29. Knaff, D.B., Ferredoxin and ferredoxin-dependent enzymes, in *Oxygenic Photosynthesis: The Light Reactions*, Ort, D.R. and Yocum, C.F., Eds., Kluwer Academic Publishers, Dordrecht, The Netherlands, 1996, p. 333.
30. Kurisu, G., Nishiyama, D., Kusunoki, M., Fujikawa, S., Datoh, M., Hanke, G. T., Hase, T., and Teshima, K. A. A structural basis of *Equisetum arvense* ferredoxin isoform II producing an alternative electron transfer with ferredoxin-$NADP^+$ reductase. *J. Biol. Chem.,* 280, 2275, 2005.
31. Hirasawa, M. et al., Ferredoxin-thioredoxin reductase: properties of its complex with ferredoxin, *Biochim. Biophys. Acta*, 935, 1, 1988.
32. Roessler, P. and Lien, S., Anionic modulation of the catalytic activity of hydrogenase from *Chlamydomonas reinhardtii*, *Arch. Biochem. Biophys.*, 213, 37, 1982.
33. Happe, T. and Naber, J.D., Isolation, characterization and N-terminal amino acid sequence of hydrogenase from green alga *Chlamydomonas reinhardtii*, *Eur. J. Biochem.*, 214, 475, 1993.
34. King, P.W. et al., Functional studies of [FeFe]-hydrogenase maturation in an *Escherichia coli* biosynthetic system, *J. Bacteriol.*, 188, 2163, 2006.
35. Tamagnini, P. et al., Hydrogenase in *Nostoc* sp. strain PCC 73102, a strain lacking a bidirectional enzyme, *Appl. Enviorn. Microbiol.*, 63, 1801, 1997.
36. Tamagnini, P. et al., Diversity of cyanobacterial hydrogenase, a molecular biology approach, *Curr. Microbiol.*, 40, 356, 2000.
37. Tamagnini, P. et al., Hydrogenases and hydrogen metabolism of cyanobacteria, *Microbiol. Mol. Biol. Rev.*, 66, 1, 2002.
38. Happe, T., Schüz, K., and Böhme, H., Transcriptional and mutational analysis of the uptake hydrogenase of the filamentous cyanobacterium *Anabaena variabilis* ATCC 29413, *J. Bacteriol.*, 182, 1624, 2000.
39. Masukawa, H., Mochimaru, M., and Sakurai, H., Disruption of uptake hydrogenase gene, but not of the bidirectional hydrogenase gene, leads to enhanced photobiological hydrogen production by the nitrogen-fixing cyanobacterium *Anabaena* sp. 7210, *Appl. Microbiol. Biotechnol.*, 58, 618, 2002.
40. Schmitz, O. and Bothe, H., $NAD(P)^+$-dependent hydrogenase activity in extracts from the cyanobacterium *Anacystis nidulans*, *FEMS Microbiol. Lett.*, 135, 97, 1996.
41. Antal, T.K. et al., The dependence of algal H_2 production on photosystem II and O_2 consumption activities in sulfur-deprived *Chlamydomonas reinhardtii* cells, *Biochim. Biophys. Acta*, 1607, 153, 2003.
42. Lee, J.W. and Greenbaum, E., A new oxygen sensitivity and its potential application in photosynthetic H_2 production, *Appl. Biochem. Bioeng.*, 105–108, 303, 2003.
43. De Vitry, C. et al., The chloroplast Rieske iron-sulfur protein: at the crossroad of electron transport and signal transduction, *J. Biol. Chem.*, 279, 44621, 2004.
44. Polle, J.E.W. et al., Truncated chlorophyll antenna size of the photosystems: a practical method to improve microalgal productivity and hydrogen production in mass culture, *Int. J. Hydrogen Energy*, 27, 2002, 1257.
45. Melis, A., Niedhardt, J., and Benemann, J.R., *Dunaliella salina* (Chlorophyta) with small chlorophyll antenna sizes exhibit higher photosynthetic productivities and photon use efficiencies than normally pigmented cells, *J. Appl. Phycol.*, 10, 515, 1999.
46. Jolley, C. et al., Structure of plant photosystem I revealed by theoretical modeling, *J. Biol. Chem.*, 280, 33627, 2005.
47. Zouni, A. et al., Crystal structure of photosystem II from *Synechococcus elongatus* at 3.8 angstrom resolution, *Nature*, 409, 739, 2001.

48. Fujita, Y. and Murakami, A., Regulation of electron transport composition in cyanobacterial photosynthetic system: stoichiometry among PSI and PSII complexes and their light harvesting antenna and *Cyt* b_6-f complex, *Plant Cell Physiol.*, 28, 1547, 1987.
49. Kondo, T. et al., Enhancement of hydrogen production by a photosynthetic bacterium mutant with reduced pigment, *J. Biosci. Bioeng.*, 93, 145, 2002.
50. Nakajima, Y. and Ueda, T., Improvement of photosynthesis in dense microalgal suspension by reduction of light harvesting pigments, *J. Appl. Phycol.*, 9, 503, 1997.
51. Nakajima, Y. and Ueda, T. Improvement of microalgal photosynthetic productivity by reducing the content of light harvesting pigment. *J. Appl. Phycol.* 11, 195, 1999.
52. Nakajima, Y. and Ueda, T., The improvement of marine microalgal productivity by reducing the light-harvesting pigment, *J. Appl. Phycol.*, 12, 285, 2000.
53. Nakajima, Y., Tsuzuki, M., and Ueda, R., Improved productivity by reduction of the content of light-harvesting pigment in *Chlamydomonas perigranulata*, *J. App. Phycol.*, 13, 95, 2001.
54. Polle, J.E.W. et al., Photosynthetic apparatus organization and function in wild type and a Chl b-less mutant of *Chlamydomonas reinhardtii*. Dependence on carbon source, *Planta*, 211, 335, 2000.
55. Polle, J.E.W., Niyogi, K.K., and Melis, A. Absence of lutein, violaxanthin and neoxanthin affects the functional chlorophyll antenna size of photosystem-II but not that of photosystem-I in the green algae *Chlamydomonas reinhardtii*, *Plant Cell Physiol.*, 42, 482, 2001.
56. Polle, J.E.W., Kanakagiri, S., and Melis, A., *tla1*, a DNA insertional transformant of the green alga *Chlamydomonas reinhardtii* with a truncated light-harvesting chlorophyll antenna size, *Planta*, 217, 49, 2003.
57. Wykoff, D.D. et al., The regulation of photosynthetic electron-transport during nutrient deprivation in *Chlamydomonas reinhardtii*, *Plant Physiol.*, 177, 129, 1998.
58. Ghirardi, M.L. et al., Sustained photobiological hydrogen gas production upon reversible inactivation of oxygen evolution in the green alga *Chlamydomonas reinhardtii*, *Trends Biotechnol.*, 18, 506, 2000.
59. Fedorov, A.S. et al., Continuous hydrogen photoproduction by *Chlamydomonas reinhardtii* using a novel two-stage, sulfate-limited chemostat system, *Appl. Biochem. Biotechnol.*, 121–124, 403, 2005.
60. Ghirardi, M.L. et al., Development of algal systems for hydrogen photoproduction: addressing the hydrogenase oxygen-sensitivity problem, in *Artificial Photosynthesis*, Collings, C., Ed., Wiley–VCH Verlag, Weinheim, Germany, 2005, p. 213.
61. Laurinavichene, T.V. et al., Demonstration of sustained hydrogen photoproduction by immobilized, sulfur-deprived *Chlamydomonas reinhardtii* cells, *Int. J. Hydrogen Energy*, 31, 659, 2006.
62. Kruse, O. et al., Improved photobiological H_2 production in engineered green algal cells, *J. Biol. Chem.*, 280, 34170, 2005.
63. Kosourov, S., Seibert, M., and Ghirardi, M.L., Effects of extracellular pH on the metabolic pathways in sulfur-deprived, H_2-producing *Chlamydomonas reinhardtii* cultures, *Plant Cell Physiol.*, 44, 146, 2003.
64. Grossman, A. and Takahashi, H., Macronutrient utilization by photosynthetic eukaryotes and the fabric of interactions, *Annu. Rev. Plant Physiol. Plant Mol. Biol.*, 52, 163, 2001.
65. Chen, H.-C. et al., *SulP*, a nuclear gene encoding a putative chloroplast-targeted sulfate permease in *Chlamydomonas reinhardtii*, *Planta*, 218, 98, 2003.
66. Chen, H.-C. and Melis, A., Localization and function of SulP, a nuclear-encoded chloroplast sulfate permease in *Chlamydomonas reinhardtii*, *Planta*, 220, 198, 2004.
67. Melis, A. and Chen, H.C., Chloroplast sulfate transport in green algae genes, proteins and effects, *Photosynt. Res.*, 86, 299, 2005.

68. Tsygankov, A. et al., Hydrogen photoproduction under continuous illumination by sulfur-deprived, synchronous *Chlamydomonas reinhardtii* cultures, *Int. J. Hydrogen Energy*, 27, 1239, 2002.
69. Hallenbeck, P.C. and Benemann, J.R., Biological hydrogen production: fundamentals and limiting processes, *Int. J. Hydrogen Energy*, 27, 1185, 2002.
70. Fouchard, S. et al., Autotrophic and mixotrophic hydrogen photoproduction in sulfur-deprived *Chlamydomonas* cells, *Appl. Environ. Microbiol.*, 71, 6199, 2005.
71. Tsygankov, A.A. et al., Hydrogen production by sulfur-deprived *Chlamydomonas reinhardtii* under photoautotrophic conditions, *Int. J. Hydrogen Energy,* 31, 1574, 2006.
72. Guan, Y. et al., Two-stage photo-biological production of hydrogen by marine green alga *Platymonas subcordiformis*, *Biochem. Eng. J.*, 19, 69, 2004.
73. Guan, Y. et al., Significant enhancement fo photobiological H_2 evolution by carbonylcyanide m-chlorophenylhydrazone in the marine green alga *Platymonas subcordiformis*, *Biotechnol. Lett.*, 26, 1031, 2004.
74. Torzillo, G. and Vonshak, A., Biotechnology for algal mass cultivation, in *Recent Advances in Marine Biotechnology*, Fingerman, M. and Nagabhushanam, R., Eds., Vol. 9, Science Publishers, Inc., Enfield, NH, 2003, p. 45.
75. Amos, W.A., *Updated Cost Analysis of Photobiological Hydrogen Production from Chlamydomonas reinhardtii Green Algae*, NREL/MP-560-35593, National Renewable Energy Laboratory, Golden, CO, 2004.
76. Akkerman, I. et al., Photobiological hydrogen production: photochemical efficiency and bioreactor design, in *Bio-methane and Bio-hydrogen*, Reith, J.H., Wijfels, R.H., and Barten, H., Eds., Dutch Biological Hydrogen Foundation, Petten, The Netherlands, 2003, chap. 6.
77. Palz, O. and Scheibenbogen, K., Photobioreactors: design and performance with respect to light energy input, in *Advances in Biochemical Engineering/Biotechnology*, Scheper, T., Ed., 59, Springer-Verlag, Berlin, 1998, p. 123.
78. Casamajor, A.B. and Parsons, R.E., *Design Guide for Shallow Ponds*, Section 2, UCRL-52385, Rev. 1, Lawrence Livermore Laboratory, Livermore, CA, January 1979.
79. Pruett, M.L., Solar Power and Energy Storage System, U.S. Patent 6,374,614, 2002.
80. Platt, E.A. and Wood, R.I., *Engineering Feasibility of a 150 kW Irrigation Pumping Plant Using Shallow Solar Ponds*, UCRL-52397, Lawrence Livermore Laboratory, Livermore, CA, 1978.
81. Matsunaga, T. et al. Glutamate production from CO_2 by marine cyanobacterium, *Synechococcus* sp. Using a novel biosolar reactor employing light-diffusing optical fibers, *Appl. Biochem. Biotechnol.*, 28/29, 157, 1991.
82. Gorden, J.M., Tailoring optical systems to optimized photobioreactors, *Int. J. Hydrogen Energy*, 27, 1175, 2002.
83. Suh, I.S. and Lee, C.-G., Photobioreactor engineering: design and performance, *Biotech. Bioproc. Eng.*, 8, 313, 2003.
84. Blake, D.M. and Kennedy, C.E., *Hydrogen Reactor Development and Design for Photofermentation and Photolytic Processes: Reactor Materials Testing*, Milestone Report, AOP 3.1.5, Subtask 3.1.5.1, National Renewable Energy Laboratory, Golden, CO, 2004. (For a copy, e-mail D. Blake at dan_blake@nrel.gov.)
85. Farrah, M., Ultraviolet Aging of Transparent Plastic Coverplates for Solar Energy Equipment, M.S. thesis, University of Lowell, Lowell, MA, 1983.
86. Anon., *Materials in Solar Thermal Collectors: Identification of New Types of Transparent Polymeric Materials*, Report 1996-11-08, IEA Solar Heating and Cooling Programme, Brussels, Belgium, 1996.
87. Raman, R., Mantel, S., Davidson, J., Wu, C., and Jorgensen, G., A review of polymer materials for solar water heating systems, *Trans. ASME*, 122, 92–100, 2000.

88. Anon., *Solar Hot Water Heating Systems: Identification of Plastic Materials for Low Cost Glazings*, PolyNEW, Inc., Subcontract TAR-9-29449-02 Report, National Renewable Energy Laboratory, Golden, CO, 2002.
89. Davidson, J.H., Mantell, S.C., and Jorgensen, G.J., Status of the development of polymeric solar water heating systems, in *Advances in Solar Energy*, Goswami, Y., Ed., Vol. 15, ASES, Inc., Boulder, CO, 2003, p. 149.
90. Czanderna, A., Masterson, K., and Thomas, T., *Silver/Glass Mirrors for Solar Thermal Applications*, SERI/SP-271-2293, SERI, Golden, CO, 1985.
91. Jorgensen, G. and Rangaprasad, G., *Ultraviolet Reflector Materials for Solar Detoxification of Hazardous Waste*, SERI/TP-257-4418, SERI, Golden, CO, 1991.
92. Watt, A.S. and Mann, M.K., *Evaluation of the Cost of Manufacturing a Housing Unit for Photoelectrochemical Hydrogen Production*, Milestone Report, U.S. DOE, Hydrogen Program, National Renewable Energy Laboratory, Golden, CO, 1999. (For a copy, e-mail D. Blake at dan_blake@nrel.gov.)
93. Blake, D.M. and Kennedy, C.E., *Hydrogen Reactor Development & Design for Photofermentation and Photolytic Processes: Reactor Materials Testing*, Milestone Report, AOP 3.1.5, Subtask 3.1.5.1, National Renewable Energy Laboratory, Golden, CO, 2005. (For a copy, e-mail D. Blake at dan_blake@nrel.gov.)
94. Massey, L.K., *Permeability Properties of Plastics and Elastomers: A Guide to Packaging and Barrier Materials*, 2nd ed., Plastic Design Library/William Andrew Publishing, Norwich, NY, 2003, Appendix II.
95. Langowski, H.-C., Flexible barrier materials for technical applications, *Vacuum Technology & Coatings*, January 2004, p. 39.
96. Spath, P.L. and Amos, W.A., *Assessment of Natural Gas Splitting with a Concentrating Solar Reactor for Hydrogen Production*, NREL/TP-510-31949, National Renewable Energy Laboratory, Golden, CO, 2002.

[illegible]

6 Dense Membranes for Hydrogen Separation and Purification*

U. (Balu) Balachandran, T. H. Lee, and S. E. Dorris

CONTENTS

6.1 INTRODUCTION

The U.S. Department of Energy's Office of Fossil Energy sponsors a wide variety of research, development, and demonstration programs aimed at maximizing the use of vast domestic fossil resources and ensuring a fuel-diverse energy sector while responding to global environmental concerns. Development of cost-effective, membrane-based reactor and separation technologies is of significant interest for applications in advanced fossil-based power and fuel technologies. Because concerns over global climate change are driving nations to reduce CO_2 emissions, hydrogen is considered the fuel of choice for the electric power and transportation industries. In his 2003 State of the Union address, President Bush announced a Hydrogen Fuel Initiative to develop hydrogen production and distribution technologies for powering fuel cell vehicles and stationary fuel cell power sources. The goal of this initiative is to lower the cost of hydrogen enough to make fuel cell cars cost-competitive with conventional gasoline-powered vehicles by 2010, and to advance the methods of producing hydrogen from renewable resources, nuclear energy, and coal.

As part of the effort to devise cost-effective, efficient processes for producing and utilizing hydrogen, Argonne National Laboratory (ANL) is developing dense, hydrogen-permeable membranes for separating hydrogen from mixed gases at commercially significant fluxes under industrially relevant operating conditions. Of particular interest is the separation of hydrogen from product streams that are generated

* Work supported by the U.S. Department of Energy, Office of Fossil Energy, National Energy Technology Laboratory's Hydrogen and Gasification Technologies Program, under Contract W-31-109-Eng-38.

during coal gasification, methane partial oxidation, reforming, and water–gas shift reactions. Because the membrane will separate hydrogen without using electrodes or an external power supply (i.e., its operation will be nongalvanic), it requires materials that exhibit suitable electronic and protonic conductivities as well as high hydrogen diffusivity and solubility. Good mechanical properties will also be necessary to withstand operating stresses, and to maximize the hydrogen flux and maintain high hydrogen selectivity, the fabricated materials must be thin and dense.

The ANL membrane development effort focused initially on $BaCe_{0.8}Y_{0.2}O_{3-\delta}$ (BCY), because it is a mixed proton–electron conductor whose high total electrical conductivity[1,2] suggested that it might yield a high hydrogen flux without using electrodes or electrical circuitry. Despite having a high total electrical conductivity, its electronic component of conductivity was insufficient to support a high nongalvanic hydrogen flux.[3,4] To increase the electronic conductivity and thereby the hydrogen flux, we developed various cermet (i.e., ceramic–metal composite) membranes, in which a metal powder is dispersed in a ceramic matrix.[5,6] In these cermets, the metal enhances the hydrogen permeability of the ceramic phase by increasing the electronic conductivity of the composite. If the metal has a high hydrogen permeability, it may also provide an additional transport path for the hydrogen.

The ANL-1, -2, or -3 cermet membranes are classified on the basis of the hydrogen transport properties of the metal and matrix phases. In ANL-1 membranes, a metal with low hydrogen permeability is distributed in the hydrogen-permeable matrix of BCY. ANL-2 membranes also have a matrix of hydrogen-permeable BCY, but they contain a hydrogen transport metal, i.e., a metal with high hydrogen permeability. In ANL-3 membranes, a metal with high hydrogen permeability is dispersed in a ceramic matrix of low hydrogen permeability, e.g., Al_2O_3, ZrO_2, or $BaTiO_3$. Each membrane is identified by a number representing the class of membrane, as described above, and a letter indicating a specific combination of metal and matrix phases. For example, ANL-3a is an ANL-3 membrane that contains a specific metal in an Al_2O_3 matrix, whereas ANL-3b contains a different combination of hydrogen transport metal and ceramic. A letter is not included when general comments are made about an entire class of membranes, e.g., ANL-3 membranes.

Hydrogen permeation through ANL-1a is higher than that in monolithic BCY because the metal in ANL-1a increases its electronic conductivity. Because the metal in ANL-1a has low hydrogen permeability, only a small part of the hydrogen diffuses through the metal phase. ANL-2a membranes, in which a hydrogen transport metal replaces the metal of ANL-1a, give a still higher hydrogen flux. The metal in ANL-2a facilitates hydrogen diffusion by increasing the electronic conductivity and by providing an alternative path for hydrogen diffusion. Although BCY and the metal phase both contribute to hydrogen permeation through ANL-2 membranes, most of the hydrogen diffuses through the metal phase.[7] Because BCY contributes relatively little to the hydrogen flux, has poor mechanical properties, and is chemically unstable under some conditions of interest,[8] ANL-3 membranes were developed. These membranes contain a hydrogen transport metal in a thermodynamically stable ceramic matrix with superior mechanical properties, e.g., Al_2O_3 or ZrO_2. They have high hydrogen permeability, improved strength, and greater chemical stability compared to ANL-2 membranes, and have given the highest hydrogen flux (20

cm^3(STP)/min-cm^2) to date for an ANL membrane. In this chapter, the hydrogen flux rates of ANL-1, -2, and -3 membranes are compared. Also, because the membranes will contact gas streams that contain CO, CO_2, and H_2S, their chemical stability is an important issue. We thus report the hydrogen flux of ANL-2 and -3 membranes vs. time during exposure to feed gases containing various levels of H_2S, CO_2, and CO.

6.2 EXPERIMENTAL

Preparation of BCY powders for ANL-1 and -2 membranes is described in detail elsewhere.[5] BCY and metal powders were mixed together to prepare powders for ANL-1a and -2a membranes. Powders for ANL-3 membranes were prepared by mixing one of two hydrogen transport metals with ceramic powders that are reported to be poor proton conductors.[8] All membranes contained 40 vol% metal, except where otherwise noted. The powder mixtures were pressed uniaxially to prepare disks (≈22 mm in diameter and ≈2 mm thick) for sintering. Cermet membranes were sintered in either air or 4% H_2/balance He, N_2, or Ar in the temperature range of 1,350 to 1,420°C.

A sintered disk was polished with 600-grit SiC paper to obtain the desired thickness and produce faces that were flat and parallel to one another. The polished disk was then affixed to an Al_2O_3 tube by a procedure described elsewhere.[9] A seal formed after heating to 950°C, when spring-loaded rods squeezed a gold ring between the membrane and the Al_2O_3 tube. One side of the sample was purged with 4% H_2/balance He during sealing, while the other side was purged with 100 ppm H_2/balance N_2. The leakage rate following this procedure was typically <10% of the total permeation flux. The hydrogen flux through the membranes was measured as a function of time and temperature during exposure to various feed gases.

The flow rate of sweep gas (100 ppm H_2/balance N_2) during permeation measurements was controlled with an MKS mass flow controller and was measured using a Humonics Field-Cal 570 flow calibrator. The sweep gas was analyzed with a Hewlett-Packard 6890 gas chromatograph. Feed gases included dry or wet 4% H_2/balance He, 100% H_2, and simulated syngas (66% H_2, 33% CO, and 1% CO_2). For wet feed gas, 4% H_2/balance He was bubbled through a water bath at room temperature to give ≈0.03 atm H_2O; for the dry condition, 4% H_2/balance He was introduced directly into the furnace from the gas cylinder. The stability of ANL-3e membranes in H_2S-containing atmospheres was determined by measuring the hydrogen flux vs. time in atmospheres with progressively higher H_2S concentrations. Gas mixtures for these tests were prepared using mass flow controllers to blend ultra high purity (UHP) He with H_2 that contained a known H_2S concentration. The compositions of the gas mixtures and the H_2S-containing gas used to prepare them are given in table 6.1.

6.3 RESULTS

Figure 6.1 compares the hydrogen fluxes for ANL-1a, -2a, and -3b membranes using a feed gas of 4% H_2/balance He. To compensate for slight differences in membrane thickness, the measured fluxes were normalized to a thickness of 0.50 mm. The actual thickness of each membrane is shown in the inset of figure 6.1. The flux of ANL-1a,

TABLE 6.1
Compositions of Gas Mixtures Used to Test Stability of ANL-3e Membranes

Composition of Gas Mixture	Gases Used to Prepare Mixture
51 ppm H2S/19.2% H2/balance He	250 ppm H2S/balance H2 and UHP He
97 ppm H2S/19.4% H2/balance He	500 ppm H2S/balance H2 and UHP He
400 ppm H2S/79.8% H2/balance He	500 ppm H2S/balance H2 and UHP He
2922 ppm H2S/19.2% H2/balance He	1.5% H2S/balance H2 and UHP He

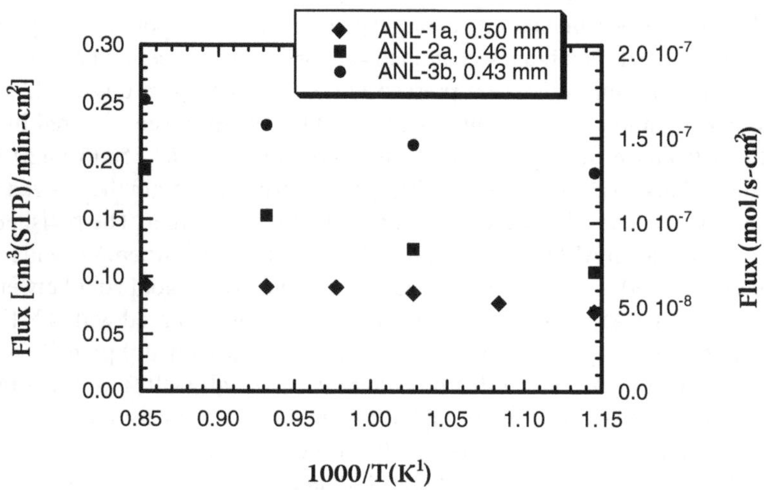

FIGURE 6.1 Hydrogen flux through ANL-1a, -2a, and -3b membranes using wet 4% H_2/balance He as the feed gas. Measured fluxes were normalized to a thickness of 0.50 mm to compensate for differences in membrane thickness. The actual thickness of each membrane is given in the inset.

while the lowest for these membranes, is significantly higher than that of BCY alone, because ANL-1a contains a continuous metal phase. Without this continuous metal phase, the electronic conductivity of BCY is too low to support a significant non-galvanic hydrogen flux.[5] Like ANL-1a, the ANL-2a membrane contains a matrix of BCY, but the metal phase of ANL-1a with low hydrogen permeability is replaced by a hydrogen transport metal. The addition of this favorable path for hydrogen diffusion significantly increases the flux of ANL-2a relative to ANL-1a, especially at higher temperatures. Comparing the properties of the ANL-2a and -3b membranes also highlights the importance of the metal phase. Although ANL-2a contains a hydrogen-permeable matrix, whereas ANL-3b does not, the hydrogen flux for ANL-3b was ≈35% higher at 900°C and ≈80% higher at 600°C. ANL-3b gave the highest flux for these membranes because its metal phase had the highest hydrogen permeability.

A 40-μm-thick ANL-3a membrane containing 50 vol% of a hydrogen transport metal attained the highest hydrogen flux (20 cm^3(STP)/min-cm^2) to date for an ANL membrane; however, these membranes sometimes contain interconnected porosity after sintering. To more reproducibly fabricate ANL-3 membranes without interconnected porosity, we developed ANL-3e membranes, which contain the same hydrogen transport metal as ANL-3a membranes but have a ceramic matrix that densifies more readily. Figure 6.2 shows that the hydrogen flux through an ANL-3e membrane, like that through an ANL-3a[9] or -3b membrane,[10] increases linearly with the difference in the square root of the hydrogen partial pressure in the feed and sweep gases. This behavior is characteristic of bulk-limited hydrogen diffusion through a metal[11] and is expected, because the membrane contains a hydrogen transport metal and the ceramic phase has a low hydrogen permeability. Such behavior suggests that reducing the membrane thickness may increase the hydrogen flux.

Figure 6.3 shows the hydrogen flux of ANL-3e membranes vs. the inverse of membrane thickness at 900°C using 100% H_2 as the feed gas. The hydrogen flux varies linearly with the inverse of membrane thickness, confirming that it is controlled by bulk diffusion over this thickness range (22 to 100 μm). At lesser thickness, however, interfacial reactions may become rate limiting. The highest flux for the ANL-3e membranes (19.0 cm^3(STP)/min-cm^2) was only slightly lower than that for an ANL-3a membrane (20 cm^3(STP)/min-cm^2).[12] However, if the ceramic matrix only supports the hydrogen transport metal without influencing its hydrogen permeation, the ANL-3e membrane should have had the higher flux, because it was slightly thinner (22 μm vs. 40 μm) and contained an equal concentration (50 vol%) of the same hydrogen transport metal. This apparent anomaly suggests that the ceramic phase plays a secondary role in hydrogen permeation.

The chemical stability of hydrogen separation membranes is a critical issue, because they will operate at elevated temperatures and pressures in atmospheres containing CO, CO_2, and H_2S, among other constituents. Figure 6.4 shows the hydrogen

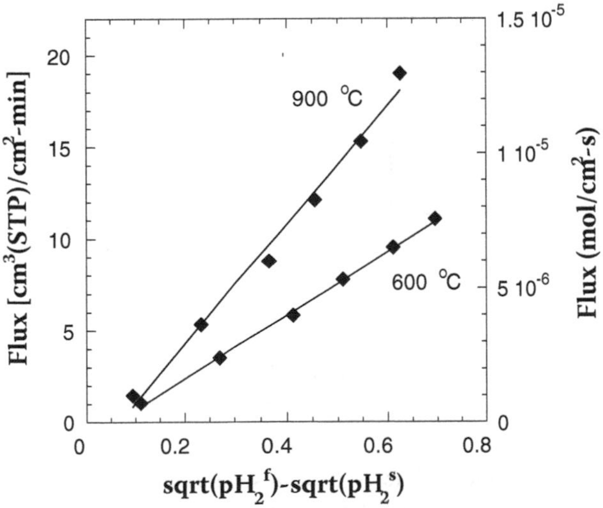

FIGURE 6.2 Hydrogen flux through 22-μm-thick ANL-3e membrane vs. the difference in the square root of hydrogen partial pressure for the feed (pH_2^f) and sweep (pH_2^s) gases at 900 and 600°C

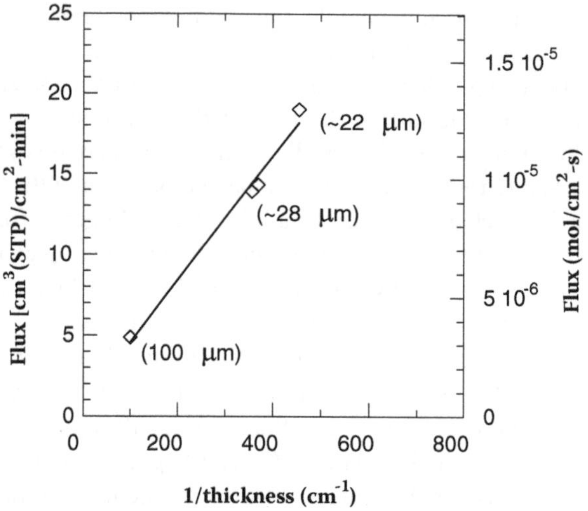

FIGURE 6.3 Hydrogen flux through ANL-3e membranes vs. the inverse of membrane thickness at 900°C using 100% H_2 as the feed gas.

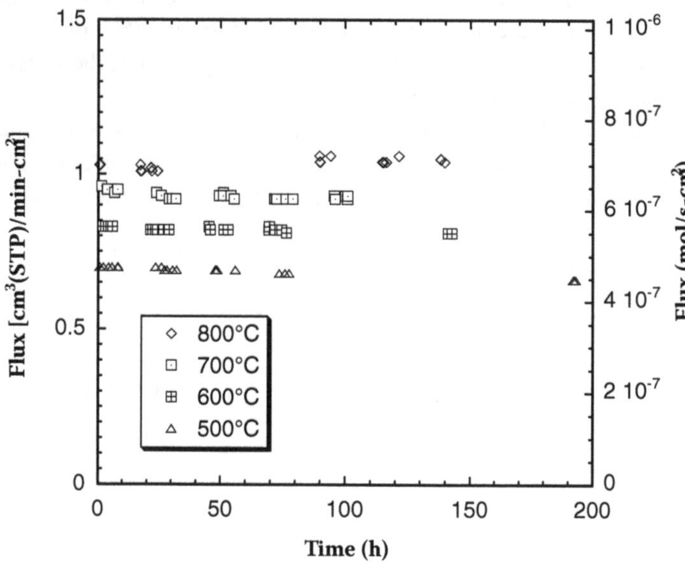

FIGURE 6.4 Hydrogen flux through ANL-3b (0.43 mm thick) vs. time in simulated syngas (66% H_2, 33% CO, and 1% CO_2) at various temperatures.

flux through an ANL-3b membrane vs. time at several temperatures in an atmosphere of simulated syngas (66% H_2, 33% CO, and 1% CO_2). A feed gas of 4% H_2/balance He was flowed before and after exposure to syngas at each temperature, and the leakage rate of hydrogen was determined by measuring the helium concentration in the sweep gas. No helium leakage was measured at any of the temperatures. Figure 6.4

shows no noticeable decrease in flux during up to 190 h of operation at each temperature. Similar tests with another ANL-3b membrane showed no decrease in the hydrogen flux during 120 h of exposure to syngas at 900°C.

The importance of a chemically stable matrix is clearly demonstrated in figure 6.5, where the hydrogen flux is plotted vs. time for ANL-2a and -3d membranes during their exposure to dry and wet syngas of composition 2.0% CH_4, 19.6% H_2, 19.6% CO, and 58.8% CO_2 (mol%). ANL-3d membranes contain the same hydrogen transport metal as ANL-3b membranes, but they have an Al_2O_3 matrix rather than the $BaTiO_3$ matrix of ANL-3b membranes; they were developed to improve the mechanical strength of thin membranes. Figure 6.5 shows that in environments with high concentrations of CO_2, the Al_2O_3 matrix of ANL-3d is chemically more stable than the BCY matrix of ANL-2a. The permeation rate through ANL-2a decreased dramatically after only several minutes, whereas the hydrogen flux through ANL-3d was stable for >3 h. Scanning electron microscopy on the ANL-2a surface after the permeation measurements showed that the BCY matrix had decomposed to form $BaCO_3$ and other phases. These results show that a chemically stable matrix such as Al_2O_3 or ZrO_2 will be required for application of the membrane in atmospheres with high CO_2 concentrations.

Figure 6.6 shows the hydrogen flux of an ANL-3e membrane (thickness = 100 μm) vs. time at several temperatures in a feed gas of 51 ppm H_2S and 19.6% H_2/balance He. For measurement of the initial (time = 0 h) hydrogen flux at each temperature, UHP H_2 and UHP He were mixed with mass flow controllers to give the same hydrogen concentration that the H_2S-containing gas would subsequently have. Next, UHP H_2 was switched to H_2 with 250 ppm H_2S, and the flux was measured vs. time. At every temperature, the hydrogen flux decreased slightly during the first hour of

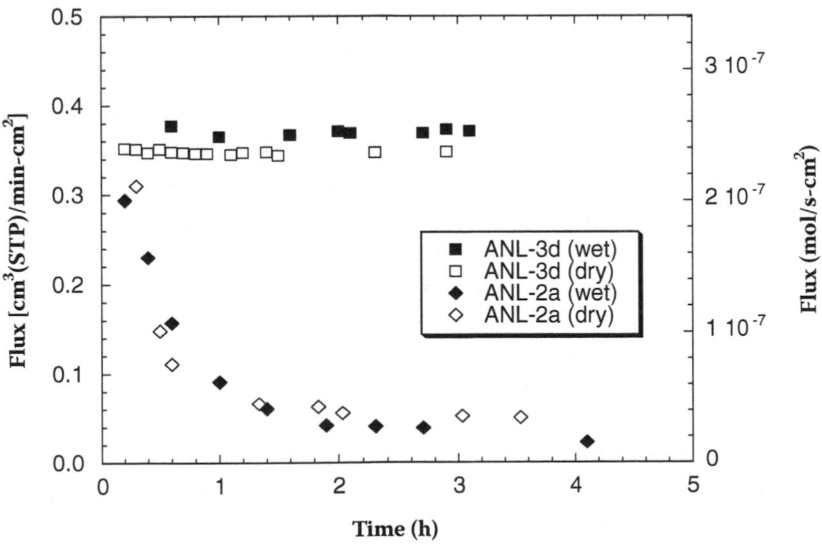

FIGURE 6.5 Hydrogen flux vs. time in feed gas of 2.0% CH_4, 19.6% H_2, 19.6% CO, and 58.8% CO_2 (mol%) for ANL-3d and ANL-2a membranes.

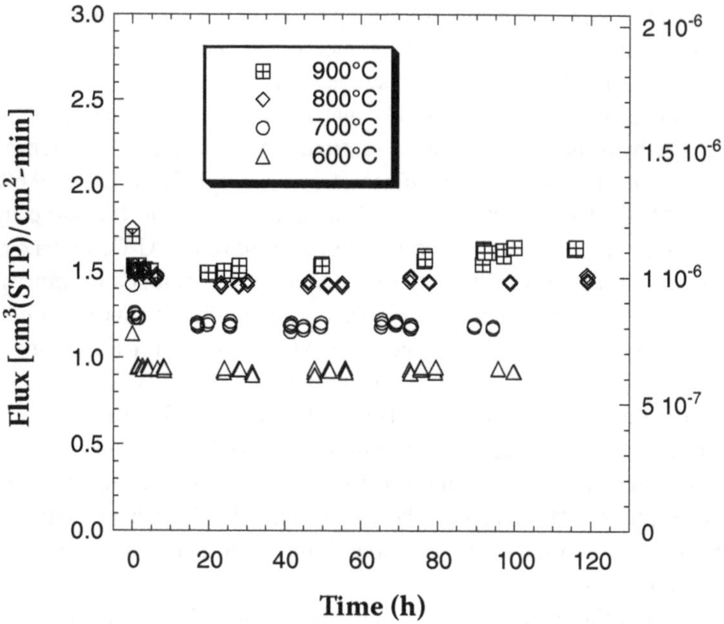

FIGURE 6.6 Hydrogen flux versus time at several temperatures for 100 mm thick ANL-3e membrane in feed gas of 51 ppm H_2S/19.6% H_2/balance He.

exposure but was stable thereafter. At 900°C, the flux actually increased slightly during longer exposures, and the final flux was only ≈3% lower than the initial flux. The reason for the initial decrease in flux is not understood at this point, but thermodynamic data indicate that a reaction between the hydrogen transport metal and H_2S is not favorable under these conditions.

Figure 6.7 shows the hydrogen flux of an ANL-3e membrane (thickness = 200 μm) vs. time at 900°C in feed gases with different H_2S concentrations. As in the earlier measurements with 51 ppm H_2S, a mixture of UHP H_2 and UHP He was used for the initial reading, then UHP H_2 was switched to an H_2S-containing gas. For measurements with a given H_2S concentration, the hydrogen concentration was constant during the initial and subsequent measurements. In the gas mixtures with 97 ppm H_2S (19.4% H_2) and 400 ppm H_2S (79.8% H_2), the hydrogen flux decreased moderately (~10%) in the first hour of exposure and then was stable, perhaps even increasing slightly with time.

The hydrogen flux of the ANL-3e membrane decreased sharply after gas containing 2,922 ppm H_2S (19.2% H_2) was introduced into the reactor. The flux decreased ≈60% during the first hour of exposure and ≈70% after 2 h of exposure, and continued to decrease at longer exposures. In addition, leakage through the sample increased with time, as indicated by the measured He concentration in the sweep gas. After 70

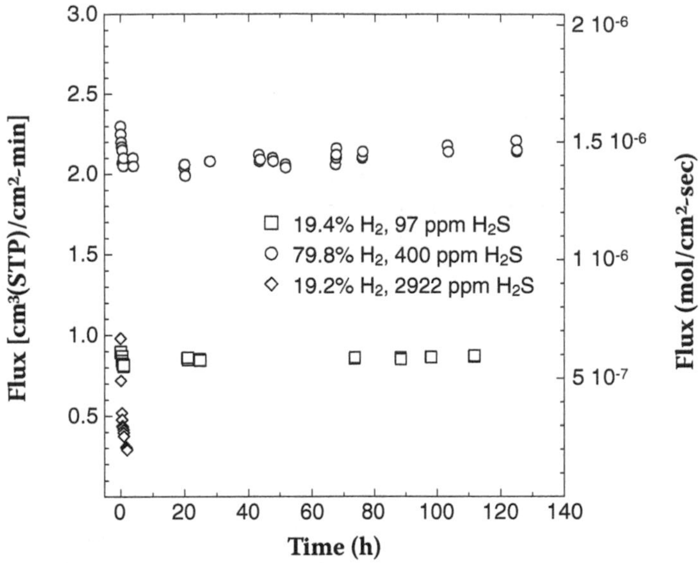

FIGURE 6.7 Hydrogen flux through 200-μm-thick ANL-3e membrane at 900°C in feed gases with a range of $H_2{}^S$ concentrations. Balance of each gas was helium.

h of exposure, the sample was cooled in He. Penetration of the sample by isopropyl alcohol, from one face to the other, showed that the sample contained interconnected porosity after the permeation test in 2,922 ppm H_2S; alcohol had not penetrated the sample before the permeation test. Also, examination of the sample by scanning electron microscopy indicated a loss of metal from the membrane surface. Thus, the sharp drop in hydrogen flux and increase in leakage rate were likely caused by loss of the hydrogen transport metal. In combination with the results for lower H_2S concentrations, this finding indicates that the stability limit for ANL-3e membranes in H_2S-containing atmospheres is in the range of 400 to 2,922 ppm H_2S.

6.4 CONCLUSIONS

We have developed cermet membranes that nongalvanically separate hydrogen from gas mixtures. The highest measured hydrogen flux was 20.0 cm^3(STP)/min-cm^2 for an ANL-3a membrane at 900°C; the flux through an ANL-3e membrane was nearly as high. The effect of hydrogen partial pressure on hydrogen flux indicates that the flux is limited by the bulk diffusion of hydrogen through the metal phase. The dependence of hydrogen flux on membrane thickness confirmed this conclusion over the range in thickness that was studied (22 to 100 μm). Although interfacial reactions are expected to become important at some smaller thickness, these results suggest that decreasing the membrane thickness will increase the hydrogen flux. The hydrogen flux of ANL-3b and -3d membranes showed no degradation in a syngas atmosphere for times up to 190 h, while the hydrogen flux through ANL-3e membranes was stable, after a small initial decrease, for up to 120 h in atmospheres containing up to 400 ppm H_2S. These results indicate that ANL-3 membranes may be suitable for long-term, practical hydrogen separation.

ACKNOWLEDGMENTS

This work was supported by the U.S. Department of Energy, Office of Fossil Energy, National Energy Technology Laboratory's Hydrogen and Gasification Technologies Program, under Contract W-31-109-Eng-38.

REFERENCES

1. Iwahara, H., Yajima, T., and Uchida, H. *Solid State Ionics*, 70/71, 1994, 267-271.
2. Iwahara, H. *Solid State Ionics*, 77, 1995, 289-298.
3. Guan, J., Dorris, S. E., Balachandran, U., and Liu, M. *Solid State Ionics*, 100, 1997, 45-52.
4. Guan, J., Dorris, S. E., Balachandran, U., and Liu, M. *J. Electrochem. Soc.*, 145, 1998, 1780-1786.
5. Guan, J., Dorris, S. E., Balachandran, U., and Liu, M. *Ceram. Trans.*, 92, 1998, 1-9.
6. Balachandran, U., Lee, T. H., and Dorris, S. E. In *Proceedings 16th Annual International Pittsburgh Coal Conf.*, Pittsburgh, PA, October 11-15, 1999.
7. Balachandran, U., Lee, T. H., Zhang, G., Dorris, S. E., Rothenberger, K. S., Howard, B. H., Morreale, B., Cugini, A. V., Siriwardane, R. V., Poston, J. A. Jr., and Fisher, E. P. In *Proceedings 26th International Technical Conference on Coal Utilization and Fuel Systems*, Clearwater, FL, March 5-8, 2001. Gaithersburg, MD: Coal Technical Association, 751-761.
8. Balachandran, U. et al. Proton-Conducting Membranes, Annual Report for FY 2001 Argonne National Laboratory (2001).
9. Balachandran, U. et al. Proton-Conducting Membranes, Annual Report for FY 2002 Argonne National Laboratory (2002).
10. Balachandran, U., Lee, T. H., Wang, S., Zhang, G., and Dorris, S. E. In *Proceedings 27th International Technical Conference on Coal Utilization and Fuel Systems*, Clearwater, FL, March 4-7, 2002.
11. Buxbaum, R. E. and Marker, T. L. *J. Memb. Sci.*, 85, 1993, 29-38.
12. Balachandran, U. et al. Proton-Conducting Membranes, Quarterly Report for October-December 2002, Argonne National Laboratory (2003).

7 Effects of Hydrogen Gas on Steel Vessels and Pipelines

Brian P. Somerday and Chris San Marchi

CONTENTS

7.1 INTRODUCTION

Carbon and low-alloy steels are common structural materials for high-pressure hydrogen gas vessels and pipelines. These steels are low cost, and a wide range of properties can be achieved through alloying, processing, and heat treatment.[1] Fabricating complex structures such as gas containment vessels and pipelines is readily accomplished with steels since these materials can be formed, welded, and heat treated in large sections.

The containment and transport of high-pressure hydrogen gas in steel structures present a particular challenge. Hydrogen gas can adsorb and dissociate on the steel surface to produce atomic hydrogen.[2,3] The subsequent dissolution and diffusion of atomic hydrogen into steels can degrade mechanical properties, a phenomenon generally referred to as hydrogen embrittlement. The manifestation of hydrogen embrittlement is enhanced susceptibility to fracture. Hydrogen reduces typical measures of fracture resistance such as tensile strength, ductility, and fracture toughness, accelerates fatigue crack propagation, and introduces additional material failure modes.[3] In particular, steel structures that do not fail under static loads in benign environments at ambient temperature may become susceptible to time-dependent crack propagation in hydrogen gas.

The objective of this chapter is to provide guidance on the application of carbon and low-alloy steels for hydrogen gas vessels and pipelines, emphasizing the variables that influence hydrogen embrittlement. Section 7.2 reviews published experience with hydrogen gas vessels and pipelines. Industrial gas and petroleum companies have successfully used carbon and low-alloy steels for hydrogen gas containment and transport, but only within certain limits of material, environmental, and mechanical conditions.[4–6] In the proposed hydrogen energy infrastructure, it is anticipated that hydrogen gas vessels and pipelines will be subjected to operating conditions that are outside the windows of experience. Thus, section 7.4 will demonstrate trends in hydrogen embrittlement susceptibility for steels as a function of important material, environmental, and mechanical variables. The metric for hydrogen embrittlement susceptibility is based on fracture mechanics properties. Fracture mechanics principles are reviewed in section 7.3.

This chapter focuses on effects of hydrogen gas on steel structures at near-ambient temperatures. For these conditions, atomic hydrogen is in solid solution in the steel lattice and can facilitate fracture through one of several broadly accepted mechanisms.[7,8] Excluded from this chapter are references to hydrogen embrittlement mechanisms that are promoted by elevated temperatures or aqueous environments. A well-known mechanism in this category is hydrogen attack, which involves a chemical reaction between atomic hydrogen and carbon in steel to form methane gas. The formation of high-pressure methane gas in internal fissures and depletion of carbon from the steel enable material failure.[3] Other mechanisms not referenced in this chapter involve the internal precipitation of high-pressure hydrogen gas.[3] Failure caused by the internal formation of methane or hydrogen gas is not considered pertinent to steel structures used in the containment and transport of high-pressure hydrogen gas.[5]

This chapter is not intended to provide detailed guidance on the design of hydrogen gas vessels and pipelines. General design approaches for structures in hydrogen gas as well as details on vessels and pipelines are available.[4,5,9,10] While this chapter emphasizes hydrogen embrittlement of steels, it does not represent a comprehensive review of the subject. The literature on hydrogen embrittlement of steels is extensive (e.g., references 11–15) and includes numerous review articles.[3,16–18] The content of this chapter does complement previous publications that address hydrogen compatibility of structural materials for hydrogen energy applications.[9,19–21] Finally, while this chapter presents some specific data to illustrate hydrogen embrittlement trends in steels, the document is not intended to serve as a data archive. Such a data compilation has been created to guide the application of materials in a hydrogen energy infrastructure.[22]

7.2 REVIEW OF HYDROGEN GAS VESSELS AND PIPELINES

This section summarizes the experience of industrial gas and petroleum companies with steel hydrogen gas vessels and pipelines. Extensive information is published in two European Industrial Gases Association (EIGA) documents, which were created to provide guidance on the design of hydrogen gas vessels and pipelines.[4,5] The document on hydrogen gas pipelines[5] was developed jointly with the Compressed Gas Association (CGA) and has been published concurrently as the CGA document G-5.6. Presentations from a workshop sponsored by the U.S. Department of Energy[6] served as additional sources of information on hydrogen piping systems. From this collective published information, the material, environmental, and mechanical conditions that have been identified by industrial gas producers and consumers to impact performance of steel hydrogen gas vessels and pipelines are reported below.

7.2.1 Hydrogen Gas Vessels

The information reported here is for cylindrical and tube-shaped steel vessels, where the primary function of the vessels is to distribute hydrogen gas.[4] Current European hydrogen gas distributors have several hundred thousand vessels in service, which supply up to 300×10^6 m^3 of hydrogen gas to customers annually. Over the past two decades, these hydrogen gas vessels have functioned safely and reliably.

Failures of hydrogen gas vessels have been encountered in Europe, particularly in the late 1970s.[4] Subsequent studies of hydrogen gas vessels led to the conclusion that failures were ultimately enabled by hydrogen-enhanced fatigue crack propagation from surface defects.

7.2.1.1 Material Conditions Affecting Vessel Steel in Hydrogen

Experience indicates that failure of hydrogen gas vessels has been governed primarily by properties of the steel, particularly strength and microstructure.[4] These variables affect the susceptibility of the steel to hydrogen embrittlement.

The published experience for reliable hydrogen gas vessels pertains to a narrow range of steel conditions.[4] Hydrogen gas vessels in Europe are fabricated from steel designated 34CrMo4. The steel composition (table 7.1) is distinguished by the alloying elements chromium and molybdenum and the concentration of carbon.

The 34CrMo4 steels are processed to produce a "quenched and tempered" microstructure. The heat treatment sequence to produce this microstructure consists

TABLE 7.1
Composition (wt%) of 34CrMo4 Steel[a]

Cr	Mo	C	Mn	Si	P[b]	S[b]	Fe	P + S[b]
0.90–1.20	0.15–0.25	0.30–0.37	0.50–0.80	0.15–0.35	0.025 max.	0.025 max.	Balance	

[a] The composition limits for 34CrMo4 vary slightly among European countries. The specification in table 7.1 is from Germany.[4] The 34CrMo4 steel composition is almost identical to either AISI 4130 or AISI 4135 steel.[47]

[b] Limits for P and S in new hydrogen gas vessels are 0.025 wt%.

of heating in the austenite phase field, rapidly cooling (quenching) to form martensite, then tempering at an intermediate temperature.[1] For hydrogen gas vessels, the heat treatment parameters are selected to produce a uniform tempered martensite microstructure and to limit tensile strength (σ_{uts}) below 950 MPa.[4]

Vessels used for hydrogen gas distribution are seamless, meaning the vessel body is fabricated without welds. Hydrogen gas vessels are ideally seamless since welding alters the desirable steel microstructure produced by quenching and tempering and introduces residual stress. Welds in high-pressure hydrogen gas vessels fabricated from low-alloy steels have contributed to hydrogen-assisted cracking.[23]

7.2.1.2 Environmental Conditions Affecting Vessel Steel in Hydrogen

The severity of hydrogen embrittlement in steel is affected by gas pressure, since this variable dictates the amount of atomic hydrogen that dissolves in steel.[17] Working pressures for steel vessels in hydrogen distribution applications are typically in the range of 20 to 30 MPa.[4]

The inner surface of hydrogen gas vessels is susceptible to localized corrosion due to impurities that can exist in the steel and hydrogen gas.[4] Interactions between localized corrosion and hydrogen embrittlement have not been specified; however, impurities in the gas and steel are known to affect hydrogen embrittlement, as described in section 7.4.

7.2.1.3 Mechanical Conditions Affecting Vessel Steel in Hydrogen

In addition to gas pressure, hydrostatic tensile stress increases the hydrogen concentration in metals.[18] This leads to high, localized concentrations of atomic hydrogen at stress risers, such as defects, thus promoting hydrogen embrittlement. Defects can form on the inner surface of hydrogen gas vessels from manufacturing or during service. One manifestation of defects that forms during service is localized corrosion pits.[4]

One of the detrimental mechanical loading conditions for steel hydrogen gas vessels is cyclic stress, which drives fatigue crack propagation.[4] Pressure cycling results from filling and emptying vessels during service. The presence of surface defects influences the mechanical conditions in the steel vessel wall. Surface defects intensify local stresses, which provide the mechanical driving force for fatigue crack propagation and concentrate atomic hydrogen in the steel. Cracks propagate by hydrogen embrittlement acting in concert with cyclic stress. After a certain number of vessel filling–emptying cycles, fatigue cracks reach a critical length. Then the cracks can extend by hydrogen embrittlement mechanisms that operate in a filled hydrogen vessel under static pressure.

7.2.2 Hydrogen Gas Pipelines

The information summarized here is for steel transmission and distribution piping systems that carry hydrogen gas. The industrial gas companies have accumulated decades of experience with hydrogen gas transmission pipelines and currently operate over 900 miles of pipeline in the United States and Europe.[6] These pipelines have been safe and reliable for specific ranges of material, environmental, and mechanical conditions.

7.2.2.1 Material Conditions Affecting Pipeline Steel in Hydrogen

Although steel pipelines have been operated safely with hydrogen gas, specific limits have been placed on properties of the steels. In particular, relatively low-strength carbon steels are specified for hydrogen gas pipelines.[5] Examples of steels that have been proven for hydrogen gas service are ASTM A106 Grade B, API 5L Grade X42, and API 5L Grade X52.[5,6] The compositions of these steels are provided in table 7.2 and table 7.3. The API 5L steels containing small amounts of niobium, vanadium, and titanium are referred to as microalloyed steels. Microalloyed X52 steel has been used extensively in hydrogen gas pipelines.[5]

Steels for hydrogen gas pipelines are processed to produce uniform, fine-grained microstructures.[5] A normalizing heat treatment can yield the desired microstructure in conventional steels. A typical normalizing heat treatment consists of heating steel in the austenite phase field followed by air cooling.[1] A more sophisticated process of hot rolling in the austenite–ferrite phase field is used to manufacture fine-grained microalloyed steels.[1]

Material strength is an important variable affecting hydrogen embrittlement of pipeline steels. One of the principles guiding selection of steel grades and processing

TABLE 7.2
Composition (wt%) of A106 Grade B Steel[a]

C	Mn	P	S	Si	Cr[b]	Cu[b]	Mo[b]	Ni[b]	V[b]	Fe
0.30 max.	0.291.06	0.035 max.	0.035 max.	0.10 max.	0.40 max.	0.40 max.	0.15 max.	0.40 max.	0.08 max.	Balance

[a] Specification is for seamless pipe.[48]

TABLE 7.3
Composition (wt%) of API 5L Steels[a]

	C	Mn	P[b]	S[b]	Nb + V + Ti	Fe
Grade X42	0.22 max.	1.30 max.	0.025 max.	0.015 max.	0.15 max.	Balance
Grade X52	0.22 max.	1.40 max.	0.025 max.	0.015 max.	0.15 max.	Balance

[a] Product Specification Level 2 composition for welded pipe.[49]

[b] Recommended maximum concentrations of P and S are 0.015 and 0.01 wt%, respectively, for modern steels in hydrogen gas service.[5]

procedures is to limit strength. The maximum tensile strength, σ_{uts}, recommended for hydrogen gas pipeline steel is 800 MPa.[5]

The properties of welds are carefully controlled to preclude hydrogen embrittlement. One of the important material characteristics governing weld properties is the carbon equivalent (CE). The CE is a weighted average of elements, where concentrations of carbon and manganese are significant factors.[5] Higher values of CE increase the propensity for martensite formation during welding. Nontempered martensite is the phase most vulnerable to hydrogen embrittlement in steels.[9,21] Although low values of CE are specified to prevent martensite formation in welds,[5] these regions are often still harder than the surrounding pipeline base metal. The higher hardness makes welds more susceptible to hydrogen embrittlement. The maximum tensile strength for welds is also recommended as 800 MPa.

7.2.2.2 Environmental Conditions Affecting Pipeline Steel in Hydrogen

Similar to hydrogen gas vessels, the hydrogen embrittlement susceptibility of pipeline steels depends on gas pressure. Industrial gas companies have operated steel hydrogen pipelines at gas pressures up to 13 MPa.[6]

Hydrogen gas pipelines are subject to corrosion on the external surface. While corrosion damage has created leaks in hydrogen gas pipelines,[5,6] interactions between corrosion and hydrogen gas embrittlement have not been cited as concerns for pipelines.

7.2.2.3 Mechanical Conditions Affecting Pipeline Steel in Hydrogen

Hydrogen gas transmission pipelines are operated at near constant pressure[5,6]; therefore, cracking due to hydrogen embrittlement must be driven by static mechanical forces. Cyclic loading, which can drive fatigue crack propagation aided by hydrogen embrittlement, has not been a concern for hydrogen gas transmission pipelines.[5] Experience from the petroleum industry, however, has demonstrated that hydrogen-assisted fatigue is possible with hydrogen gas distribution piping.[6]

Defects can form on the inner and outer surfaces of steel pipelines from several sources, including welds, corrosion, and third-party damage.[5,6] Welds are of particular concern since steel pipelines can require two different welds: longitudinal (seam) welds to manufacture sections of pipeline and girth welds to assemble the pipeline system. These welds are inspected to detect the presence of defects. Similar to hydrogen gas vessels, defects in pipeline walls intensify stresses locally, creating more severe mechanical conditions for crack extension and concentrating atomic hydrogen in the steel.

7.3 IMPORTANCE OF FRACTURE MECHANICS

Experience has revealed that defects can form on the surfaces of both hydrogen gas vessels and pipelines.[4,5] Since elevated stresses arise near defects in pressurized vessels and pipelines, establishing design parameters based on average wall stresses and material tensile data (i.e., strength and ductility) can be nonconservative. The design of structures containing defects is more reliably conducted using fracture mechanics methods. The application of fracture mechanics to structures exposed to hydrogen gas has been well documented.[3,7,9,10]

Fracture mechanics methods are commonly implemented in materials testing protocols. Fracture mechanics-based material properties are needed for engineering purposes, i.e., design of defect-tolerant structures, but scientific studies of materials often measure these properties as well. Laboratory fracture mechanics specimens impose severe mechanical conditions for fracture, and these conditions can promote fracture phenomena that are not revealed by other testing methods. For this reason, fracture mechanics-based materials tests are appealing for assessing hydrogen embrittlement. This section gives brief background information on fracture mechanics applied to structures and materials in hydrogen gas.

The average wall stress and the local stress near defects are related through the linear elastic stress intensity factor (K). The magnitude of the local stress is proportional to the stress intensity factor, K, according to the following relationship:[24,25]

$$\sigma_y = \frac{K}{\sqrt{2\pi x}} \tag{7.1}$$

where σ_y is the local tensile stress normal to the crack plane and x is the distance in the crack plane ahead of the crack tip. The stress intensity factor, K, is proportional to the wall stress and structural dimensions, viz.:[24,25]

$$K = \beta\sigma_w\sqrt{\pi a} \tag{7.2}$$

where σ_w is the wall stress, the parameter β is a function of both defect geometry and structure geometry, and a is the defect depth.

Design parameters of structures containing defects can be established through the stress intensity factor, K. The failure criterion for structures that contain defects and are subjected to static or monotonically increasing loads is as follows:

$$K \geq K_c \tag{7.3}$$

where K is the applied stress intensity factor and K_c is the critical value of stress intensity factor for propagation of the defect. The K_c value is a property of the structural material and can depend on variables such as the service environment. Combining equations 7.2 and 7.3, the following relationship can be established:

$$\beta\sigma_w\sqrt{\pi a} \geq K_c \tag{7.4}$$

Equation 7.4 is the essential relationship for design of structures containing defects. Assuming K_c is known for the structural material and service environment, equation 7.4 can be used in the following manner:

- If the structure dimensions and defect depth are known, the maximum wall stress can be calculated.
- If the structure dimensions and wall stress are known, the maximum defect depth can be calculated.
- If the wall stress and defect depth are known, the structural dimensions can be calculated.

The failure criterion in equation 7.4 pertains to structures subjected to static or monotonically increasing loads. Extension of a defect under these loading conditions is sustained as long as equation 7.4 is satisfied. Defects can also extend by fatigue crack propagation when the structure is loaded under cyclic stresses. The rate of fatigue crack propagation is proportional to the stress intensity factor range, i.e.:[24]

$$\frac{da}{dN} = C\Delta K^n \tag{7.5}$$

where da/dN is the increment of crack extension per load cycle, C and n are material- and environment-dependent parameters, and ΔK is the stress intensity factor range. The stress intensity factor range, ΔK, is defined as $(K_{max} - K_{min})$, where K_{max} and K_{min} are the maximum and minimum values of K, respectively, in the load cycle. K_{max} and K_{min} are calculated from equation 7.2. The relationship in equation 7.5 is relevant for fatigue crack propagation at K_{max} values less than K_c, but does not describe crack propagation in the lowest range of ΔK.

It must be noted that the fracture mechanics framework described above only applies when plastic deformation of the material is limited. Substantial plastic deformation may accompany propagation of existing defects in structures fabricated from relatively low-strength materials, e.g., carbon steels. In these cases, the linear elastic stress intensity factor, K, does not accurately apply in structural design. Alternately, elastic-plastic fracture mechanics methods may apply.[24]

The hydrogen embrittlement susceptibility of structural steels can be quantified using fracture mechanics–based material properties. The critical values of stress intensity factor for propagation of a defect under static and monotonically increasing loads in hydrogen gas are referred to as K_{TH} and K_{IH}, respectively,[7] in this chapter. For cyclic loading, the material response is given by the da/dN vs. ΔK relationship measured in hydrogen gas. Enhanced hydrogen embrittlement is indicated by lower values of K_{TH} and K_{IH} but higher values of da/dN. Fracture mechanics properties of materials in hydrogen gas are typically measured under controlled laboratory conditions using standardized testing techniques.[26–28] These properties provide consistent, conservative indices of hydrogen embrittlement susceptibility.

7.4 VESSELS AND PIPELINES IN HYDROGEN ENERGY APPLICATIONS

An open question is whether steels currently used in hydrogen gas vessels and pipelines can be employed for similar applications in the hydrogen energy infrastructure.

The answer depends on several factors, including structural design constraints as well as steel properties. The information in section 7.2 demonstrates that steels are suitable structural materials provided hydrogen gas vessels and pipelines are operated within certain limits. In the proposed hydrogen energy infrastructure, it is anticipated that hydrogen gas vessels and pipelines will be subjected to service conditions that are outside the windows of experience. For example, hydrogen gas will likely be stored and transported at pressures that exceed those in current industrial gas and petroleum industry applications. The objective of this section is to provide insight into possible limitations on steel properties by illustrating trends in hydrogen embrittlement susceptibility as a function of important material, environmental, and mechanical variables.

The hydrogen embrittlement data in this section are for structural steels that are similar to those used in current hydrogen gas vessels and pipelines. In particular, data were selected for steels having compositions, microstructures, and tensile strengths that are germane to steels in hydrogen gas vessels and pipelines. In some cases, data are presented for steels having properties that deviate substantially from those used in gas vessels and pipelines. These cases are noted in the text, but the data trends still provide important insights. Fracture mechanics data were selected to demonstrate hydrogen embrittlement trends, since these data pertain to structures containing defects and provide conservative indices of fracture susceptibility in hydrogen gas.

Much of the data demonstrate that caution must be exercised in extending current steels to operating conditions outside the windows of experience. However, other data suggest that the hydrogen embrittlement resistance of steels can be improved.

7.4.1 Effect of Gas Pressure

Steels become more susceptible to hydrogen embrittlement as the materials are exposed to higher gas pressures. Thermodynamic equilibrium between hydrogen gas and dissolved atomic hydrogen is expressed by the general form of Sievert's law:[17]

$$C = S\sqrt{f} \tag{7.6}$$

where C is the concentration of dissolved atomic hydrogen, the fugacity, f, of the hydrogen gas is related to the pressure (and temperature) of the system, and the solubility, S, of atomic hydrogen in the steel is a temperature-dependent material property. equation 7.6 shows that as fugacity (pressure) increases, the quantity of atomic hydrogen dissolved in the steel increases; consequently, embrittlement becomes more severe. This trend is illustrated from K_{TH}, K_{IH}, and da/dN data. Figure 7.1 shows data for both low-alloy steels (K_{TH}) and carbon steels (K_{IH}), where critical K values decrease as hydrogen gas pressure increases for both types of steel.[10,29] Data for a low-alloy steel in figure 7.2 demonstrate that da/dN measured at a fixed stress intensity factor range, ΔK, continuously increases as hydrogen gas pressure increases.[30] Finally, figure 7.3 shows that increasing hydrogen gas pressure also accelerates da/dN in a carbon steel, but only at lower ΔK values.[31]

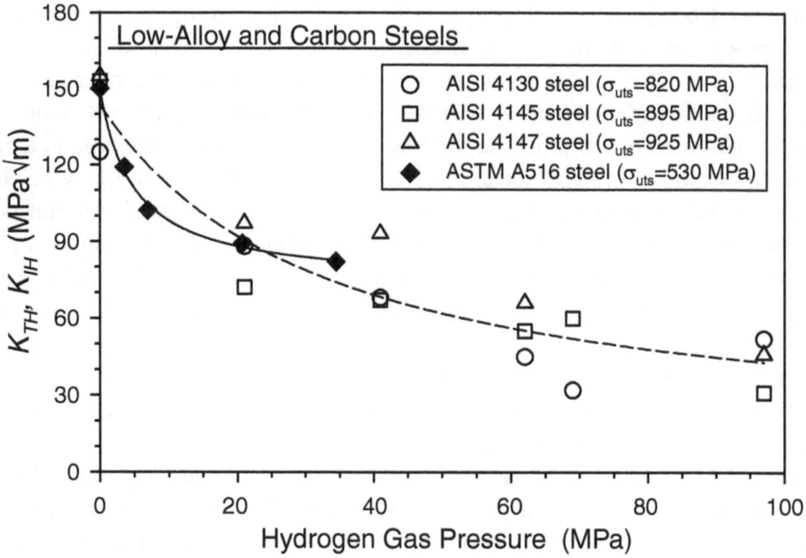

FIGURE 7.1 Effect of gas pressure on critical stress intensity factor for crack extension in hydrogen gas (K_{TH} or K_{IH}).[10,29] The low-alloy steels (open symbols) were tested under static loading, while the carbon steel (filled symbols) was tested under rising displacement loading. Data points at zero pressure represent fracture toughness measurements in air, i.e., K_{Ic}.

The data in figure 7.1 through figure 7.3 indicate that steel vessels and pipelines in hydrogen economy applications (i.e., at high hydrogen gas pressure) could be more vulnerable to hydrogen embrittlement than estimated from current experience. The quantities of hydrogen needed for a hydrogen-based economy suggest that gas could be stored and transported at pressures that exceed current limits. The American Society of Mechanical Engineers (ASME) is developing standards for hydrogen gas vessels with working pressures up to 100 MPa.[32] Current hydrogen gas vessels, however, have maximum working pressures in the range of 20 to 30 MPa.[4] Figure 7.1 and figure 7.2 demonstrate that vessels fabricated from low-alloy steels become increasingly more susceptible to hydrogen embrittlement as pressures increase above 30 MPa. Current hydrogen gas pipelines are operated at pressures up to 13 MPa.[6] Figure 7.1 and figure 7.3 indicate that enhanced hydrogen embrittlement susceptibility must be considered for pipelines operating above 13 MPa.

7.4.2 Effect of Gas Impurities

Hydrogen gas embrittlement in steels can be altered by the presence of low concentrations of other gases in the environment. Certain gases such as oxygen can impede the adsorption of hydrogen gas on steel surfaces. Consequently, the kinetics of atomic hydrogen dissolution in steel can be greatly reduced, and the apparent hydrogen embrittlement determined from short-term testing is mitigated.[2,3] Sulfur-bearing gases such as hydrogen sulfide can have the opposite effect: the presence of these gases exacerbates hydrogen embrittlement.[33,34]

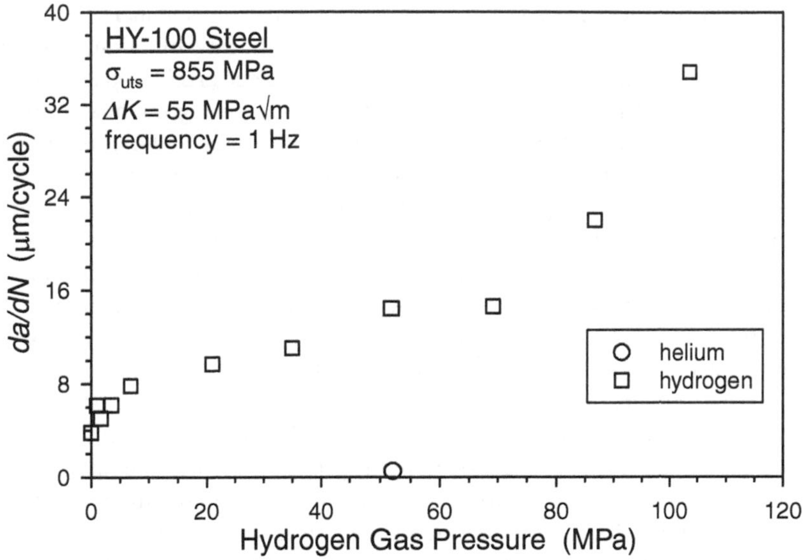

FIGURE 7.2 Effect of hydrogen gas pressure on fatigue crack growth rate (*da*/*dN*) at constant stress intensity factor range (ΔK) in a low-alloy steel.[30]

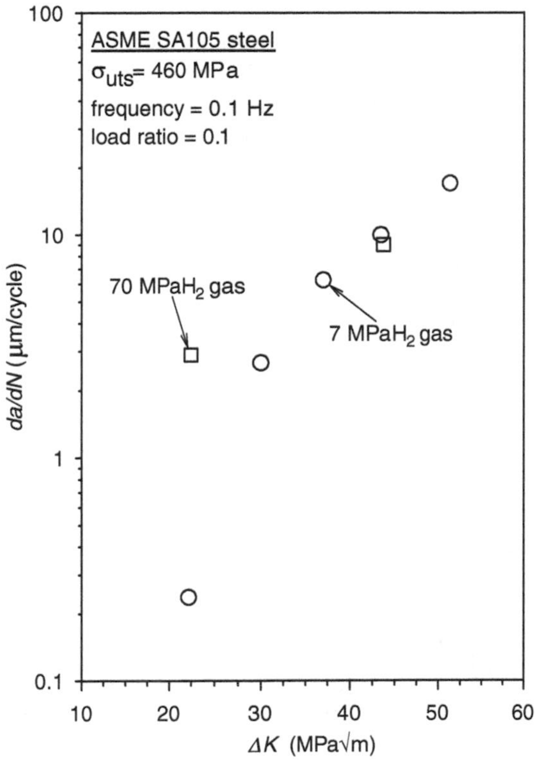

FIGURE 7.3 Effect of hydrogen gas pressure on fatigue crack growth rate (*da*/*dN*) vs. stress intensity factor range (ΔK) relationships for a carbon steel.[31]

The effect of various gas additives on hydrogen embrittlement in a low-alloy steel is illustrated in figure 7.4.[35] The data in figure 7.4 show the ratio of fatigue crack propagation rate in hydrogen gas–containing additives to fatigue crack propagation rate in hydrogen gas only. A ratio near 1.0 indicates that fatigue crack growth rates are equal in the two environments. The data demonstrate that oxygen and carbon monoxide gases in low concentrations can mitigate hydrogen embrittlement, while gases such as methyl mercaptan and hydrogen sulfide can compound hydrogen embrittlement.

The data in figure 7.4 are effective in demonstrating the potential impact of a wide range of gas additives on hydrogen embrittlement for a single steel; however, some further comments are needed. The low-alloy steel represented in figure 7.4 was not heat treated by quenching and tempering; however, the data trends are expected to apply to steel hydrogen vessels. Additionally, some studies confirm results from figure 7.4, e.g., effects of oxygen and hydrogen sulfide[33,34,36,37]; other studies, however, report conflicting results. For example, figure 7.4 shows that sulfur dioxide has no effect on fatigue crack propagation in hydrogen gas, but other studies have found that this gas species inhibits hydrogen embrittlement.[38] Finally, the measurements represented in figure 7.4 were conducted for specific gas concentrations at a high load cycle frequency (i.e., 5 Hz), but such variables impact how severely gas additives affect hydrogen embrittlement.[39] Despite these caveats, the data in figure 7.4 highlight the importance of trace gas constituents on environmental effects for steels in hydrogen gas.

The presence of nonintentional gas additives must be considered for hydrogen embrittlement of vessels and pipelines in the hydrogen energy infrastructure. The effect of gas impurities on hydrogen embrittlement may depend on the absolute partial pressure of the trace gas.[39] Increasing the operating pressure of vessels and

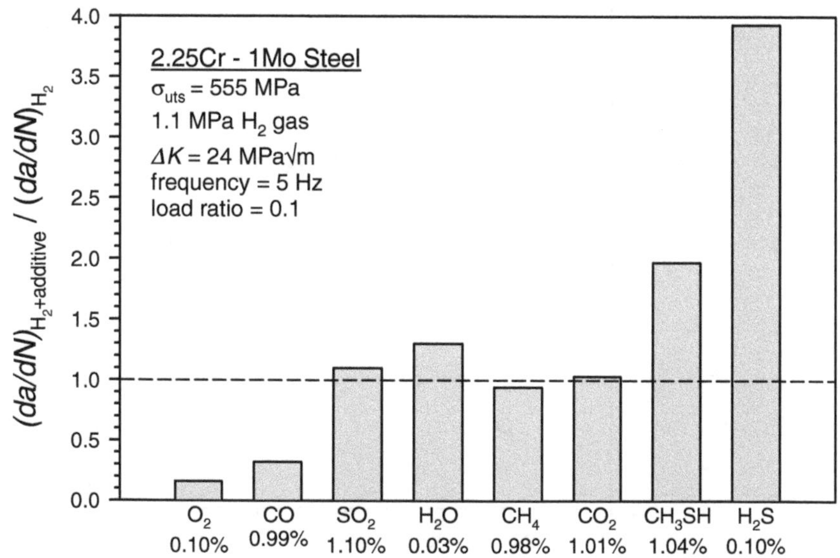

FIGURE 7.4 Effect of gas additives on the fatigue crack growth rate (*da/dN*) at constant stress intensity factor range (ΔK) for a low-alloy steel in hydrogen gas.[35]

pipelines will elevate partial pressures of impurities in hydrogen gas and potentially their role in hydrogen embrittlement.

Caution must be exercised in trying to exploit gas additives to control hydrogen embrittlement. While the data in figure 7.4 suggest that gas additives such as oxygen could be employed to mitigate hydrogen embrittlement, the mechanistic role of gas additives must be considered. For example, oxygen is reported to impede the kinetics of atomic hydrogen uptake in metals such as steels, but over long periods steels may dissolve sufficient hydrogen to suffer embrittlement. Therefore, gas additives that affect hydrogen uptake kinetics may impact manifestations of hydrogen embrittlement that operate at short timescales (e.g., fatigue loading) but not longer timescales (e.g., static loading).

7.4.3 Effect of Steel Strength

Hydrogen embrittlement in steels generally becomes more severe as material strength increases. This behavior arises because the magnitude of stress amplification near defects is proportional to material strength. These high stresses combined with the resulting enhanced hydrogen dissolution increase susceptibility to hydrogen embrittlement. The impact of material strength on hydrogen embrittlement is exemplified by the K_{TH} data in figure 7.5.[10] Values of K_{TH} measured for low-alloy steels in hydrogen gas decrease as tensile strength, σ_{uts}, increases. A similar trend is expected for carbon steels.

Numerous studies have reported hydrogen embrittlement data trends similar to those in figure 7.5.[40-43] However, some exceptions have been found in the literature.

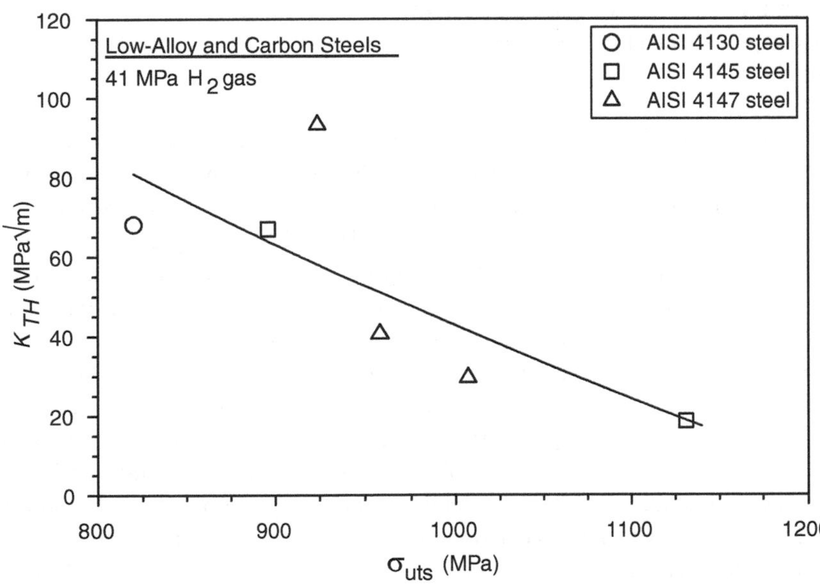

FIGURE 7.5 Effect of tensile strength (σ_{uts}) on critical stress intensity factor for crack extension in hydrogen gas (K_{TH}).[10] Data are for low-alloy steels tested under static loading.

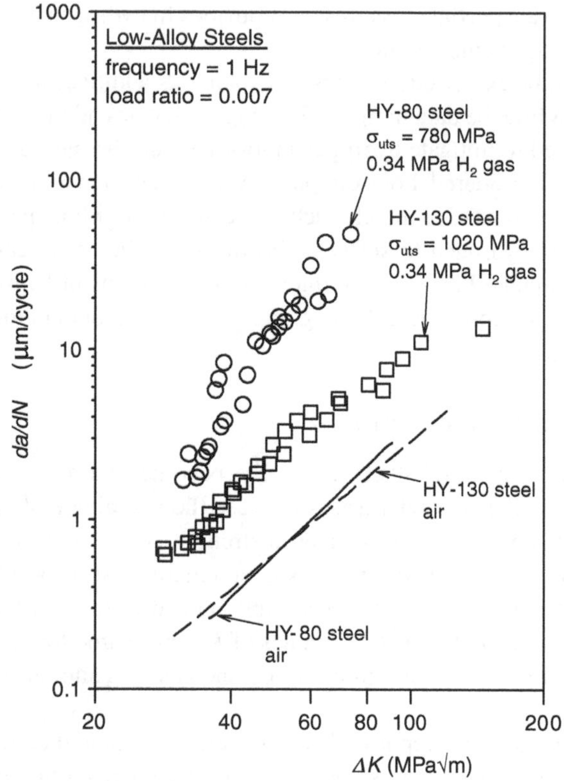

FIGURE 7.6 Fatigue crack propagation rate (*da/dN*) vs. stress intensity factor range (ΔK) relationships measured in low-pressure hydrogen gas for two low-alloy steels with different tensile strengths.[44]

An example is provided in figure 7.6, which shows fatigue crack propagation rate, *da/dN*, vs. stress intensity factor range, ΔK, plots for two low-alloy steels exposed to low-pressure hydrogen gas.[44] Crack propagation rates for the lower-strength steel (HY-80) exceed those in the higher-strength steel (HY-130) during exposure to hydrogen gas. The reason for the inconsistent hydrogen embrittlement trends portrayed in figure 7.5 and figure 7.6 has not been determined; however, it is important to note that data in the two figures were generated under two different loading formats. The K_{TH} data reflect crack growth under static loading, while the *da/dN* data pertain to fatigue crack growth under cyclic loading. Hydrogen-assisted crack growth under static loading is likely governed by crack tip stress, but hydrogen-assisted fatigue crack growth involves cyclic plastic strain. Crack propagation under these two modes of loading could be influenced by material strength differently. Additionally, fatigue crack growth rates can depend on the path of cracking through the steel microstructure. The difference in crack growth rates for HY-80 and HY-130 steels in figure 7.6 could reflect effects of crack path and not solely material strength. The data in figure 7.6 represent tests conducted in low-pressure hydrogen gas, but similar behavior is expected at higher gas pressure.

The effect of tensile strength on hydrogen embrittlement is important for vessels and pipelines in the hydrogen energy infrastructure, where high-strength materials may be attractive. Increasing the operating pressures of hydrogen gas vessels and pipelines could motivate the use of higher-strength steels. With increased gas pressure, the wall thickness of gas vessels and pipelines must increase to meet design stress requirements. However, with higher-strength steels, thinner walls can be used while maintaining the design stress. The data in figure 7.5 demonstrate that steel vessels with tensile strength exceeding the current limits, i.e., 950 MPa,[4] will be more susceptible to hydrogen embrittlement under static loading. The data in figure 7.6 suggest that higher-strength steels may be less susceptible to hydrogen-assisted fatigue crack growth.

7.4.4 Effect of Steel Composition

The concentrations of common elements in steels can significantly impact hydrogen embrittlement susceptibility. A striking demonstration of the effects of manganese, silicon, phosphorus, and sulfur on hydrogen embrittlement in a low-alloy steel is given by the data in figure 7.7.[43] Values of K_{TH} are plotted vs. the sum of bulk manganese, silicon, sulfur, and phosphorus concentrations. Examination of the steel compositions associated with individual data points in figure 7.7 reveals that increases in manganese and silicon are detrimental to hydrogen embrittlement resistance, but variations in phosphorus and sulfur have little effect. Similar trends were revealed from a study that individually varied elements such as manganese,

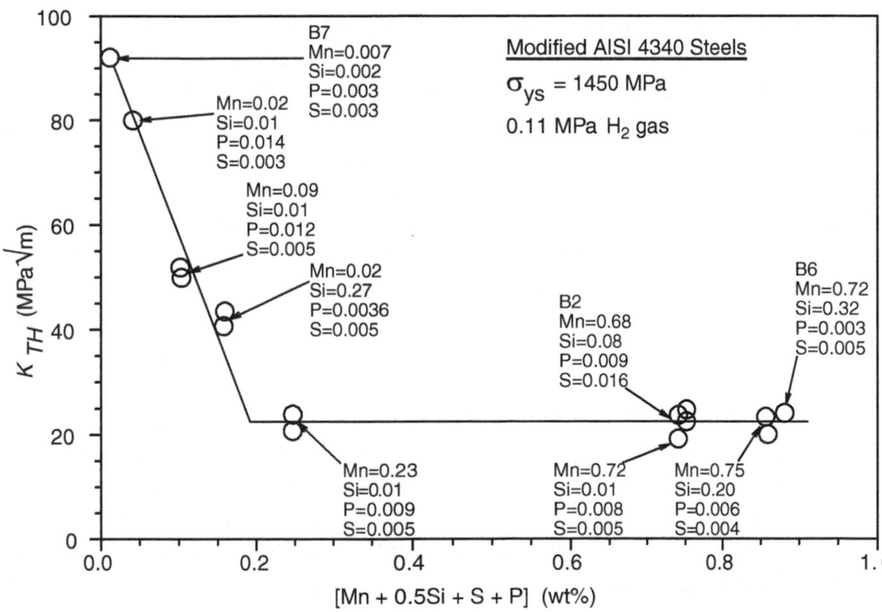

FIGURE 7.7 Effect of manganese, silicon, phosphorus, and sulfur content on critical stress intensity factor for crack extension (K_{TH}) in low-alloy steels.[43] Data are for high-strength steel tested in low-pressure hydrogen gas.

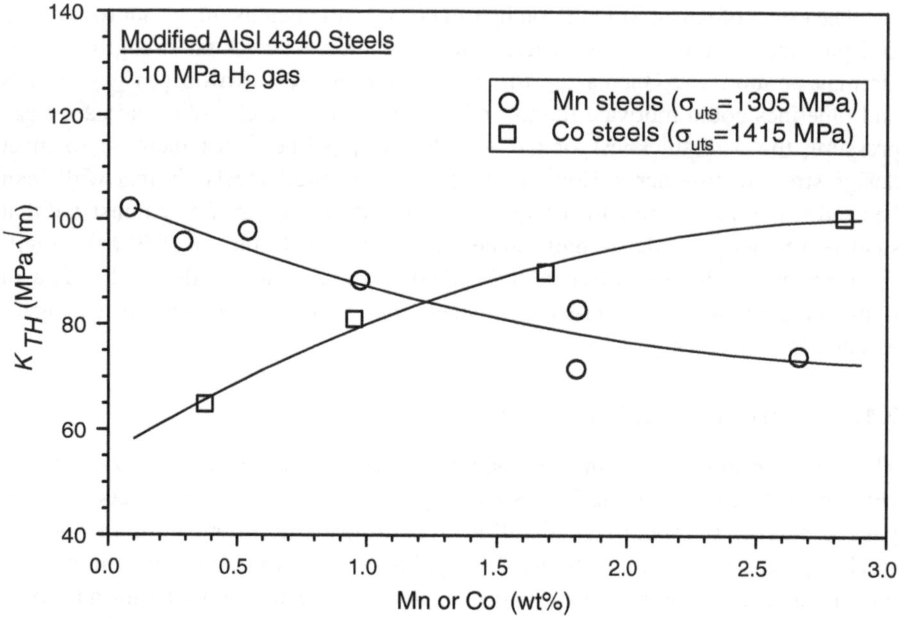

FIGURE 7.8 Effect of manganese or cobalt content on critical stress intensity factor for crack extension (K_{TH}) in low-alloy steels.[40] Data are for high-strength steel tested in low-pressure hydrogen gas.

sulfur, and phosphorus in a low-alloy steel.[40] Figure 7.8 shows that K_{TH} decreases as manganese increases from 0.07 to 2.65 wt%. Systematic variations in sulfur and phosphorus concentrations in the range 0.002 to 0.027 wt% did not affect K_{TH}. While the data indicate that variations in bulk sulfur and phosphorus in the concentration ranges examined do not alter the degree of hydrogen embrittlement, the presence of these elements is integral to the hydrogen embrittlement mechanism in low-alloy steels. While bulk compositions of sulfur and phosphorus should be minimized, the data show that additional benefit could be obtained by minimizing silicon and manganese as well. Although the low-alloy steels from Sandoz[40] and Bandyopadhyay et al.[43] had extremely high strengths and were tested in low-pressure hydrogen gas, the trends in figure 7.7 and figure 7.8 are expected to apply to lower-strength steels in high-pressure hydrogen gas.

The data in figure 7.7 and figure 7.8 apply to low-alloy steels and may not give accurate insight into behavior for carbon steels. Increasing concentrations of manganese and silicon in low-alloy steels enhances the propensity for hydrogen-assisted fracture along grain boundaries.[43] Carbon steel fracture mechanics specimens tested under rising load in hydrogen gas do not exhibit fracture along grain boundaries, but rather cracks propagate across the grains.[29] Since the role of manganese and silicon reflected in figure 7.7 and figure 7.8 is to affect fracture along grain boundaries, the data trends probably do not describe behavior in carbon steels. Data showing effects of steel composition on K_{TH} or K_{IH} measured in hydrogen gas have not been found in the literature for carbon steels.

The hydrogen embrittlement resistance of low-alloy steels used in hydrogen gas vessels cannot be substantially altered by varying concentrations of elements such as manganese and silicon within the allowable composition ranges. Table 7.1 shows that the allowable composition ranges for manganese and silicon in 34CrMo4 steel are 0.50 to 0.80 wt% and 0.15 to 0.35 wt%, respectively. The data in figure 7.7 indicate that K_{TH} noticeably improves only for manganese and silicon levels well below the lower limits in the 34CrMo4 steel composition ranges.

Altering composition may be one avenue to improve the hydrogen embrittlement resistance of steels. Vessels and pipelines in the hydrogen energy infrastructure will likely be subjected to higher gas pressures and may need to be fabricated from higher-strength steels. Increasing either hydrogen gas pressure or steel strength will degrade resistance to hydrogen embrittlement. However, manufacturing steels with much lower manganese and silicon concentrations may balance the loss in hydrogen embrittlement resistance associated with increasing gas pressure or steel strength. Other data suggest that alloying elements not typically in the specifications for low-alloy steels could improve hydrogen embrittlement resistance. For example, data in figure 7.8 show that additions of cobalt to a low-alloy steel with high tensile strength significantly increase K_{TH} values measured in low-pressure hydrogen gas.

7.4.5 Effect of Welds

Welding carbon and low-alloy steels can create residual stress and cause undesirable microstructure changes, e.g., formation of martensite, both of which make steel more vulnerable to hydrogen embrittlement.[9,21,23] Both the fusion zone and heat-affected zone regions of the weld can have microstructures that vary from the base metal.

Limited data show that both welding practice and location of defects can dictate the hydrogen embrittlement susceptibility of a weld. A study on microalloyed steel API 5L Grade X60 examined weld joints that were fabricated using either one or two weld passes.[45] Fracture mechanics specimens were extracted from the base metal, fusion zone, and heat-affected zone and tested in 7-MPa hydrogen gas. Results showed that K_{IH} values measured in the weld fusion zones were similar to values in the base metal, i.e., K_{IH} was approximately 100 MPa$\sqrt{m}$ in each region. In contrast, the heat-affected zones were more susceptible to hydrogen embrittlement, and K_{IH} was difficult to measure. The heat-affected zone in the two-pass weld was most susceptible.

Vessels and pipelines in the hydrogen energy infrastructure will be fabricated similar to current structures, where vessels are seamless and pipelines can be fabricated with both longitudinal welds and girth welds. Variables such as hydrogen gas pressure affect welds in a fashion similar to that of base metals, so the effect of increased gas pressure must be considered for hydrogen embrittlement of welds. Perhaps most important is the possibility of using steels that are outside the window of experience for hydrogen gas pipelines. Although hydrogen embrittlement at welds in current hydrogen gas pipelines has not been reported, it is acknowledged that the strength and microstructure of welds must be controlled to avoid hydrogen embrittlement.[5] The effect of alloy composition and welding practice on weld properties must be understood for any new steels used for hydrogen gas pipelines.

7.4.6 Effect of Mechanical Loading

Hydrogen embrittlement in steels can be manifested under different modes of mechanical loading, i.e., static, monotonically increasing, or cyclic. The severity of hydrogen embrittlement can depend on the specific mode of loading, e.g., static vs. monotonically increasing, as well as variations in one type of loading.

Carbon and low-alloy steels having relatively low tensile strengths resist hydrogen embrittlement under static loads, but these alloys are susceptible under monotonically increasing loads. The carbon steel A516 exhibits hydrogen embrittlement when tests are conducted in hydrogen gas under rising displacement loading (figure 7.1).[29] However, cracks do not propagate in A516 steel when fracture mechanics specimens are statically loaded at $K = 82$ MPa√m in 70-MPa hydrogen gas.[10]

Variations in the rate of monotonic loading as well as the frequency and mean load for cyclic loading affect hydrogen embrittlement. Slow loading rates enhance hydrogen embrittlement, as demonstrated in figure 7.9 for a low-alloy steel.[33] These K_{IH} measurements are for a high-strength steel tested in low-pressure hydrogen gas, but similar trends are expected for low-strength steels in high-pressure gas. Figure 7.10 shows that low load cycling frequencies increase fatigue crack growth rates for a carbon steel tested in hydrogen gas.[31] A similar effect of load cycle frequency on fatigue crack growth rate was measured for a low-alloy steel in hydrogen gas.[35] Finally, figure 7.11 shows that fatigue crack growth rates in hydrogen gas do not depend on load ratio (i.e., K_{min}/K_{max}) for values up to 0.4.[46] However, over this range of load ratios, the difference in crack growth rates measured in hydrogen gas vs.

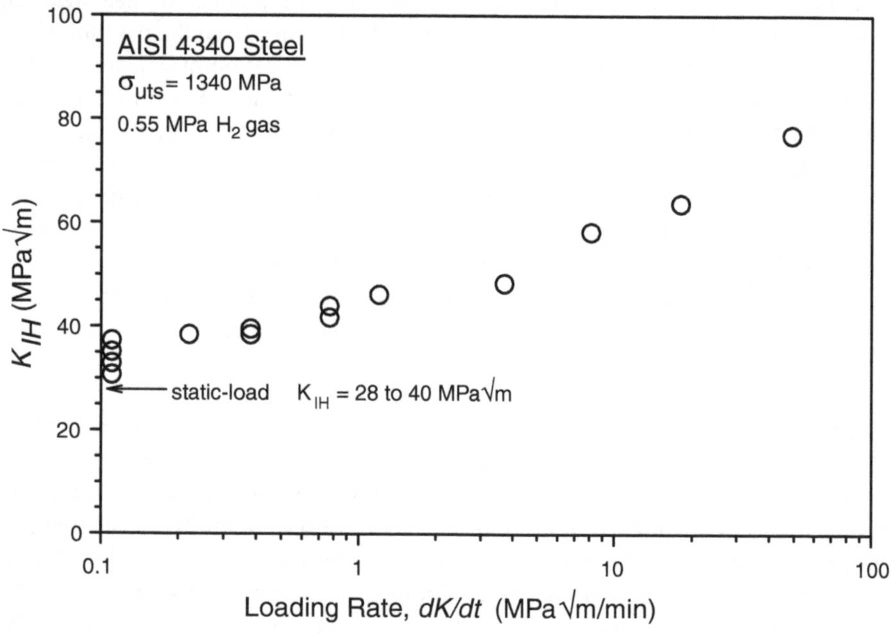

FIGURE 7.9 Effect of loading rate (dK/dt) on critical stress intensity factor for crack extension (K_{IH}) in a low-alloy steel.[33] Data are for high-strength steel tested in low-pressure hydrogen gas.

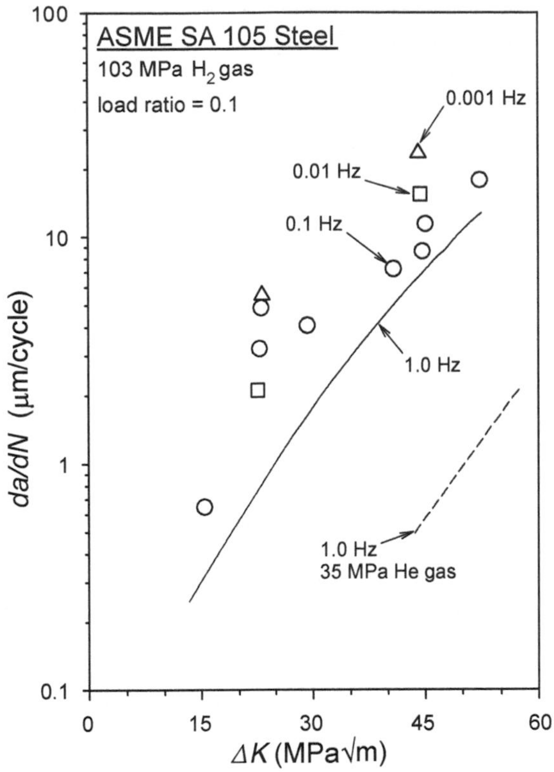

FIGURE 7.10 Effect of load cycle frequency on fatigue crack growth rate (*da*/*dN*) vs. stress intensity factor range (ΔK) relationships for a carbon steel.[31]

nitrogen gas diminishes. Crack growth rates in hydrogen gas increase at higher load ratios in figure 7.11 because K_{max} approaches K_{IH} for the steel. Fatigue crack growth rates in hydrogen gas were also found to be independent of load ratio for the carbon steel ASME SA 105.[31] Similar effects of load cycle frequency and mean load on fatigue crack growth rates in hydrogen gas are expected for low-alloy steels.

Hydrogen vessels and pipelines in current applications are subjected to a variety of loading modes during service, including static, monotonically increasing, and cyclic. Vessels and pipelines in the hydrogen energy infrastructure are expected to experience these same modes of loading. At issue is whether operating conditions needed to support the hydrogen economy will cause substantial changes in variables such as loading rate and frequency, as well as mean loads. For example, the in-line compressors needed for pipelines in the hydrogen energy infrastructure could alter the frequency and amplitude of pressure fluctuations compared to current pipelines. In addition, hydrogen gas vessels could be filled and emptied more frequently in the hydrogen economy. The data in figure 7.9 and figure 7.10 suggest that higher loading rates and frequencies mitigate hydrogen embrittlement in structural steels. However, actual duty cycles involve sequences of active and static loads that are more complex than the uniform loading conditions used in laboratory tests. Hydrogen embrittlement

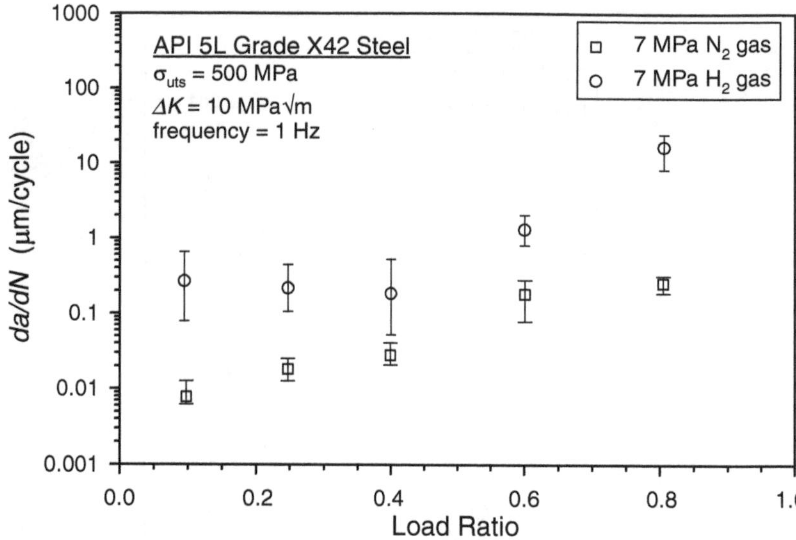

FIGURE 7.11 Effect of load ratio (ratio of minimum load to maximum load) on fatigue crack growth rate (*da*/*dN*) at fixed stress intensity factor range (ΔK) in hydrogen gas for a carbon steel.[46]

data generated under loading conditions that mimic real duty cycles are needed to better understand the impact of mechanical loading variables on hydrogen gas vessels and pipelines.

7.5 CONCLUSION

Experience with steel vessels and pipelines in the industrial gas and petroleum industries demonstrates that these structures can be operated safely with hydrogen gas, although the experience is limited to certain ranges of material, environmental, and mechanical variables. Gas pressures in vessels and pipelines for the hydrogen economy are certain to exceed the limit in current applications. Data consistently show that steels are more susceptible to hydrogen embrittlement at higher gas pressures. As operating pressures increase, designs will demand higher-strength materials. Most data indicate that steels are more vulnerable to hydrogen embrittlement when strength increases. The effects of other variables, such as gas impurities, welds, and mechanical loading on hydrogen embrittlement of steel vessels and pipelines in the hydrogen economy are not as certain. Hydrogen embrittlement resistance of steels could be improved through production of low-manganese and low-silicon steels. Data for high-strength steels in low-pressure hydrogen gas show that composition has a dramatic effect on hydrogen embrittlement; however, this trend has not been demonstrated for lower-strength steels in high-pressure hydrogen gas.

Although hydrogen embrittlement is more severe at high gas pressures and in high-strength steels, structures can still be designed with steels under these conditions by using fracture mechanics. Provided material data are available for steels in high-pressure hydrogen gas, the limiting crack depth, wall stress, and structure dimensions can be defined using fracture mechanics.

ACKNOWLEDGMENTS

This work was supported by the U.S. Department of Energy under Contract DE-AC04-94AL85000.

REFERENCES

1. Krauss, G., *Steels: Heat Treatment and Processing Principles*, ASM International, Materials Park, OH, 1990.
2. Nelson, H.G., Testing for hydrogen environment embrittlement: primary and secondary influences, in *Hydrogen Embrittlement Testing*, ASTM STP 543, ASTM, Philadelphia, 1974, pp. 152–169.
3. Nelson, H.G., Hydrogen embrittlement, in *Treatise on Materials Science and Technology: Embrittlement of Engineering Alloys*, Vol. 25, Briant, C.L. and Banerji, S.K., Eds., Academic Press, New York, 1983, pp. 275–359.
4. *Hydrogen Cylinders and Transport Vessels*, IGC 100/03/E, European Industrial Gases Association, Brussels, 2003.
5. *Hydrogen Transportation Pipelines*, IGC 121/04/E, European Industrial Gases Association, Brussels, 2004.
6. Hydrogen Pipeline Working Group Workshop, U.S. Department of Energy, Augusta, GA, 2005 (www.eere.energy.gov/hydrogenandfuelcells/wkshp_hydro_pipe.html).
7. Gangloff, R.P., Hydrogen assisted cracking of high strength alloys, in *Comprehensive Structural Integrity*, Vol. 6, Milne, I., Ritchie, R.O., and Karihaloo, B., Eds., Elsevier Science, New York, 2003, pp. 31–101.
8. Birnbaum, H.K., Robertson, I.M., Sofronis, P., and Teter, D., Mechanisms of hydrogen related fracture: a review, in *Second International Conference on Corrosion-Deformation Interactions*, Magnin, T., Ed., The Institute of Materials, London, 1997, pp. 172–195.
9. Thompson, A.W., Materials for hydrogen service, in *Hydrogen: Its Technology and Implications*, Vol. II, Cox, K.E. and Williamson, K.D., Eds., CRC Press, Cleveland, OH, 1977, pp. 85–124.
10. Loginow, A.W. and Phelps, E.H., Steels for seamless hydrogen pressure vessels, *Corrosion*, 31, 404–412, 1975.
11. Thompson, A.W. and Bernstein, I.M., Eds., *Effect of Hydrogen on Behavior of Materials*, The Metallurgical Society of AIME, Warrendale, PA, 1976.
12. Bernstein, I.M. and Thompson, A.W., Eds., *Hydrogen Effects in Metals*, The Metallurgical Society of AIME, Warrendale, PA, 1981.
13. Moody, N.R. and Thompson, A.W., Eds., *Hydrogen Effects on Material Behavior*, TMS, Warrendale PA, 1990.
14. Thompson, A.W. and Moody, N.R., Eds., *Hydrogen Effects in Materials*, TMS, Warrendale, PA, 1996.
15. Moody, N.R., Thompson, A.W., Ricker, R.E., Was, G.S., and Jones, R.H., Eds., *Hydrogen Effects on Material Behavior and Corrosion Deformation Interactions*, TMS, Warrendale, PA, 2003.
16. Thompson, A.W. and Bernstein, I.M., The role of metallurgical variables in hydrogen-assisted environmental fracture, in *Advances in Corrosion Science and Technology*, Vol. 7, Fontana, M.G. and Staehle, R.W., Eds., Plenum Press, New York, 1980, pp. 53–175.

17. Hirth, J.P., Effects of hydrogen on the properties of iron and steel, *Metallurgical Transactions*, 11A, 861–890, 1980.
18. Moody, N.R., Robinson, S.L., and Garrison, W.M., Hydrogen effects on the properties and fracture modes of iron-based alloys, *Res Mechanica*, 30, 143–206, 1990.
19. Swisher, J.H., Hydrogen compatibility of structural materials for energy-related applications, in *Effect of Hydrogen on Behavior of Materials*, Thompson, A.W. and Bernstein, I.M., Eds., The Metallurgical Society of AIME, Warrendale, PA, 1976, pp. 558–577.
20. Thompson, A.W., Structural materials use in a hydrogen energy economy, *International Journal of Hydrogen Energy*, 2, 299–307, 1977.
21. Thompson, A.W. and Bernstein, I.M., Selection of structural materials for hydrogen pipelines and storage vessels, *International Journal of Hydrogen Energy*, 2, 163–173, 1977.
22. SanMarchi, C. and Somerday, B.P., *Technical Reference for Hydrogen Compatibility of Materials*, Sandia National Laboratories, Livermore, CA, 2007 (www.ca.sandia.gov/matlsTechRef).
23. Laws, J.S., Frick, V., and McConnell, J., *Hydrogen Gas Pressure Vessel Problems in the M-1 Facilities*, NASA CR-1305, NASA, Washington, DC, 1969.
24. Anderson, T.L., *Fracture Mechanics: Fundamentals and Applications*, 2nd ed., CRC Press, New York, 1995.
25. Liu, A., Summary of stress-intensity factors, in *ASM Handbook: Fatigue and Fracture*, Vol. 19, Lampman, S.R., Ed., ASM International, Materials Park, OH, 1996, pp. 980–1000.
26. *Standard Test Method for Measurement of Fatigue Crack Growth Rates*, Standard E 647-05, ASTM International, West Conshohocken, PA, 2005.
27. *Standard Test Method: Laboratory Testing of Metals for Resistance to Sulfide Stress Cracking and Stress Corrosion Cracking in H_2S Environments*, Standard TM0177-96, NACE International, Houston, 1996.
28. *Standard Test Method for Determining Threshold Stress Intensity Factor for Environment-Assisted Cracking of Metallic Materials*, Standard E 1681-03, ASTM International, West Conshohocken, PA, 2003.
29. Robinson, S.L. and Stoltz, R.E., Toughness losses and fracture behavior of low strength carbon-manganese steels in hydrogen, in *Hydrogen Effects in Metals*, Bernstein, I.M. and Thompson, A.W., Eds., American Institute of Mining, Metallurgical, and Petroleum Engineers, New York, 1981, pp. 987–995.
30. Walter, R.J. and Chandler, W.T., *Influence of Gaseous Hydrogen on Metals Final Report*, NASA-CR-124410, NASA, Marshall Space Flight Center, AL, 1973.
31. Walter, R.J. and Chandler, W.T., Cyclic-load crack growth in ASME SA-105 grade II steel in high-pressure hydrogen at ambient temperature, in *Effect of Hydrogen on Behavior of Materials*, Thompson, A.W. and Bernstein, I.M., Eds., The Metallurgical Society of AIME, Warrendale, PA, 1976, pp. 273–286.
32. *Hydrogen Standardization Interim Report for Tanks, Piping, and Pipelines*, ASME, New York, 2005.
33. Clark, W.G. and Landes, J.D., An evaluation of rising load K_{Iscc} testing, in *Stress Corrosion: New Approaches*, ASTM STP 610, ASTM, Philadelphia, 1976, pp. 108–127.
34. Clark, W.G., Effect of temperature and pressure on hydrogen cracking in high strength type 4340 steel, *Journal of Materials for Energy Systems*, 1, 33–40, 1979.
35. Fukuyama, S. and Yokogawa, K., Prevention of hydrogen environmental assisted crack growth of 2.25Cr-1Mo steel by gaseous inhibitors, in *Pressure Vessel Technology*, Vol. 2, Verband der Technischen Uberwachungs-Vereine, Essen, Germany, 1992, pp. 914–923.

36. Hancock, G.G. and Johnson, H.H., Hydrogen, oxygen, and subcritical crack growth in a high-strength steel, *Transactions of the Metallurgical Society of AIME*, 236, 513–516, 1966.
37. Nakamura, M. and Furubayashi, E., Crack propagation of high strength steels in oxygen-doped hydrogen gas, *Transactions of the Japan Institute of Metals*, 28, 957–965, 1987.
38. Liu, H.W., Hu, Y.-L., and Ficalora, P.J., The control of catalytic poisoning and stress corrosion cracking, *Engineering Fracture Mechanics*, 5, 281–292, 1973.
39. Chandler, W.T. and Walter, R.J., Testing to determine the effect of high-pressure hydrogen environments on the mechanical properties of metals, in *Hydrogen Embrittlement Testing*, ASTM STP 543, ASTM, Philadelphia, PA, 1974, pp. 170–197.
40. Sandoz, G., A unified theory for some effects of hydrogen source, alloying elements, and potential on crack growth in martensitic AISI 4340 steel, *Metallurgical Transactions*, 3, 1169–1176, 1972.
41. Nelson, H.G. and Williams, D.P., Quantitative observations of hydrogen-induced, slow crack growth in a low alloy steel, in *Stress Corrosion Cracking and Hydrogen Embrittlement of Iron Base Alloys*, Staehle, R.W., Hochmann, J., McCright, R.D., and Slater, J.E., Eds., NACE, Houston, TX, 1977, pp. 390–404.
42. Hinotani, S., Terasaki, F., and Takahashi, K., Hydrogen embrittlement of high strength steels in high pressure hydrogen gas at ambient temperature, *Tetsu-To-Hagane*, 64, 899–905, 1978.
43. Bandyopadhyay, N., Kameda, J., and McMahon, C.J., Hydrogen-induced cracking in 4340-type steel: effects of composition, yield strength, and H_2 pressure, *Metallurgical Transactions*, 14A, 881–888, 1983.
44. Clark, W.G., The effect of hydrogen gas on the fatigue crack growth rate behavior of HY-80 and HY-130 steels, in *Hydrogen in Metals*, Bernstein, I.M. and Thompson, A.W., Eds., ASM, Metals Park, OH, 1974, pp. 149–164.
45. Hoover, W.R., Robinson, S.L., Stoltz, R.E., and Spingarn, J.R., *Hydrogen Compatibility of Structural Materials for Energy Storage and Transmission Final Report*, SAND81-8006, Sandia National Laboratories, Livermore, CA, 1981.
46. Cialone, H.J. and Holbrook, J.H., Effects of gaseous hydrogen on fatigue crack growth in pipeline steel, *Metallurgical Transactions*, 16A, 115–122, 1985.
47. *Metals & Alloys in the Unified Numbering System*, 10th ed., SAE International, Warrendale, PA, 2004.
48. *Standard Specification for Seamless Carbon Steel Pipe for High-Temperature Service*, A 106/A 106M-04b, ASTM International, West Conshohocken, PA, 2004.
49. *Specification for Line Pipe*, API 5L, American Petroleum Institute, Washington, DC, 1999.

8 Hydrogen Permeation Barrier Coatings

C.H. Henager, Jr.

CONTENTS

8.1 INTRODUCTION

Gaseous hydrogen, H_2, has many physical properties that allow it to move rapidly into and through materials, which causes problems in keeping hydrogen from materials that are sensitive to hydrogen-induced degradation. Hydrogen molecules are the smallest diatomic molecules, with a molecular radius of about 37×10^{-12} m, and the hydrogen atom is smaller still. Since it is small and light, it is easily transported within materials by diffusion processes. The process of hydrogen entering and transporting through a material is generally known as permeation, and this section reviews the development of hydrogen permeation barriers and barrier coatings for the upcoming hydrogen economy.

8.2 BACKGROUND

Hydrogen permeation is defined as the transport of hydrogen as dissociated hydrogen atoms[1] and has units of moles of hydrogen gas per square meter per second (mol m^{-2} sec^{-1}), which is the permeation rate. Known as Richardson's law this relation can be expressed as

$$J = \frac{DK}{d}(P_{high}^{1/2} - P_{low}^{1/2}) \tag{8.1}$$

where J is the permeation rate, D is the diffusion coefficient of hydrogen in the material, and K is Sievert's constant for the material, which determines the hydrogen solubility.

The product of D and K is referred to as Φ, the permeation coefficient or permeability of the material. Sievert's law gives the solubility in terms of the pressure as

$$c_H = KP_H^{1/2} \tag{8.2}$$

and equation 8.1 can be expressed as

$$J = \frac{\Phi\, \Delta P_H^{1/2}}{d} \tag{8.3}$$

where Φ is the material permeability for hydrogen and ΔP_H is the hydrogen pressure difference across the thickness d of the given material.

Both D and K, therefore Φ, are temperature dependent and have associated activation energies such that permeation is much higher at elevated temperatures for all materials than at low temperatures. For many materials, these permeation constants are too high for a given application and a permeation barrier must be considered.

8.3 HISTORICAL OVERVIEW

The concept of a hydrogen barrier seems to have arisen due to two separate issues in technology. One is the concern regarding hydrogen embrittlement of steels,[2,3] and the other is the required low permeation rate of tritium in the conceptual designs for fusion power that involve deuterium–tritium plasmas and tritium breeding blankets. These two technological areas have made advances in hydrogen barrier development that greatly impact current knowledge and state of the art. By far the most important advances and studies relevant to steels and metallic alloys have originated from the fusion energy materials community, and this will become apparent as this section looks more closely at specific barrier coatings and coating technologies.

8.4 HYDROGEN BARRIER COATINGS

The principal concern for hydrogen barrier coatings is to be able to prevent hydrogen ingress into a material that could be damaged or degraded due to hydrogen uptake. Data for hydrogen solubility and diffusivity are readily available for many materials and, for most metals, are significant in terms of amount of hydrogen and mobility of hydrogen. For example, a common stainless steel, 316SS, will dissolve about 20 parts per million hydrogen atoms at room temperature, and a 1-cm-thick piece of this steel will have a hydrogen permeation rate of about 6×10^{-13} moles of H_2 per square meter of steel per second (moles-H_2 m^{-2} sec^{-1}) for a hydrogen pressure of 5,000 psi, which amounts to about 3.5×10^{11} atoms of hydrogen per square meter of this 1-cm-thick steel plate. The solubility of the steel and its permeation rate increase exponentially with temperature, and a solubility of several thousands parts per million can exist at 800°C in 316SS, and the permeation rate has increased by eight orders of magnitude over room temperature.[1] This is quite remarkable, but of serious consequence since many steels and other alloys can be embrittled by this much hydrogen.[2] For fusion energy concerns this is manifested in a tritium inventory in materials that are unacceptably high, or permeation through metallic components into flowing coolants that are too high.[4] *

Therefore, hydrogen barriers have been developed and evolved from these twin concerns: (1) reducing hydrogen uptake into materials to prevent degradation and (2) preventing tritium permeation through materials to reduce radioactive transport to increase public health and safety for fusion power plants. Most hydrogen barriers have been conceived as external coatings on existing metallic alloys in order to prevent hydrogen uptake into the metal. In some cases, exposure treatments can be designed to produce an external scale on the metal or alloy that also serves this purpose. In general, the most effective approach seems to be the application of a suitable external coating having low hydrogen solubility and slow hydrogen transport. Some materials possess intrinsically low hydrogen permeation, such as gold or tungsten as pure metal examples, while many oxides, carbides, and nitrides possess low hydrogen permeation and have other desirable properties, such as high-temperature utility and corrosion protection. Typical coating-related issues, such as thermal expansion mismatch, coating defects, and inferior mechanical properties, must be dealt with for ceramic-based barrier coatings.

8.4.1 External Coatings

This class of coatings comprises those barrier coatings that are applied externally to a material in order to prevent hydrogen permeation, and are designed specifically for that purpose. In the next section we will consider barriers that may evolve or be developed on the surface of a material using a natural process, such as oxidation. For this section, however, we have two main cases, one for which a barrier coating is conceived and applied as an external coating using some physical or chemical deposition process, including electrochemical processes using liquids. We then consider the second case, where a coating is applied and then postprocessed to produce a hydrogen barrier layer.

The most widely used permeation barrier coating is aluminum oxide, or alumina, since it possesses one of the lowest hydrogen permeation rates of any material and one that is many orders of magnitude lower than most metals.[5] Table 8.1 lists, in decreasing order of hydrogen permeability, the permeability of a variety of materials at 500°C (773K) in units of moles-H_2 m^{-1} sec^{-1} $Pa^{-0.5}$, which are moles of hydrogen gas per meter thickness of material per second per square root of hydrogen pressure in Pa. Several ceramic coatings are listed and make good barrier coatings, but alumina seems to be superior. The work of Roberts et al.[5] showed that permeation was controlled by dissolution and transport through the grains of dense sintered alumina. Hydrogen transport and solubility are greatly reduced in alumina compared to metals.[6,7] More recent data mainly confirm the earlier work, but all the data for alumina are at high temperatures and must be extrapolated for our purposes.[7]

* Interested readers are referred to references 1 to 4 for more information regarding embrittlement and hydrogen permeation.

TABLE 8.1
Hydrogen Permeability of Various Materials

Material	Permeability	Reference	Comments
Vanadium	2.9×10^{-8}	[8]	Extremely sensitive to surface oxides
Niobium	7.5×10^{-9}	[8]	
Titanium	7.5×10^{-9}	[8]	Stable hydrides
Iron	1.8×10^{-10}	[8]	Iron
Ni	1.2×10^{-10}	[8]	
Ferritic steels	3×10^{-11}	[8, 9]	Tritium
Inconel 600	2.8×10^{-11}	[8]	
Austenitic steels	0.7 to 1.2×10^{-11}	[8]	Large data compilation
Molybdenum	1.2×10^{-11}	[8]	Variable
Titanium carbide	~1 to 8×10^{-15}	[10, 11]	External PVD coating
Tungsten	4.3×10^{-15}	[8]	Extrapolated from high T tests
Titanium carbide/titanium nitride layered	~7×10^{-16}	[10]	External PVD coating
Aluminum oxide	~9×10^{-17}	[7, 10, 11]	Extrapolated form high T tests
Beta-SiC	~1×10^{-20}	[10]	Extrapolated

As noted, the majority of research into hydrogen permeation barriers has been performed in the fusion materials research community concerned with tritium fate and transport in structures and breeding blankets. Since the work of Roberts et al.[5] showing the excellent permeation resistance of alumina, one thread of research has concentrated on ceramic coatings on fusion alloys, either 316 stainless steels or ferritic-martensitic alloys, such as MANET,* MANET-II, etc. Mühlratzer et al.[12] discuss a 1,000-fold reduction† in hydrogen permeation through Hastelloy-X that is coated with CVD alumina, although this work also points out the difficulties with forming a dense external ceramic coating on a metal substrate. Film defects, cracks, and spallation are issues that must be addressed for external permeation barrier coatings. This becomes apparent when either improved coating methods are used or there is a better mechanical and thermal property match between the coating and the substrate. For example, thin alumina deposited on amorphous tungsten oxide, WO_3, by a filtered vacuum arc method can reduce hydrogen permeation by a factor greater than 3,000 even for only thicknesses of 500 nm.[13] However, plasma-sprayed coatings of alumina on steels are not very effective permeation barriers since the films are highly defected.[14] Table 8.1 suggests that much higher permeation reductions are possible relative to steels provided the coatings are dense and relatively defect-free, which remains the principal reason that PRF values of only 1,000 to 10,000 are reported.

Another class of external coatings that has been investigated is TiC and TiN, alone and in combination as a composite film.[4,11,15–17] Although TiC and TiN have

* Designation for European ferritic-martensitic Nb-rich steel DIN 1.4914.

† A permeation reduction factor (PRF) is often quoted for comparison between treated and untreated metals. For this case, a PRF of 1,000 would be given.

low intrinsic permeabilities relative to steels, vapor-deposited films on steels do not reach this potential and often exhibit activation energies for permeation equal to that of the steel, which indicates permeation is controlled by defects in the coating rather than through the coating material.[16] Thus, permeation reduction factors of about 10 or so have been realized, but not the full potential of these materials as barriers. The large tritium permeation reduction reported by Shan et al.[17] for TiN and TiC/TiN coatings on stainless steel, which was on the order to 10^5 to 10^6, does not appear to be reasonable, and their measurement technique is questionable.[4]

The choice of permeation barrier materials selected here for coatings is reasonable based on permeation resistance, but external coatings of ceramics on metals are difficult to perfect since many metals have high thermal expansion coefficients and most ceramics have low ones. This causes large thermal stresses in the coatings to develop, which leads to defect formation in the coatings and lowers permeation resistance. A better technique, which will be discussed in the next section, relies on the formation of intrinsic oxide films on the surface of the metals, either by direct oxidation or by alloying followed by oxidation.[4]

8.4.2 Grown-On Oxide Films

Since oxides appear to have intrinsic low hydrogen permeabilities, one avenue of barrier development has pursued the direct oxidation of suitable alloys and (the proven more successful) aluminization of steels with subsequent alumina surface film formation. Direct oxidation relies on the intrinsic oxide layers that can be grown on alloys to provide hydrogen barrier capabilities. Such films typically consist of Fe-Cr-Al mixed oxides, depending on the steel, and several researchers have oxidized Fecralloy for this very reason.[18–21] In general, some concerns with oxidized steels are that spallation of the typical Fe- and Cr-containing oxides prevents excellent hydrogen barrier formation. Fecralloy, however, forms an alumina film that is more adherent and a better hydrogen barrier. A permeation reduction of 1,000 has been determined in Fecralloy steels.[21] Grown-on coatings perform better than plasma-sprayed Fe-Cr-Al coatings.[20]

Aluminizing steels, as noted above, has produced the best hydrogen barriers on steels and other compatible alloys.[22–33] Interestingly, the process of aluminization or aluminizing is flexible and robust such that a wide variety of techniques can be used, and this adds to the success of this method. Steels are aluminized by the in-diffusion of aluminum from the surface via the melt or vapor phase. In a relatively short time, on the order of hours, a 50-micron-thick aluminide alloy layer can be produced on the surface of a given steel and subsequently oxidized to form an alumina layer on top of hard aluminide intermetallic layers. In figure 8.1 the phase diagram for FeAl is shown to illustrate the varied layers that form during this diffusion–reaction process.

Although the diffusion–reaction method may vary, the results of aluminization remain similar and depend on the steel composition and on the temperature of the reaction processing. Methods that have been used for steels include hot dipping,[24,27–29,34–37] plasma spraying,[14,20,37,38] pack aluminizing, which is a form of chemical vapor deposition (CVD),[22,25,27,33] vacuum evaporation,[39] and polymer slurry methods.[40] Each of these methods has advantages and disadvantages compared to the others

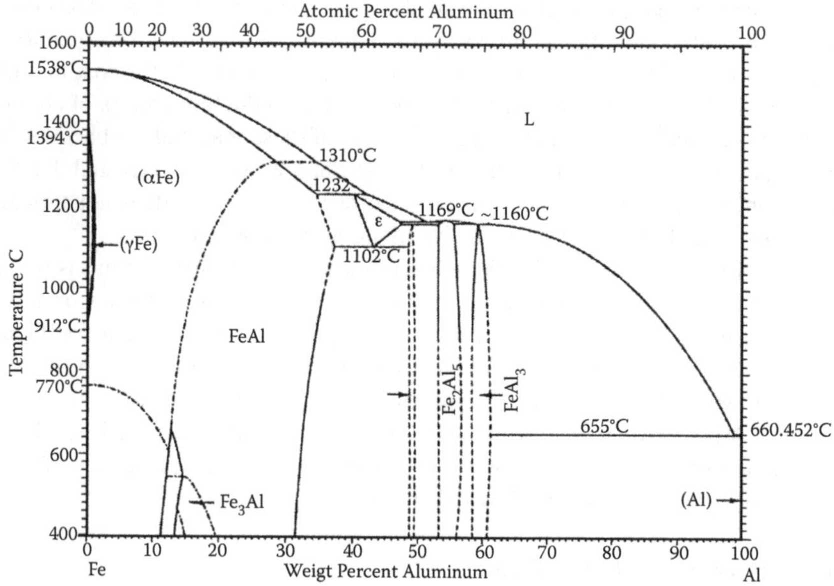

FIGURE 8.1 AlFe phase diagram showing intermetallic phases, such as $FeAl_3$, Fe_2Al_5, $FeAl_2$, and FeAl, which can form as separate layers in aluminized steel.

and has been used to create permeation barrier coatings on steels. One consideration is the temperature required for the process relative to heat treatment temperatures of steels. The treatment temperature of the MANET-II ferritic-martensitic steel limits the process temperature to about 750°C, which in turn limits the amount of aluminum and depth of diffusion of aluminum in the surface. For this case, hot dipping was chosen as a preferred method.[41]

Permeation reduction factors of up to 10,000, or 10^4, have been realized with the best coatings based on aluminized steels. Ferritic-martensitic steels that were aluminized had the Fe_2Al_5 phase predominant in the layer sequence, while a 316L steel had $FeAl_3$ and $FeAl_2$ as the main aluminide phases.[25] The best permeation barrier resulted from an external alumina film of about 1 micron in thickness grown on the aluminide layers.[25]

The vacuum evaporation process and polymer slurry process are quite new relative to the others and have the potential to provide more control in the processing, in the case of the vacuum evaporation technique, or greatly reduce the cost and environmental concerns of pack aluminizing with the polymer slurry methods. The vacuum evaporation process allows one to diffuse other elements than Al into the steels or to deposit FeAl coatings directly onto the surface of the steels, with Al diffusion occurring to help bond the deposited coating.[39]

This process is referred to as enclosed vacuum evaporation (EVE) coating technology and is applicable to a variety of coating and substrate materials, with a unique capability of producing smooth and uniform coatings on the inner surface of small-diameter, high aspect ratio cylindrical components or other confined geometries.

The technique has been used to deposit reproducible coatings on the inner surface of tubes as small as 10 mm in diameter and in lengths up to 3.8 m. For larger-diameter tubes or pipes in which radiant heating of the substrate from the source filament is impractical, separate resistive or inductive heating of the substrate to the desired temperature is used. Figure 8.2 shows the inner surface of a steel tube coated with a FeAl alloy for hydrogen barrier testing. The deposition is rapid, and substrate temperature rise can be controlled to avoid de-tempering alloys.

The polymer slurry method for aluminizing steel surfaces is straightforward and simple, and should be low cost since the raw materials and processing steps are also low cost. Aluminum flake of 1 to 2 microns in diameter is blended with a preceramic polysiloxane polymer and heated in air or nitrogen to 700 to 800°C for several hours to allow the aluminum to diffuse into the steel and to allow an external Si-Al-O film to form from reactions between the siloxane backbone and the Al. As with all aluminizing reaction–diffusion coatings, a series of aluminide layers form on the surface of the steel, as shown in figure 8.3. The outermost layer of alumina is the hydrogen permeation barrier, while the aluminum-rich layers provide additional aluminum for alumina formation in oxidizing environments, as required to maintain the external oxide layer.

The advantages of aluminizing steels go beyond hydrogen barrier formation, however, as such surface treatments also provide additional corrosion protection. The fusion materials community continues to study these processing methods and may continue to be the main driving force for research in this area until hydrogen infrastructure issues become more important.[27]

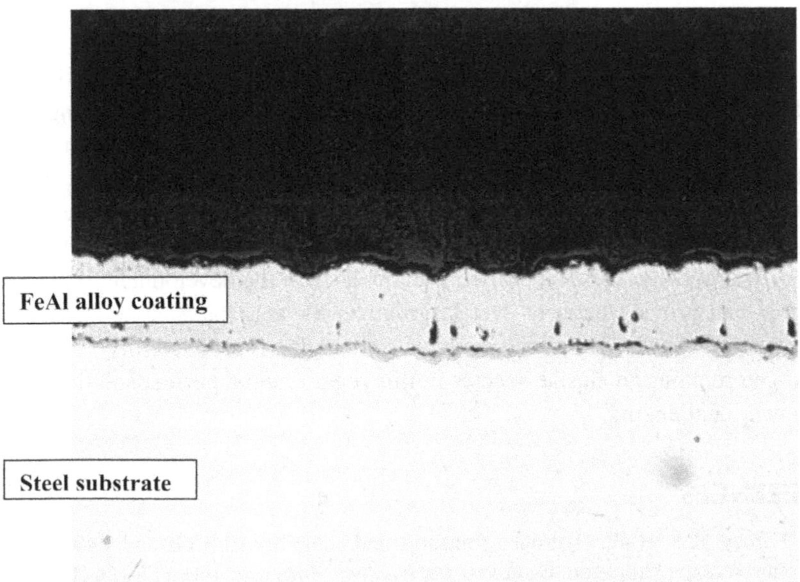

FIGURE 8.2 FeAl coating on the inner diameter of a 316SS tube that was deposited using the EVE technique.

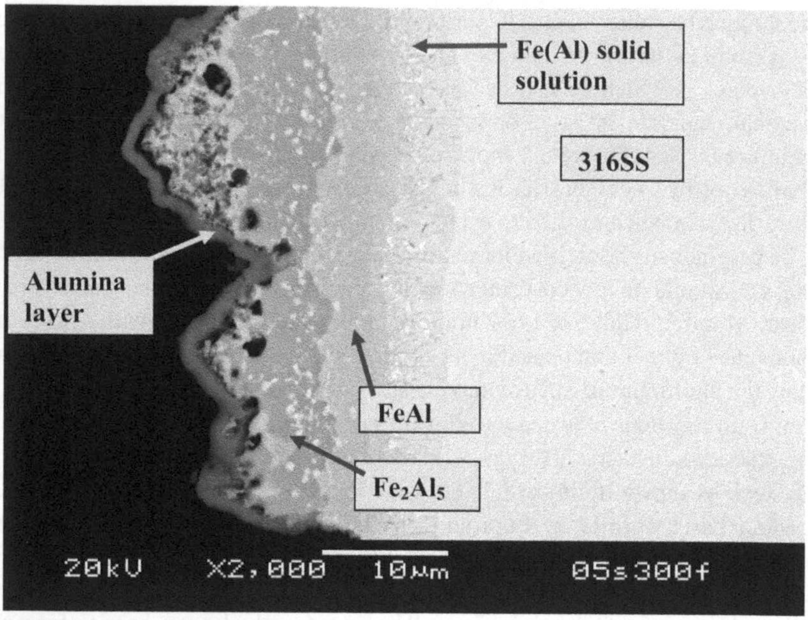

FIGURE 8.3 A typical aluminized steel surface following the polymer slurry method at 800°C in air. Any of the methods mentioned in this section about aluminized films will produce similar reaction layers. Ease of processing and cost may dictate which method is preferred for a given application.

8.5 SUMMARY

The best hydrogen barrier coatings have been fabricated using aluminized steels produced by a variety of methods, including pack aluminizing, hot dipping, vacuum evaporation, or polymer slurry techniques. Permeation reduction factors of up to 10^4 have been realized in this manner. Titanium-based coatings offer an alternative choice to Al but are not as permeation resistant as the alumina-based methods, and are not as reproducible in fabrication. Much work remains to be done in the general area of hydrogen permeation barriers, particularly in the development of new methods that can provide barriers over large areas for anticipated hydrogen economy infrastructure needs. Low-cost methods and better reproducibility are required. Hydrogen remains an elusive species in this regard, and a perfect solution is apparently very challenging.

REFERENCES

1. Forcey, K.S. et al., Hydrogen transport and solubility in 316L and 1.4914 steels for fusion reactor applications, *Journal of Nuclear Materials*, 160, 117–124 (1988).
2. Gibala, R. and R.F. Hehemann, Eds., *Hydrogen Embrittlement and Stress Corrosion Cracking*, ASM, Metals Park, OH, 1984, p. 324.
3. Honeycombe, R.W.K., Steels: microstructure and properties, in *Metallurgy and Materials Science*, R.W.K. Honeycombe and P. Hancock, Eds., London: Edward Arnold, 1981.

4. Hollenberg, G.W. et al., Tritium/hydrogen barrier development, *Fusion Engineering and Design*, 28, 190–208 (1995).
5. Roberts, R.M. et al., Hydrogen permeability of sintered aluminum oxide, *Journal of the American Ceramic Society*, 62, 495 (1979).
6. Yu, G.T. and S.K. Yen, Determination of the diffusion coefficient of proton in CVD gamma aluminum oxide thin films, *Surface and Coatings Technology*, 166, 195 (2003).
7. Serra, E. et al., Hydrogen permeation measurements on alumina, *Journal of the American Ceramic Society*, 88, 15 (2005).
8. Brimhall, J.L., E.P. Simonen, and R.H. Jones, *Data Base on Permeation, Diffusion, and Concentration of Hydrogen Isotopes in Fusion Reactor Materials*, Fusion Reactor Materials Semiannual Progress Report, DOE/ER-0313/16, 1994.
9. Forcey, K.S. et al., Hydrogen transport and solubility in 316L and 1.4914 steels for fusion reactor applications, *Journal of Nuclear Materials*, 160, 117 (1988).
10. Hollenberg, G.W. et al., Tritium/hydrogen barrier development, *Fusion Engineering and Design*, 28, 190 (1995).
11. Perujo, A. and K.S. Forcey, Tritium permeation barriers for fusion technology, *Fusion Engineering and Design*, 28, 252 (1995).
12. Mühlratzer, A., H. Zeilinger, and H.G. Esser, Development of protective coatings to reduce hydrogen and tritium permeation, *Nuclear Technology*, 66, 570 (1984).
13. Yamada-Takamura, Y. et al., Hydrogen permeation barrier performance characterization of vapor deposited amorphous aluminum oxide films using coloration of tungsten oxide, *Surface and Coatings Technology*, 153, 114 (2002).
14. Song, R.G., Hydrogen permeation resistance of plasma-sprayed Al_2O_3 and Al_2O_3-13wt% TiO_2 ceramic coatings on austenitic stainless steel, *Surface and Coatings Technology*, 168, 191 (2003).
15. Tazhibaeva, I.L. et al., Hydrogen permeation through steels and alloys with different protective coatings, *Fusion Engineering and Design*, 51/52, 199 (2000).
16. Forcey, K.S. et al., Formation of tritium permeation barriers by CVD, *Journal of Nuclear Materials*, 200, 417 (1993).
17. Shan, C. et al., Behaviour of diffusion and permeation of tritium through 316L stainless steel with coating of TiC and TiN + TiC, *Journal of Nuclear Materials*, 191–194, 221 (1992).
18. Van Deventer, E.H. and V.A. Maroni, Hydrogen permeation characteristics of some Fe-Cr-Al alloys, *Journal of Nuclear Materials*, 113, 65 (1983).
19. Earwaker, L.G. et al., Influence on hydrogen permeation through steel of surface oxide layers and their characterisation using nuclear reactions, *IEEE Transactions on Nuclear Science*, NS-28, 1848 (1980).
20. Fazio, C. et al., Investigation on the suitability of plasma sprayed Fe-Cr-Al coatings as tritium permeation barrier, *Journal of Nuclear Materials*, 273, 233 (1999).
21. Forcey, K.S., D.K. Ross, and L.G. Earwalker, Investigation of the effectiveness of oxidised Fecralloy as a containment for tritium in fusion reactors, *Zeitschrift fur Physikalische Chemie Neue Folge*, 143, 213 (1985).
22. Shen, J.-N. et al., Effect of alumina film prepared by pack cementation aluminizing and thermal oxidation treatment of stainless steel on hydrogen permeation, *Yuanzineng Kexue Jishu/Atomic Energy Science and Technology*, 39, 73 (2005).
23. Aiello, A. et al., Hydrogen permeation through tritium permeation barrier in Pb-17Li, *Fusion Engineering and Design*, 58/59, 737 (2001).
24. Glasbrenner, H., A. Perujo, and E. Serra, Hydrogen permeation behavior of hot-dip aluminized MANET steel, *Fusion Technology*, 28, 1159 (1995).
25. Forcey, K.S., D.K. Ross, and C.H. Wu, Formation of hydrogen permeation barriers on steels by aluminising, *Journal of Nuclear Materials*, 182, 36 (1991).

26. Fukai, T. and K. Matsumoto, Surface modification effects on hydrogen permeation in high-temperature, high-pressure, hydrogen-hydrogen sulfide environments, *Corrosion* (Houston), 50, 522 (1994).
27. Konys, J. et al., Status of tritium permeation barrier development in the EU, *Fusion Science and Technology*, 47, 844 (2005).
28. Glasbrenner, H. et al., Corrosion behaviour of Al based tritium permeation barriers in flowing Pb-17Li, *Journal of Nuclear Materials*, 307–311, 1360 (2002).
29. Glasbrenner, H. et al., Development of a Tritium Permeation Barrier on F82H-mod, in *Sheets and on MANET Tubes by Hot Dip Aluminising and Subsequent Heat Treatment*, Forschungszentrum Karlsruhe GmbH, Karlsruhe, Germany, 1998, p. 30.
30. Glasbrenner, H. et al., *The Formation of Aluminide Coatings on MANET Stainless Steel as Tritium Permeation Barrier by Using a New Test Facility*, Vol. 2, Elsevier, Lisbon, 1997, p. 1423.
31. Iordanova, I., K.S. Forcey, and M. Surtchev, Structure and composition of aluminized layers and RF-sputtered alumina coatings on high chromium martensitic steel, *Materials Science Forum*, 321–324, 422 (2000).
32. Iordanova, I., K.S. Forcey, and M. Surtchev, X-ray and ion beam investigation of alumina coatings applied on DIN1.4914 martensitic steel, *Nuclear Instruments and Methods in Physics Research, Section B: Beam Interactions with Materials and Atoms*, 173, 351 (2001).
33. Forcey, K.S. et al., Use of aluminising on 316L austenitic and 1.4914 martensitic steels for the reduction of tritium leakage from the NET blanket, *Journal of Nuclear Materials*, 161, 108 (1989).
34. Yao, Z.Y. et al., Hot dipping aluminized coating as hydrogen permeation barrier, *Acta Metallurgica Sinica*, 14, 435 (2001).
35. Aiello, A., A. Ciampichetti, and G. Benamati, An overview on tritium permeation barrier development for WCLL blanket concept, *Journal of Nuclear Materials*, 329–333, 1398 (2004).
36. Aiello, A. et al., Qualification of tritium permeation barriers in liquid Pb-17Li, *Fusion Engineering and Design*, 69, 245 (2003).
37. Benamati, G. et al., Development of tritium permeation barriers on Al base in Europe, *Journal of Nuclear Materials*, 271/272, 391 (1999).
38. Perujo, A., K.S. Forcey, and T. Sample, Reduction of deuterium permeation through DIN 1.4914 stainless steel (MANET) by plasma-spray deposited aluminum, *Journal of Nuclear Materials*, 207, 86 (1993).
39. Knowles, S.D. et al., Method of Coating the Interior Surface of Hollow Objects, U.S. Patent 6,866,886, *2005*.
40. C.H. Henager, Jr., Low-cost aluminide coatings using polymer slurries, personal communication, 2006.
41. Serra, E., H. Glasbrenner, and A. Perujo, Hot-dip aluminium deposit as a permeation barrier for MANET steel, *Fusion Engineering and Design*, 41, 149 (1998).

9 Reversible Hydrides for On-Board Hydrogen Storage

G. J. Thomas

CONTENTS

9.1 INTRODUCTION

In concept, reversible hydrides offer a direct means of storing hydrogen on-board fuel cell vehicles and would be compatible with a hydrogen-based transportation fuel infrastructure. A tank, or perhaps more accurately a storage system, containing an appropriate hydride material would remain fixed on a vehicle and could be refueled simply by applying an overpressure of hydrogen gas. Once filled, the hydrogen gas would remain at the equilibrium pressure for the particular hydride material, changing only with temperature changes induced in the storage tank. When hydrogen was needed, it would be released endothermically, using the waste heat from the fuel cell (or internal combustion engine [ICE]) to supply the required energy. This approach offers certain advantages over high-pressure compressed gas tanks and cryogenic liquid hydrogen systems—it is inherently stable with regard to hydrogen release, it can operate at a low or moderate gas pressure, and it could eliminate some of the energy costs of compression or liquefaction. It also has the potential to achieve volumetric hydrogen densities much higher than those of compressed gas and even liquid hydrogen.

In practice, however, the use of hydrides for on-board hydrogen storage is much more complicated than is described above, and a number of issues arise when one attempts to choose a material and design a storage system. These issues arise because (1) many hydride materials do not meet minimal on-board storage requirements for weight and volume density, and (2) there are some fundamental material properties that determine system performance that are competing against one another; that is, they affect system requirements in opposing directions. Thus, a hydride-based storage system will likely be a design with numerous trade-offs in terms of capacity, kinetics, and thermal requirements. Furthermore, there are a host of other issues beyond capacity and kinetic performance, such as hydride–dehydride cycling-induced changes or degradation in performance, response of the material to impurities in the incoming hydrogen gas, evolution of impurities from the bed affecting the fuel cell, and, importantly, cost of the material and its impact on the fuel supply system cost.

The development of storage materials with properties that can encompass all of the required performance attributes for on-board hydrogen storage will be an extremely challenging task and likely require a multidisciplinary approach. In 2003, the U.S. Department of Energy (DOE) launched a concerted effort to develop high-capacity materials that have the potential to meet the hydrogen storage system performance targets established by the DOE and FreedomCAR and Fuel Partnership[1] a government/industry collaboration. This hydrogen storage initiative has spawned a considerable level of effort over the last few years, and it is in this arena that this chapter will focus.

There have been many review articles on metal hydrides, and a few of the more recent ones are referenced here.[2–10] This review will attempt to cover only fairly recent studies on high hydrogen capacity hydrides that (1) have been demonstrated to be reversible or, at the least, partially reversible and (b) have the potential for exhibiting other properties (e.g., kinetics, operating temperatures) suitable for on-board hydrogen storage applications. This area of materials research and development is very active at the present time, so that it is likely that not all of the relevant work will be included. The author apologizes for any omissions.

9.2 HYDRIDE PROPERTIES AND HYDROGEN CAPACITY

Hydrides can be loosely categorized by their chemical binding—metallic, covalent, ionic, or complex—between the host elements and hydrogen. Intermetallic alloys form a large class of hydrides, generally with metallic bonds, that can be further subcategorized by the ratio of the alloying constituents A and B. Thus, for example, one refers to $LaNi_5H_6$ as an AB5 hydride. An online database of hydride properties, hydpark,[11] is largely organized along these lines. From an on-board hydrogen storage perspective, however, it is the nature of the chemical bond that is key because it determines the thermodynamic stability of the hydride, the hydrogen stoichiometry of the material, and the mechanisms for hydrogen absorption and release.

In 2001, Schlapbach and Zuttel published a paper on hydrogen storage[4] and included a plot of volumetric and gravimetric hydrogen densities in a variety of materials. Figure 9.1 shows a similar plot that also includes corresponding values for compressed gas and liquid hydrogen, as well as the DOE and FreedomCAR and Fuel Partnership targets for on-board storage systems. One can see that there are a number of materials that contain hydrogen concentrations well above the system targets, with some having more than twice the density of liquid hydrogen. In addition to mixing material properties with system properties, the plot also includes many different material types, such as solid reversible hydrides, liquid and solid nonreversible chemical systems, and, for comparison, a few liquid and gaseous fuels (e.g., octane, methane, etc.) plotted in terms of their hydrogen content.

One can notice some trends in material properties from the plot. First, the intermetallic hydrides (plotted in bright red), such as $LaNi_5H_6$, generally have low gravimetric hydrogen density and are clustered toward the left-hand side of the graph. But these materials are also relatively dense, and so they can have high volumetric hydrogen densities, often greater than 100 g H_2/l. A few elemental hydrides (also plotted in bright red), such as MgH_2 and TiH_2, are also shown. Although they may exhibit high hydrogen densities, they tend to be heavy as well and, in many cases, have strong covalent hydrogen bonds resulting in more stable structures. Their higher stability means that higher temperatures are required to release the hydrogen. For example, MgH_2 must be heated above 300°C in order to release hydrogen at a significant rate.

The highest-capacity materials (shown in blue on the plot) lie in the upper-right-hand quadrant and above the system target lines. This chapter will be limited to a discussion of these materials only because they are the materials that are of greatest interest for hydrogen storage applications. Also, the chapter will not cover historical developments and will largely be concerned with research published within the last several years. One additional limitation in scope is that only material properties related to the thermally induced release of hydrogen will be described. Other hydrogen release reactions, such as hydrolysis, that generate an oxide or hydroxide by-product that must be processed off-board will not be discussed.

The solid, reversible hydride materials plotted in blue in figure 9.1 generally contain an Al–H complex anion (alanates), a B–H complex anion (borohydrides), or N–H groups (amides, imides). The plot also includes some nonreversible materials, such as ammonia borane, NH_3BH_3, that have very high hydrogen capacities that can be released by thermolysis, but must be regenerated through a chemical process. This means that the spent fuel must be removed and processed externally. These "chemical hydride" materials will also not be discussed in this chapter.

In intermetallic systems, hydrogen absorption (desorption) is relatively straightforward and occurs through (1) molecular dissociation (recombination) at the metal surface and (2) atomistic diffusion of the hydrogen through the solid. Grochala and Edwards[8] refer to these materials as interstitial hydrides since the hydrogen resides in the interstices of the metal lattice. In contrast, hydrogen dissociation in complex hydrides generally occurs through the formation of intermediate compounds. The reverse processes, the re-formation of the hydride phases, as well as hydrogen transport mechanisms, are generally not well understood in these materials. An additional issue with the high hydrogen capacity materials is that they are often quite stable and require high temperatures for hydrogen release.

The desired form of a reversible hydride reaction for on-board storage may be written as

$$M_XH_Y + \text{heat} \equiv XM(s) + Y/2H_2(g) \tag{9.1}$$

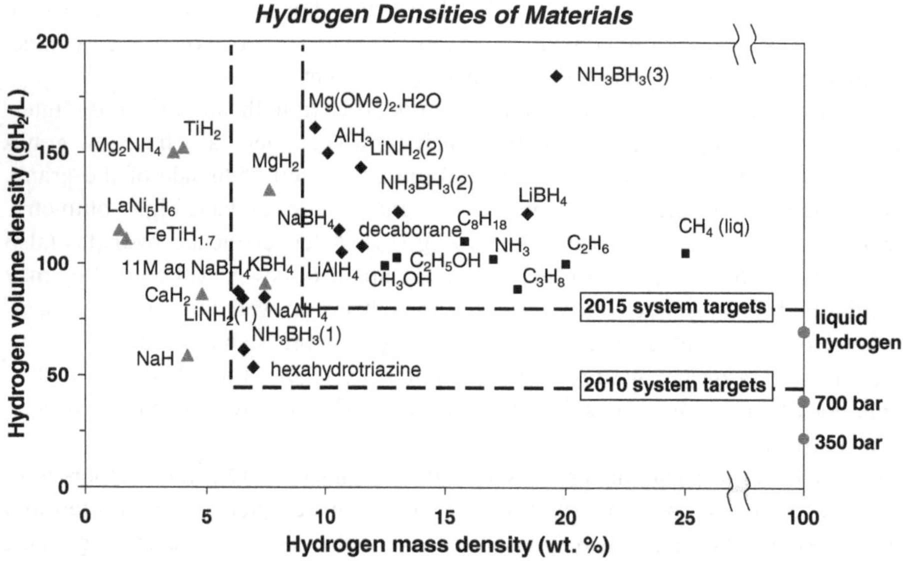

FIGURE 9.1 Plot of hydrogen weight fraction and hydrogen volume density for some representative hydrogen storage materials. For comparison, the current 2010 and 2015 DOE/FreedomCAR and Fuel Partnership targets for system weight and system volume densities are indicated by the dashed lines. The densities of compressed hydrogen at ambient temperature and liquid hydrogen at 20K are also shown.

where M is a single element or combination of elements. The heat term on the left side of the reaction indicates that the dissociation of the hydride coupled with the release of hydrogen is endothermic, and conversely, formation of the hydride by reacting the element(s) with gaseous hydrogen is exothermic. From the perspective of a vehicular hydrogen storage system then, the material remains stable on-board the vehicle at low or moderate hydrogen pressures until heat is applied. The preferred source of this heat is waste heat from the fuel cell or ICE, so that there would be no energy penalty for releasing the hydrogen. Ideally then, the operating temperature range of the storage material should lie within the operating temperature range of the fuel cell or ICE coolant loop.

It should also be noted from equation 9.1 that heat must be dissipated when refueling the tank (recharging the spent hydride), and for refueling rates equivalent to filling conventional gas tanks, the cooling power could, for the more stable hydride materials, exceed the capacity of on-board coolant systems. In these cases, there would be an energy cost borne by the off-board refueling facility.

The thermodynamic parameters of the reaction quantify the energy requirements. The Gibbs free energy of formation per mole of hydride at constant temperature, T, is

$$\Delta G_f = \Delta H_f - T\Delta S \tag{9.2}$$

where ΔH_f is the formation enthalpy of the hydride and ΔS is the change in entropy of the system when the hydride is formed. When ΔG_f is negative, the reaction is favored and heat is released as the hydride is formed.

For an ideal gas, the hydrogen overpressure, P, in equilibrium with the hydride at temperature T, can then be expressed in the form[12]

$$RT\ln(P/P^o) = \Delta G_f = -\Delta H_f + T\Delta S \tag{9.3}$$

where R is the gas constant and P^o is the pressure at standard conditions, that is, 1 atm of pressure. The enthalpy is expressed as the formation energy per mole of hydrogen. The entropy change is the difference between the entropy of hydrogen gas and the configurational and vibrational entropy of the hydrogen in the solid, and is generally considered to have roughly the same value for most hydrides. One can then readily see from the equation that the more stable a hydride is (larger ΔH_f), the lower the equilibrium pressure is at a given temperature. The equation also shows that a plot of lnP vs. 1/T for a given hydride is a straight line with a slope of $\Delta H_f/R$, the familiar van't Hoff plot. Graphically, the Y-intercept at $1/T \rightarrow 0$ corresponds to an equilibrium pressure at infinite temperature.

Current research toward developing high hydrogen capacity materials for on-board storage applications is largely concentrated on reducing the energy requirements, either by modifying the material to reduce the enthalpy of formation, ΔH_f, of the hydride phase, or through altering the reaction pathway to hydrogen dissociation or recombination. This stems largely from the on-board need to use the waste heat from an "engine" (e.g., a fuel cell or ICE) to supply the required energy for hydrogen release from the hydride. Roughly speaking, this means that the "operating window" in temperature and pressure for a hydride storage system lies between room temperature and 100°C, and ~1 and 100 bars. An examination of the hydpark database[11] indicates that the bulk of experimental values for ΔS range from about 95 to 130 J/mol H_2. A simple calculation, then, using equation 9.3 would show that ΔH_f should be in the range of about 20 to 40 kJ/mol H_2. The lower enthalpy values would actually be preferred in order to reduce the cooling power requirements during rehydriding. Even lower ΔH_f materials, e.g., ~15 kJ/mol H_2, could be used with higher-pressure containers. However, materials with formation enthalpies higher than the upper limit could not be used as storage materials because the operating temperatures would be too high for on-board systems as they are currently envisioned.

Concurrently, the hydrogen kinetics of absorption and release must also be improved to meet minimal performance standards through, for example, the development of effective catalysts or the formation of very small particle sizes (e.g., nanoscale materials). Of course, an added requirement for small particle sizes is that their small dimensions be maintained through repeated hydride–dehydride cycling. The kinetic requirements for hydrogen release in a storage material are dictated by the needs of the vehicle driver and the fuel cell power. For example, a 100-kW peak power fuel cell with about 42% fuel efficiency would need 2 g H_2/sec to produce full power when demanded by the vehicle driver. Furthermore, this hydrogen delivery rate must be available through nearly all of the range of the hydride composition.

This is a particularly severe requirement for a material when it is nearly depleted (the fuel tank is nearly empty).

Hydrogen transport in the high-capacity hydrides appears to be through the movement of heavy atoms, complexes, or lattice defects rather than hydrogen atoms, with correspondingly higher activation energies for diffusion than for interstitial hydrides. In Ti- and Zr-doped $NaAlH_4$ and Na_3AlH_6, for example, Sandrock et al.,[13] Luo and Gross,[14] and Kiyobayashi et al.[15] all measured activation energies for hydrogen release in the range of 80 to 100 kJ/mol. These high enthalpy values translate to slower diffusion rates, and hence slower release rates, at the desired operating temperatures. As an example, an increase in the activation energy for migration of ~25 kJ/mol reduces the diffusivity by a factor of 10^{-4}. Since the diffusion distance is proportional to (diffusivity × time)$^{1/2}$, a diffusing species would then take 100 times longer to travel the same distance out of a particle. Hence, one approach considered to mitigate the slower transport rates is to reduce the particle size of the material. For the above example, an equivalent release rate with the slower diffusivity could be achieved by reducing the hydride particle size, for example, from 10 to 0.1 μm. Of course, on an absolute scale, particles may have to be in the nanosize range to exhibit sufficiently fast kinetics.

In summary, the ideal reversible hydride would simultaneously have all of the desired properties discussed above, as well as additional properties, such as good stability through multiple hydride–dehydride cycles. The values derived in the preceding text are compiled in table 9.1. It must be emphasized that these values do not represent a comprehensive analysis, nor do they reflect the DOE/FreedomCAR and Fuel Partnership hydrogen storage system targets.[1] They are intended only to highlight some of the hydride material properties from the perspective of hydrogen storage system needs.

The hydride weight and volume densities were simply chosen so that a reasonably designed storage system could meet or exceed the current 2010 FreedomCAR and Fuel Partnership system targets. The actual energy densities for a specific system would depend not only on the hydrogen weight and volume densities of the hydride, but also on a number of other factors, including system design, the other materials used to fabricate the system components, thermal requirements (including the hydride materials' enthalpy and thermal conductivity), the packing density of the hydride, and the maximum operating pressure of the system (also dependent on the enthalpy). The hydrogen release rate estimate was based on a 5-kg hydrogen system capacity.

TABLE 9.1
Material Properties for Reversible Hydrides in Hydrogen Storage Systems

Gravimetric hydrogen density	>9 wt% H2
Volumetric hydrogen density	>60 g H2/l
Enthalpy of hydride formation	15–40 kJ/mol H2
Hydrogen release rate	>1 wt% H2/min

In the following sections, specific materials will be discussed in more detail, and recent work on Al–H-based, B–H-based, and N-based complex hydrides will be summarized.

9.3 ALANATES

As in most discussions on alanates, that is, materials containing the AlH_4- complex, it is appropriate to start by referencing the work of Bogdanovic and Schwickardi where they showed that $NaAlH_4$ could be made fully reversible through the addition of a small amount (~2 mol%) of Ti.[16] The Ti addition also improved the kinetics of hydrogen release. Since the alanates generally have high hydrogen capacities, but were previously considered to be nonreversible, this work immediately spawned a large amount of research on this class of materials, which continues to the present.[17–54] Figure 9.2 shows the hydrogen weight fraction (based on their chemical formulae) for ternary hydrides based on the AlH_4- anion. One can see that hydrogen contents tend to be much higher than in intermetallic hydrides, and that higher-valency cations, although generally increasing the molecular weight of the compound, can still have fairly high hydrogen capacity. The plot is also limited to alanates with a single cation element. It should be mentioned that some of these alanates are not stable at room temperature and are difficult to synthesize. Also, alanates can be synthesized with mixed cations, forming quarternary or even higher hydride compounds.

Since hydrogen release from alanates involves dissociation of the compound and the formation of other reaction products, not all of the hydrogen is typically

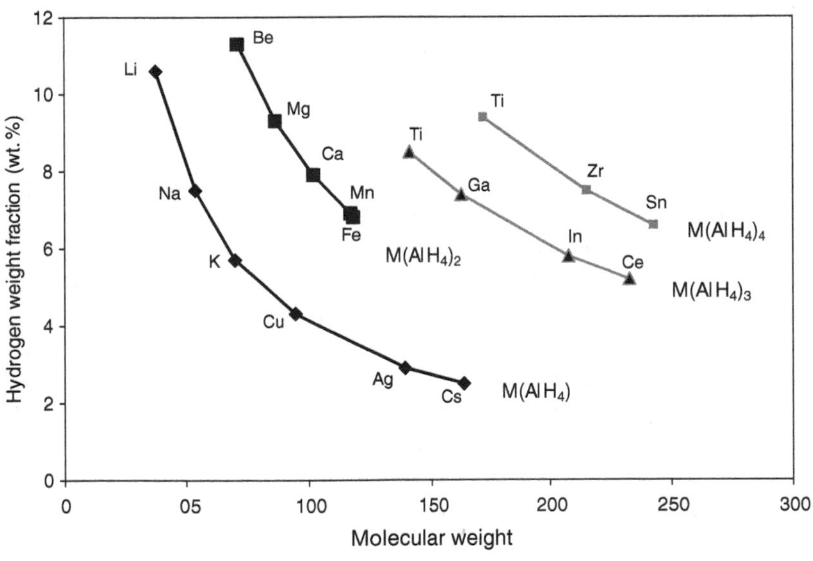

FIGURE 9.2 Trend of hydrogen weight fraction for various alanate compounds and cations as a function of the molecular weight of the compound. The trend is also representative of other complex hydrides, such as borohydrides.

available. This can be seen in more detail using a description of the Na–Al–H system as an example.

$NaAlH_4$ decomposes to Na and Al, releasing its hydrogen through the following reaction chain:

$$NaAlH_4 \text{ (+ Ti)} \equiv 1/3\ Na_3AlH_6 + 2/3\ Al + H_2 \qquad \Delta H = 37 \text{ kJ/mol } H_2$$
$$1/3\ Na_3AlH_6 \text{ (+ Ti)} \equiv NaH + Al + \tfrac{1}{2} H_2 \qquad \Delta H = 47 \text{ kJ/mol } H_2$$
$$NaH \equiv Na + \tfrac{1}{2} H_2 \qquad \Delta H = 112 \text{ kJ/mol } H_2$$

All of the reactions are endothermic, and each has a different enthalpy, as shown above. The last reaction, the decomposition of NaH, would not be of much value in a hydrogen storage application because the large enthalpy would require a very high temperature to release the hydrogen (>400°C). Hence, the full formula hydrogen weight fraction of 7.4 wt% for the tetrahydride would not be available in a practical system. Rather, only the first two reactions would be used. These have a combined theoretical yield of about 5.6 wt% H_2, the hydrogen capacity value usually quoted for this material. Experimentally, even lower hydrogen capacities are observed due to kinetic limitations, impurities, and a reduction of the available material associated with the Ti doping process.

The reactions also point out the complexities associated with these materials. As mentioned earlier, kinetic measurements yielding the activation energies for hydrogen release find much higher values than typically found for atomistic hydrogen diffusion and are indicative of the transport of heavier metal atoms or even M–H complexes.[14–16] Secondly, it is remarkable that the addition of a small amount of Ti would cause the reverse reactions to occur in the solid state. *In situ* X-ray diffraction (XRD) measurements during desorption[22] exhibit sharp diffraction lines corresponding to metallic Al, indicating that the Al has clustered into relatively large crystallites. How then do the complexes form, dissociate from the metal clusters, and recombine with the Na under the modest conditions of temperature (100 to 200°C) and hydrogen overpressure (<100 bars)? This phenomenon is even more surprising in light of the high pressure and temperature required to form pure alane, AlH_3.[55,56] Finally, it should be mentioned that in spite of concerted experimental[13–15,22–28] and theoretical[29–31,36] efforts, the role of Ti in enhancing reaction kinetics in both the forward and reverse directions has not been definitively determined, and the mechanisms of hydrogen transport during release and rehydrogenation remain unclear.

Work has also continued toward development of more effective catalysts or dopants for improving the low-temperature kinetics of hydrogen release and absorption in Na alanates. Bogdanovic et al.[19] extended their earlier work to include a comprehensive survey of other precursor materials. Zidan et al.[18] found Zr to be an effective catalyst for one of the decomposition reactions and further showed that including both Ti and Zr catalysts improved the overall desorption kinetics of Na alanate. Others extended this approach of using co-dopants to consider Ti, Zr, Fe combinations,[32] adding graphite[33] and other catalytic complexes.[34,35] Although some improvements have been reported with these alternative additives or catalysts, the overall performance of the sodium alanate materials has not been significantly improved over the original Ti dopant initially reported by Bogdanovic and Schwickardi.[16]

Considerable work has been directed toward other alanate compounds in an effort to find materials with improved reversible properties (capacity, enthalpy, kinetics) over Na alanate.[37–46,52,53] A very large number of options are available to materials researchers for consideration. In addition to the single-element cations shown in figure 9.2, there are many combinations that form stable compounds. Since the Gibbs free energy of a particular alanate is affected by the electronic binding between the cations and the Al–H complex, hydride properties can, in principle, be modified to improve performance. This area has been explored in some detail through experimental and theoretical studies, and due to the very large number of candidates, combinatorial or rapid screening methods have been employed. Studies using a combination of modeling and rapid experimental synthesis and measurement are described here.

Sachtler et al.[47] studied the material phase space of Na-Li-Mg/AlH_4 using an 8-reactor system to measure the reversible hydrogen content in $NaAlH_4$, $LiAlH_4$, $Mg(AlH4)_2$, and about 18 binary and ternary mixtures. The materials were fabricated by using a modified milling procedure, in one case starting with the alanates and in a second case starting with mixtures of the hydrides NaH, LiH, and MgH_2 along with small (~100-nm) Al particles. After the materials were mixed, XRD characterization showed no new phases. An initial thermal desorption was then made, followed by a single rehydriding condition (87 bars H_2 at 125°C for 12 h). With both starting materials (hydrides and alanates) under the conditions applied, they found that the highest-capacity material was simply $NaAlH_4$. No higher-capacity material was produced. The experimental results agree roughly with first-principles calculations in that some of the experimental mixture stoichiometries proved to be unstable (positive enthalpies).

The brief description above points out the complexities in exploring potential hydride materials, particularly with regard to rapid screening methods. Some material combinations are inherently unstable and will not form. On the other hand, stable compounds may require unique synthesis conditions of, for example, temperature or hydrogen overpressure. Other compounds may even need a solution-based synthesis approach. Theoretical estimates of material stability, if accurate, can be of great value to experimental efforts by eliminating unstable candidates and identifying promising ones. The question of hydrogen reversibility is another issue. Different compounds are likely to require different hydrogen pressures and temperatures to rehydride, as indicated in equation 9.2. Rapid screening methods typically employ a single condition to all of the material combinations at a given time. Another variable is the potential role of catalysts, as exemplified by the importance of a small amount of Ti in $NaAlH_4$.

One of the most aggressive and comprehensive combinatorial study efforts on alanates is due to Opalka and co-workers at UTRC, and their partners (Albemarle Corp., Questek LLC, SRNL, and IFE, Norway).[48–51] They have been exploring the quarternary or higher phase space consisting of alkaline metal hydrides, alkaline earth hydrides, transition metal hydrides, alane, and molecular hydrogen. Their work employs a combination of first-principles theoretical modeling, thermodynamic modeling, three different synthesis routes (solid-state processing, molten-state processing, and solution-based processing) for forming stable compounds, and analytical methods to measure and characterize materials.

Initial modeling studies surveyed a material phase space that partially overlapped that of Sachtler et al.,[47] that is, quarternary phases of Na-Li-Mg-Al-H, but they also included Ti as a partial substitute with Mg. Over 200 phases were considered. Consistent with the previously discussed work, only a few promising candidates were identified.

9.4 BOROHYDRIDES

Borohydrides, materials with the tetraborohydride complex, BH_4^-, offer the potential for even higher gravimetric capacity than alanates because of the lower molecular weight of boron. They also have high hydrogen densities. This can be seen in table 9.2, which lists hydrogen weight and volume capacities for a few borohydride compounds. However, the chemistry is quite different in this case, and generally speaking, many of these compounds are relatively stable, with reversibility an issue.

The series of alkali borohydrides from Li through Cs was studied theoretically by Vajeeston et al. using density-functional theory (DFT).[57] They found the stable structures for each of the compounds, laying the ground work for further work in this area. Experimentally, however, little progress has been made from the standpoint of reversible hydrides.[58–64]

$LiBH_4$ was studied by Zuttel et al.[58] as a potentially new hydrogen storage material. They found that they could release about 13.5 wt% hydrogen from the compound starting at 200°C using an SiO_2 catalyst. The remaining 4.5 wt% hydrogen apparently stays in the form of LiH, an extremely stable hydride. They were not successful in forming the borohydride from the elements with hydrogen pressures up to 150 bars at temperatures up to 650°C. In later work, Ohba et al. performed first-principles calculations[59] and predicted that a monoclinic phase, $Li_2B_{12}H_{12}$, was the stable intermediate phase formed in the decomposition of Li borohydride. The proposed reaction of $LiBH_4$ to this intermediate phase releases about 10 wt% H_2. Subsequent decomposition of the phase to LiH and B would result in the release of an additional

TABLE 9.2
Hydrogen Weight and Volume Densities in Representative Borohydrides

Material	Formula Weight Density	Formula Volume Density
LiBH4	18.4 wt% H2	122 H2 g/l
NaBH4	10.6 wt% H2	114 H2 g/l
KBH4	7.5 wt% H2	83 H2 g/l
Mg(BH4)2	14.8 wt% H2	110 H2 g/l
Ca(BH4)2	11.6 wt% H2	107 H2 g/l
Zn(BH4)2	8.4 wt% H2	—

3.8 wt% H_2. The quantity of desorbed hydrogen and the calculated enthalpies for these reactions are in good agreement with the experimental results.

The decomposition of $Mg(BH_4)_2$ has been extensively studied using *in situ* XRD techniques coupled with residual gas analysis (RGA) measurements of the gas release.[60] They report that the borohydride decomposed starting at ~300°C, releasing ~9 wt% H_2 by ~350°C. No ordered phase was detected by the XRD between these two temperatures, indicative of an amorphous phase or phases. Above 350°C, MgH_2 apparently recrystallized and was detected by the XRD. This phase subsequently released the additional hydrogen as the temperature was increased further. Initial attempts to recharge the spent material indicated that only the Mg rehydrided to form MgH_2.

Nakamori et al.[61] have been studying a wide range of potential borohydrides. They have examined the stability of borohydrides with transition metal cations and have shown a correlation between the borohydride stability and the electronegativity of the cation element. Based on their analysis, they formed a number of borohydrides by reacting $LiBH_4$ with chlorides of the various elements and found that the hydrogen desorption temperature decreased in those compounds formed with a higher electronegativity element.

The interaction of chlorides with borohydrides has also been studied by Jensen et al.[62] They have looked at stabilizing the transition metal borohydrides $Zr(BH_4)_2$ and $Zn(BH_4)_2$, which have very high vapor pressures at the temperatures needed for dehydrogenation, by partial substitutions of alkali metal cations. The substitution is accomplished by reacting the borohydride with chlorides of the alkali metals. The substitution has the added advantage of increasing the formula weight fraction over the transition metal hydride alone. Their results, although promising for specific substitiutions, also show that diborane, B_2H_6, can be formed during the decomposition of borohydrides. Diborane would be an unwelcome by-product in any storage system because of its toxicity and its potential poisoning of fuel cell catalysts. The level of diborane production was found to be significantly lower in Mn borohydride.[63]

9.5 DESTABILIZED BOROHYDRIDES

The high heats of formation of complex hydrides, particularly the borohydrides, and the limitations inherent with alloy substitutions have lead to a somewhat different approach—reacting a borohydride with another compound to form a dehydrogenated compound with a lower reaction energy than for the enthalpy of the hydride alone. This was shown by Reilly and Wiswall in the late 1960s.[65] More recently, Vajo and Skeith[66,67] have applied this method to Li borohydride. $LiBH_4$, ball milled with MgH_2 as a destabilizing additive, was studied by Vajo and Skeith.[66] In this work, 2 to 3 mol% of $TiCl_3$ was also included in the mixtures. They found that the effective enthalpy was reduced significantly, from ~69 kJ/mol H_2 for the borohydride decomposition alone (to LiH and H_2) down to ~45 kJ/mol H_2 when the borohydride reacted with MgH_2 to form MgB_2 and LiH. The lower enthalpy value results in an equilibrium hydrogen overpressure of 1 bar at ~200°C, as compared to ~400°C for the borohydride.

This striking result, however, is also accompanied by some side effects. As with other techniques that may improve the thermodynamics, there is an accompanying loss in capacity. The reversible hydrogen content for the $LiBH_4$–MgH_2 couple was found to be in the range of 8 to 10 wt%, rather than 18.4 wt% for the formula value, or the 13.6 wt% hydrogen released when the borohydride decomposes to LiH, even though this hydrogen capacity, coupled with the lower enthalpy, indicates a material that is approaching storage system requirements.

Other Mg-based destabilizing additives with Li borohydride have been examined.[68,69] MgF_2, MgS, and MgSe were mixed and reacted with $LiBH_4$ over multiple cycles. Essentially the same reactions as with MgH_2 were found, with the formation of MgB_2 and LiF, Li_2S, or Li_2Se, respectively, with the Mg flouride, sulfide, and selenide. These reactions all reduced the enthalpy, but produced a lower hydrogen yield than with MgH_2. As before, only partial reversibility was demonstrated and the hydrogen capacity was reduced in subsequent cycles. Additional destabilized systems that have been studied include $MgH_2/NaBH_4$.[70]

The potential number of reaction combinations for destabilizing hydrides is very large, and ideally, only the most promising combinations should be explored experimentally. As with the alanate materials, rapid screening in this case proved to be best performed using computational methods. Alapati et al.[71] used first-principles calculations of the reaction enthalpies to screen over 100 destabilization reactions. Using the criterion that the desired enthalpy should be in the range of 30 to 60 kJ/mol H_2, they found only five reaction schemes that appeared promising. The reaction schemes included amides as well as borohydrides. It is expected that experimental work will follow on their recommendations.

Perhaps the greatest obstacles to overcome with this method are the very slow rates of hydrogen absorption and release. The hydrogen kinetics are now limited by the reaction rates for forming the destabilized compound, and generally speaking, chemical reaction rates for forming compounds in the solid state are slower by orders of magnitude than interstitial hydrogen transport in metal hydrides. This effect is exacerbated by the lower operating temperatures of the destabilized systems. Since it is unlikely that the diffusivities of the reacting species can be improved much, current efforts to overcome this severe limitation are focused on reducing the transport distances by forming very small particles.[69] This also implies that small particle sizes must be maintained over multiple hydride–dehydride cycles. Thus, the use of carbon scaffolds, where the materials are imbedded into nanoporous carbon structures, such as carbon aerogels, is being explored.[69] Work is continuing in this area of hydride development.

9.6 NITROGEN SYSTEMS

Nitrogen also forms compounds that can have relatively large hydrogen contents. In fact, one of the highest hydrogen density materials is ammonia borane, NH_3BH_3, with about 19.5 wt% H_2 (by formula). It is currently under study as a potential chemical hydride, that is, as a nonreversible material that would be recharged with hydrogen in a chemical process off-board a vehicle.[72]

Most stable solids in this category of materials are generally metal amides containing the NH_2 radical. With a lower hydrogen stoichiometry than the alanates or borohydrides, their overall capacities are somewhat lower. However, they readily form with lightweight elements and are generally reacted with other hydrides. Thus, they have the potential to contain comparable amounts of recoverable hydrogen.

In 2002, Chen et al. reported that Li amide, $LiNH_2$, and LiH could be reacted to form an imide with partial release of hydrogen.[73,74] The reaction is

$$LiNH_2 + LiH \equiv Li_2NiH + H_2$$

The reaction yielded a reversible hydrogen capacity of 6.5 wt%. If the imide were subsequently decomposed, the overall hydrogen capacity of the amide–hydride pair would be 11.5 wt%. As with other systems, however, this total capacity has not been achieved reversibly. Furthermore, the formation enthalpy and hydrogen transport kinetics of this system require high temperatures (~350°C) for hydrogen release at reasonable rates. Some improvement in hydrogen release kinetics was achieved by incorporating Ti catalysts.[75]

Researchers have explored alternative reaction pairs. One such system is Li amide with Mg hydride. Luo and Ronnebro[76] reported a reversible hydrogen capacity of just over 5 wt% at 200°C with this pair, an improvement in operating temperature over the previous LiH system, but with some loss in available hydrogen. Further studies showed that, initially, $LiNH_2$ and MgH_2 exothermically react to partially form $Mg(NH_2)_2$ and LiH. The Mg amide and LiH then react to form an imide, $MgLi_2(NH)_2$, to release some of the hydrogen endothermically.[77,78] Only small improvements in the kinetics of hydrogen release were found with catalyst additions.[79]

Other combinations of hydrides with amides have been considered. Nakamori et al.[80–83] and others[84,85] have looked at mixtures of Li amide with Li alanate and with Li borohydride. DFT calculations[86,87] suggest that these systems will behave in a fashion similar to that of the destabilized borohydrides described above; that is, the reaction products in the dehydrogenated state should have lower enthalpies than the borohydride or the alanate alone. The potential yields are higher than with LiH. For the borohydride case, the expected reaction pathway is

$$LiBH_4 + 2\ LiNH_2 \equiv Li_3BN_2 + 4\ H_2$$

which would yield 11.9 wt% hydrogen. The calculated enthalpy for forming Li_3BN_2 is 23 kJ/mol H_2, considerably less than for the pure borohydride phase (69 kJ/mol H_2). The equivalent reaction with the mixed amide–alanate would have an expected yield of 9.6 wt%. As one might expect, however, the experimental results are much more complex. Hydrogen release for the $LiBH_4$–2 $LiNH_2$ case occurred in a single peak between ~300 and ~375°C and was measured to be ~8 to 9 wt% H_2. The expected phase, Li_3BN_2, was found by XRD analysis following the decomposition. Thus, the expected reaction product was formed, and the amount of hydrogen released, although not as high as expected (about 70% of the expected level), was not unreasonable. However, the temperature was much higher than would be expected based simply on the estimated enthalpy.

Another experimental study on the same material system was performed by Pinkerton et al.,[88] who found that an intermediate compound was formed. When a 2:1 molar ratio mixture of $LiNH_2$ and $LiBH_4$ was ball milled for a sufficient length of time (300 min), or heated above about 95°C, they found a new quarternary hydride compound with the approximate composition of $Li_3BN_2H_8$. This material was stable at room temperature, but melted at ~190°C. When heated above 250°C, it released ~10 wt% hydrogen. The final product remaining was identified as a mixture of Li_3BN_2 polymorphs. Calorimetric measurements of the decomposition suggested that the dehydriding was exothermic, which implies that the hydride may not be reversible.

One final issue with nitrogen-containing material systems is the propensity to generate ammonia, NH_3, during hydrogen release. In the measurements described above, Pinkerton et al.[88] found about 2 to 3 mol% ammonia in the released gas. Measurements on the $LiNH_2$–MgH_2 system showed a concentration of about 300 to 400 ppm in the desorbed hydrogen stream at ~200°C.[89] This impacts storage performance in two ways. First, NH_3 can be a strong poisoning agent to fuel cell membranes. Current fuel purity specifications for fuel cells require NH_3 concentrations below 100 ppb in the hydrogen supply, a very stringent condition for amides or other nitrogen-based material systems. The examples cited above are orders of magnitude above the current limit. Second, the formation of NH_3 results in a loss of material available for recharging. Although this might be a small effect over a single charge–discharge cycle, the cumulative effect over the life of a storage system might be large if ammonia or other by-products are produced during hydrogen release.

9.7 OTHER MATERIALS

The discussion above was limited to alanates, borohydrides, amides, and combinations of these materials. Other hydrides or alternative approaches have also been proposed for storage applications. Zaluska et al.[90] studied lightweight lithium–beryllium hydride and showed a reversible hydrogen capacity of over 8 wt%. They also showed that the hydride may be usable down to ~150°C. Although these results are rather promising, it is unlikely that any beryllium-containing compound would be considered for vehicular use because of the toxicity of this element, even though the hydride may be quite stable.

Rather than modifying the chemical composition of the material, an alternative approach to reducing the hydride formation enthalpy was recently reported by Wagemans et al.[91] Their quantum chemical calculations showed that the stability of magnesium hydride, MgH_2, was reduced for sufficiently small (less than ~1.3 nm) Mg clusters. Mg hydride is normally quite stable, with an equilibrium pressure of 0.1 MPa at around 300°C. At the nanoscale, however, the calculations indicated that the hydride decomposition temperature could be lowered significantly, for example, down to about 200°C for a crystallite size of 0.9 nm. Of course, at this size range, kinetics would not be expected to be an issue; however, preventing the growth of such small particles would be the biggest obstacle toward this nanoscale approach. Perhaps embedding the Mg clusters into a scaffold material (e.g., a carbon aerogel), as proposed for improving the kinetics for destabilized hydrides, may be a potential solution.

We have seen that material development efforts for hydrogen storage are largely focused on reducing the enthalpy of formation, ΔH_f, of stable hydrides. From an engineering point of view, that is, managing the thermal requirements of the material in a storage system, the lower the enthalpy change, the better. But from a material point of view, the material becomes progressively less stable as the formation enthalpy approaches zero and the equilibrium hydrogen overpressure increases exponentially. The reverse reaction may then be difficult, requiring high pressure, or the hydride may be considered nonreversible. As an example, alane, AlH_3, with a hydrogen capacity of 10 wt%, a formation enthalpy of ~7 to 11 kJ/mol (depending on the phase), reasonable release kinetics, and high volume density,[56] has many of the ideal properties needed in a storage material. But the hydride is sufficiently unstable that it can decompose at moderate temperatures. (Also, extremely high pressures or a chemical process is required to rehydride the Al.) Similarly, some of the alanates shown in figure 9.2, such as $LiAlH_4$, $Mg(AlH_4)_2$, or $Ca(AlH_4)_2$, have been found to be metastable at moderate temperatures. Recently, Graetz and Reilly[92] have made a unique suggestion regarding the use of these materials. They have considered such metastable materials to be kinetically stabilized; that is, although they can decompose at moderate temperatures, they may be usable as potential storage materials because the reaction rates of decomposition may be sufficiently slow at the ambient temperatures likely to be encountered in storage systems.

9.8 SUMMARY

It is readily apparent that research and development activities in the field of reversible hydrides have greatly increased over the last few years due, in large part, to the increased level of government funding in the U.S., Europe, and Asia for hydrogen storage materials. Industrial R&D has significantly increased as well. Correspondingly, much progress has been achieved, particularly toward understanding the complexity and diversity of complex hydrides and their behavior. Many unique and innovative concepts have been explored. In contrast to earlier years, little or no studies have been tailored toward intermetallic and covalent hydrides, except for the continuing efforts at improving materials for NiMH batteries.

REFERENCES

1. Storage system targets can be accessed at http://www1.eere.energy.gov/hydrogenandfuelcells/storage/current_technology.html.
2. Sandrock, G., Hydrogen-metal systems, in *Hydrogen Energy System, Utilization of Hydrogen and Future Aspects*, Y. Yurum, ed., NATO ASI Series, The Netherlands: Kluwer Publishing, Dordrecht.
3. Sandrock, G., Applications of hydrides, in *Hydrogen Energy System, Utilization of Hydrogen and Future Aspects*, Y. Yurum, Ed., NATO ASI Series, The Netherlands: Kluwer Publishing, Dordrecht.
4. Schlapbach, L. and Zuttel, A., *Nature*, 414, 353, 2001.
5. Bowman, R.C. Jr. and Fultz, B., *MRS Bulletin*, September 2002, 688.
6. Bogdanovic, B. and Sandrock, G., *MRS Bulletin*, September 2002, 712.
7. Akiba, E. and Okada, M., *MRS Bulletin*, September 2002, 699.

8. Grochala, W. and Edwards, P., *Chem. Rev.*, 104, 1283, 2004.
9. Chandra, D., Reilly, J.J., and R. Chellappa, R., *JOM*, February 2006, 26.
10. Fakioglu, E., Yurum, Y., and Veziroglu, T.N., *Int. J. Hydrogen Energy*, 290, 1371, 2004.
11. Sandrock, G. and Thomas, G., DOE/IEA/SNL Hydride Database, http://hydpark.ca.sandia.gov.
12. See, for example, Mueller, W.M., Blackledge, J.P., and Libowitz, G.G., *Metal Hydrides*, New York: Academic Press, 1968.
13. Sandrock, G., Gross, K., and Thomas, G., *J. Alloys Comp.*, 339, 299, 2002.
14. Luo, W. and Gross, K.J., *J. Alloys Comp.*, 385, 224, 2004.
15. Jensen, C.M., Kiyobayashi, T., Sun, D., and Sauara, A., 225th ACS National Meeting. 2003.
16. Bogdanovic, B. and Schwickardi, M. *J. Alloys Comp.*, 253/254, 1, 1997.
17. Jensen, C.M., Zidan, R., Mariels, N., Hee, A.G., and Hagen, C. *Int. J. Hydrogen Energy*, 24, 461, 1999.
18. Zidan, R., Takara, S., Hee, A.G., and Jensen, C.M., *J. Alloys Comp.*, 285, 119, 1999.
19. Bogdanovic, B., Brand, R.A., Marjanovic, A., Schwickardi, M. and Tolle, J. J., *Alloys Comp.*, 302, 36, 2000.
20. Chen, J., Kuriyama, N., Xu, Q., Takeshita, H.T., and Sakai, T., *J. Phys. Chem B*, 105, 11214, 2001.
21. Sandrock, G., Gross, K., Thomas, G., Jensen, C., Meeker, D., and Takura, S., *J. Alloys Comp.*, 330/332, 696, 2002.
22. Gross, K.J., Sandrock, G., and Thomas, G., *J. Alloys Comp.*, 330/332, 691, 2002.
23. Thomas, G.J., Gross, K., and Sandrock, G., *J. Alloys Comp.*, 330, 2002.
24. Brinks, H.W., Jensen, C.M., Srinivasan, S.S., Hauback, B.C., Blanchard, D., and Murphy, K., *J. Alloys Comp.*, 376, 215, 2004.
25. Walters, R.T. and Scogin, J.H. *J. Alloys Comp.*, 379, 135, 2004.
26. Majer, G. Stanik, E., Valiente Banuet, L.E., Grinberg, F., Kircher, O., and Fichtner, M., *J. Alloys Comp.*, 404–406, 738, 2005.
27. Leon, A., Kircher, O., Rosner, H., Decamps, B., Leroy, E., Fichtner, M., and Percheron-Guegan, A., *J. Alloys Comp.*, 414, 190, 2006.
28. Graetz, J., Ignatov, A.Y., Tyson, T.A., Reilly, J.J., and Johnson, J., in *Mater. Res. Soc. Conf. Proceedings*, 2005, p. 837.
29. Chaudhuri, S. and Muckerman, J.T., *J. Phys. Chem. B*, 109, 6952, 2005.
30. Chaudhuri, S., Graetz, J., Ignatov, A., Reilly, J.J., and Muckerman, J.T, *J. Am. Chem. Soc.*, 2006.
31. Iniguez, J. and Yildirim, T., *Phys. Rev. B*, 0604472, 2006.
32. Wang, J., Ebner, A.D., Zidan, R., and Ritter, J.A., *J. Alloys Comp.*, 391, 245, 2005.
33. Wang, J., Ebner, A.D., Prozorov, T., Zidan, R., and Ritter, J.A., *J. Alloys Comp.*, 395, 252, 2005.
34. Zaluska, A. and Zaluski, L., *J. Alloys Comp.*, 404–406, 706, 2005.
35. Resan, M., Hampton, M.D. Lomness, J.K., and Slattery, D.K., *Int. J. Hydrogen Energy*, 30, 1417, 2005.
36. Lovvik, O.M. and Opalka, S.M., *Appl. Phys. Lett.*, 88, 161917-1-3, 2006.
37. Hauback, B.C., Brinks, H.W., and Fjellvag, H., *J. Alloys Comp.*, 346, 184, 2002.
38. Brinks, H.W., Hauback, B.C., Norby, P., and Fjellvag, H., *J. Alloys Comp.*, 351, 222, 2003.
39. Fichtner, M., Fuhr, O., and Kircher, O., *J. Alloys Comp.*, 356, 418, 2003.
40. Morioka, H., Kakizaki, K., Chung, S.C., and Yamada, A., *J. Alloys Comp.*, 353, 310, 2003.
41. Lovvik, O.M. and Opalka, S.M., *Phys. Rev. B*, 69, 134117, 2004.

42. Fossdal, A., Brinks, H.W., Fichtner, M., and Hauback, B.C., *J. Alloys Comp.*, 404–406, 752, 2005.
43. Fossdal, A., Brinks, H.W., Fichtner, M., and Hauback, B.C., *J. Alloys Comp.*, 387, 47, 2005.
44. Graetz, J., Lee, Y., Reilly, J.J., Park, S., and Vogt, T., *Phys. Rev. B*, 71, 184115, 2005.
45. Resan, M., Hampton, M.D., Lomness, J.K., and Slattery, D.K., *Int. J. Hydrogen Energy*, 30, 1413, 2005.
46. Andreasen, A., Vegge, T., and Pedersen, A.S., *J. Solid State Chem.*, 178, 3672, 2005.
47. Sachtler, A. et al., U.S. DOE Hydrogen Program FY2006 Annual Progress Report.
48. Qiu, C., Opalka, S.M., Olson, G.B., and Anton, D.L., *Int. J. Mater. Res.*, 97(11), 1484-1494, 2006.
49. Opalka, S.M., Lovvik, O.M., Brinks, H.W., Saxe, P., and Hauback, B.C., *J. Am. Chem. Soc.*, in press.
50. Lovvik, O.M., Swang, O., and Opalka, S.M., *J. Mater. Res.*, 20, 3199, 2005.
51. Opalka , S.M. et al., U.S. DOE Hydrogen Program FY2006 Annual Progress Report.
52. Chung, S.C. and Morioka, H., *J. Alloys Comp.*, 372, 92, 2004.
53. Varin, R.A., Chiu, C., Czujko, T., and Wronski, Z., *J. Alloys Comp.*, 439(1-2), 302-311, 2007. 2006.
54. Xiao, X., Chen,L., Wang, X., Wang, Q., and Chen, C., *Int. J. Hydrogen Energy*, in press, 2006.
55. Sinke, G.C., Walker, L.C., Oetting, F.L., and Stull, D.R., *J. Chem. Phys.*, 47, 2759, 1967.
56. Graetz, J. and Reilly, J.J., *J. Alloys Comp.*, 2006.
57. Vajeeston, P., Ravindran, P., Kjekshus, A., and Fjellvag, H., *J. Alloys Comp.*, vol. 404, 377-383, 2005.
58. Zuttel, A., Wenger, P., Rentsch, S., Sudan, P., Mauron, Ph. and Emmerenegger, Ch., *J. Power Sources*, 118, 1, 2003.
59. Ohba, N., Miwa, K., Aoki, M., Noritake, T., Towata, S., Nakamori, Y., Orimo, S., and Zuttel, A., *Phys. Rev. B*, 74, 075110, 2006.
60. Zhao, J.-C., U.S. DOE Hydrogen Program FY2006 Annual Progress Report.
61. Nakamori, Y., Miwa, K., Li, H., Ohba, N., and Orimo, S., in *Proceedings of the MRS*, Boston, November 2006, Abstract Z2.1.
62. Jensen, C., Eliseo, J., and Severa, G., in *Proceedings of the MRS*, Boston, November 2006, Abstract Z2.3.
63. Srinivasau, S. and Jensen, C., *Proc. Mater. Res. Soc.*, 837, 101-107, 2005.
64. Varin, R.A. and Chiu, Ch., *J. Alloys Comp.*, 397, 276, 2005.
65. Reilly, J.J. and Wiswall, R.H., *Inorg. Chem.*, 6, 2220, 1967; 7, 2254, 1968.
66. Vajo, J.J. and Skeith, S.L., *J. Phys. Chem B*, 109, 3719, 2005.
67. Vajo, J.J., invited presentation at Gordon Research Conference of Hydrogen-Metal Systems, Waterville, ME, July 10–15, 2005.
68. Bowman, R.C. Jr., Hwang, S.-J., Ahn, C.C., and Vajo, J.J., *Mater. Res. Soc. Symp. Proc.*, 837, 3.6.1, 2005.
69. Vajo, J.J., U.S. DOE Hydrogen Program FY2006 Annual Progress Report.
70. Czujko, T., Varin, R.A., Wronski, Z., Zaranski, Z., and Durejko, T., *J. Alloys Comp.*, 427, 291, 2007.
71. Alapati, S.V., Johnson, J.K., and Scholl, D., *J. Phys. Chem. B.* 110(17), 8769-8776, 2006.
72. Aardahl, C., U.S. DOE Hydrogen Program FY2006 Annual Progress Report.
73. Cheng, P., Xiong, Z., Luo, J., Lin, J., and Tan, K.L., *Nature*, 420, 302, 2002.
74. Chen, P., Xiong, Z., Luo, J., Lin, J., and Tan, K.L., *J. Phys. Chem B*, 107, 10967, 2003.
75. Isobe, S., Ichikawa, T., Hanada, N., Leng, H.Y., Fichtner, M., Fuhr, O., and Fujii, H., *J. Alloys Comp.*, 404–406, 439, 2005.

76. Luo, W. and Ronnebro, E., *J. Alloys Comp.*, 404–406, 392, 2005.
77. Luo, W. and Sickafoose, S., *J. Alloys Comp.*, 407, 274, 2006.
78. Lohstroh, W. and Fichtner, M., *J. Alloys Comp.*, in press, 2006.
79. Lohstroh, W. and Fichtner, M., U.S. DOE Hydrogen Program FY2006 Annual Progress Report.
80. Nakamori, Y., Ninomiya, A., Kitahara, G., Aoki, M., Noritake, T., Miwa, K., Kojima, Y., and Orimo, S., *J. Power Sources*, 155, 447, 2006.
81. Aoki, M., Miwa, K., Noritake, T., Kitahara, G., Nakamori, Y., Orimo, S., and Towata, S., *Appl. Physics A*, 80, 1409, 2005.
82. Noritake, T., Aoki, M., Towata, S., Ninomiya, A., Nakamori, Y., and Orimo, S., *Appl. Physics A*, 83(2), 277-279, 2006.
83. Nakamori, Y., Ninomiya, A., Kitahora, G., Aoki, M., Noritake, T., Miwa, K., Kojima, Y., and Orimo, S., *J. Power Sources*, 155, 447-455, 2006.
84. Ichikawa, T., Isobe, S., N., Hanada, N., and Fujii, H., *J. Alloys Comp.*, 365, 271, 2004.
85. Hu, Y.H., Yu, N.Y., and Ruckenstein, E., *Ind. Eng. Chem. Res.*, 43, 4174, 2004.
86. Miwa, K., Ohba, N., Towata, S., Nakamori, Y., and Orimo, S., *Phys. Rev. B*, 69, 245120, 2004.
87. Orimo, S., Nakamori, Y., Kitahara, G., Miwa, K., Ohba, N., Noritake, T., and Towata, S., *Appl. Phys. A*, 79, 1765, 2004.
88. Pinkerton, F.E. et al., *J. Phys. Chem. B*, 109, 6, 2005.
89. Luo, W., U.S. DOE Hydrogen Program FY2006 Annual Progress Report.
90. Zaluska, A., Zaluska, L., and Strom-Olsen, J.O., *J. Alloys Comp.*, 307, 157, 2000.
91. Wagemans, R.W.P. et al., *J. Am. Chem. Soc.*, 127, 16675, 2005.
92. Graetz, J. and Reilly, J.J., *Scripta Material*, 2006.

10 The Electrolytes for Solid-Oxide Fuel Cells

Xiao-Dong Zhou and Prabhakar Singh

CONTENTS

10.1 INTRODUCTION

10.1.1 Fuel Cell Background

In principle, fuel cells can be considered as batteries that convert chemical energy to electricity without combustion, but instead through electrochemical reactions. The difference between fuel cells and conventional batteries is that fuel cells can run continuously as long as fuels are available for electrochemical reactions, whereas a battery needs to be recharged periodically. The concept of fuel cells was conceived by Sir William Robert Grove, known as the father of the fuel cell, who developed a

TABLE 10.1
Classification of Fuel Cells[2,12]

	AFC	DMFC	MCFC	PAFC	PEMFC	SOFC
Electrolyte	Potassium hydroxide	Polymer member	Molten carbonates	Phosphoric acid	Ion exchange membrane	Ceramic
Operating temperature	60–90°C	60–130°C	650°C	200°C	80°C	<1,000°C
Charge carrier	OH–	H+	CO32–	H+	H+	O2-
Electrodes	Metal/carbon	Pt on carbon	Ni + Cr	Pt on carbon	Pt on carbon	LSM/Ni-YSZ
ASR of the electrolyte						
Efficiency	45–60%	40%	45–60%	35–40%	40–60%	50–65%
Typical electrical power	Up to 20 kW	<10 kW	>1 MW	>50 kW	Up to 250 kW	W-MW
Possible applications	Submarines, spacecraft	Portable applications	Power stations	Power stations	Vehicles, small and stationary	Power stations

"gas voltaic battery" in 1839, later named the Grove cell.[1] Accelerated research and engineering for fuel cell development started in the 1960s due to the unique needs for the long-endurance manned space exploration missions. For space applications, in addition to less toxicity, fuel cells have the advantage over traditional batteries, as they offer significantly higher energy density (energy per equivalent unit of weight) than conventional and advanced batteries.

Fuel cells are classified primarily according to the nature of the electrolyte. Moreover, the nature of the electrolyte governs the choices of the electrodes and the operation temperatures. Shown in table 10.1 are the fuel cell technologies currently under development.[2–4] Technologies attracting attention toward development and commercialization include direct methanol (DMFC), polymer electrolyte membrane (PEMFC), solid-acid (SAFC), phosphoric acid (PAFC), alkaline (AFC), molten carbonate (MCFC), and solid-oxide (SOFC) fuel cells. This chapter is aimed at the solid-oxide fuel cells (SOFCs) and related electrolytes used for the fabrication of cells.

10.1.2 The Electrolyte for SOFCs

As discussed in table 10.1, the mobile species within a fuel cell are ions, which necessitate the electrolyte being an ionic conductor and electronic insulator. If the oxygen ions are the only charge carriers, the electron motive force (EMF) of the cell is determined from the chemical potential of oxygen (i.e., oxygen activity), which is expressed by the Nernst equation as

$$EMF = \Gamma \frac{RT}{4F} \ln\left(\frac{(pO_2)_a}{(pO_2)_b}\right) \tag{10.1}$$

where Γ is the ionic transference number (ionic conductivity/total conductivity), T is the cell operation temperature, F is the Faraday constant, $(pO_2)_a$ is the oxygen partial pressure (fugacity/activity) at the cathode side, and $(pO_2)_b$ is similarly the oxygen partial pressure (fugacity/activity) at the anode side. In the case of an open circuit (without external current flow), the EMF of the cell corresponds to the open-circuit voltage (OCV). The importance of the electrolyte to SOFCs was realized several decades ago and has been extensively studied and reviewed.[5–11]

10.1.2.1 Requirements

Long-term successful operation of the SOFCs requires that the electrolyte possess adequate chemical and structural stability over a wide range of oxygen partial pressures, from air or oxygen to humidified hydrogen or hydrocarbons. The requirements for the electrolyte used in the intermediate-temperature SOFCs (IT SOFCs) include:

1. *Conductivity*: The materials must have an ionic transference number close to unity; i.e., the electronic conductivity in the electrolyte must be sufficiently low in order to minimize internal shorting and provide high energy conversion efficiency. The electrolyte materials should also possess high oxygen ion conductivity to minimize the ohmic losses in the cell.
2. *Density*: In order to minimize molecular transport of gases across the electrolyte membrane and prevent combustion (to produce maximum power density), the electrolyte must remain gas tight during the life of the cell. This indicates that when we consider the SOFC electrolyte, the main challenge has to be related to the processing of dense, thin electrolyte layers using either the anode or cathode as the supporting structure.
3. *Stability*: The operation of SOFCs requires the cathode and the anode to be porous for gas transport; therefore, the electrolyte is exposed to both the air and the fuel at elevated temperature. The electrolyte must remain chemically phase stable in these environments, along with thermal and mechanical stability during thermal cycling. This requires matching of the thermal expansion coefficients of adjoining layers. Chemical interactions and formation of interfacial compounds between adjoining components must also be minimized or prevented to mitigate increase in cell resistance and polarization losses. It should be kept in mind that the SOFCs, currently designed for stationary applications, have a target life of 40,000 h, and hence the long-term chemical and structural stability of the electrolyte plays a crucial role.

10.1.2.2 Materials

The most commonly used electrolyte materials currently under development and deployment in SOFCs are the oxides with low-valence element substitutions, sometimes named acceptor dopant, which create oxygen vacancies through charge compensation. It is straightforward to design the oxygen ion conductors by increasing the oxygen vacancy concentration. This, however, may not be valid in many cases, as other factors, such as vacancy ordering, charge mobility, and compatibility with other cell components, must also be taken into consideration. As a result, the most commonly utilized electrolyte materials that can satisfy these requirements are Y-stabilized ZrO_2 (YSZ), with acceptor-substituted CeO_2 and $(La,Sr)(Mg,Ga)O_3$. Other interesting electrolyte materials include pseudo-perovskite structures, $Bi_4V_2O_{11}$ (MIMEVOX)[13–16] and $La_2Mo_2O_9$[17–19]; pyrochlore structures,[20–28] $(Gd,Ca)_2Ti_2O_{7-\_}$; and apatites, $La_{10-x}Sr_6O_{26}$.[29,30]

10.2 FLUORITES: ZRO_2, CEO_2, AND BI_2O_3

The fluorite class of oxides is most studied as solid-oxide electrolyte materials because of their chemical and structural stability. The fluorite lattice structure is basically face center cubic (FCC; space group, Fm^3m) with a general formula of MO_2, in which M typically is Zr or Ce, as in ZrO_2 or CeO_2. There are four MO_2 formulas in a unit cell, in which cations are in cubic closest packing with oxygen ions in all tetrahedral holes. The CeO_2 lattice is known to possess a large open-volume fraction (>35%) and, as a result, is capable of tolerating oxygen nonstoichiometry and forming solid solution with various low-valence elements. This gives the materials scientist an opportunity to alter the properties of a given base oxide by substituting different cations or varying oxygen content.

10.2.1 ZrO_2-Based Materials

Doped ZrO_2 has been extensively investigated as an oxygen ion conductor since Nernst invented a lamp with stabilized ZrO_2. Yttrium-stabilized zirconia (YSZ) is still the most interesting and practical material used as the electrolyte in SOFC. The only drawback of stabilized ZrO_2 is the low ionic conductivity in the IT regime. Two solutions that have been tried to resolve this problem are (1) to decrease the thickness of the YSZ electrolyte layer and (2) to find other acceptors to replace Y. The first solution seems to be valid only when the cell operating temperature is higher than 600°C, at which the maximum electrolyte thickness is about 5 μm, assuming the electrolyte area-specific resistance (ASR) ~ 0.1 Ω cm². Oxygen ion conductivity in doped ZrO_2 is found to be dependent on the atomic number, ionic radius, and concentration of the dopants. Additionally, the microstructural features, such as dopant segregation, impurities, and oxygen ion ordering, also play an important role in controlling the overall ionic conductivity. The maximum ionic conductivity in doped ZrO_2 is obtained when the concentration of acceptor-type dopants is close to the minimum necessary to completely stabilize the cubic fluorite-type phase. Further additions decrease the ionic conductivity due to increasing association of the oxygen vacancies and dopant cations into complex defects of low mobility. It is commonly accepted that this tendency increases with increasing difference between the host and dopant cation radii. The ionic radius for Zr^{4+} is 0.84 Å when the coordination number is 8; it is 0.87 Å for Sc^{3+} and 1.109 for Y^{3+}.[11,31] Sc-doped zirconia, therefore, is a promising candidate because it has a conductivity similar to that of doped CeO_2,

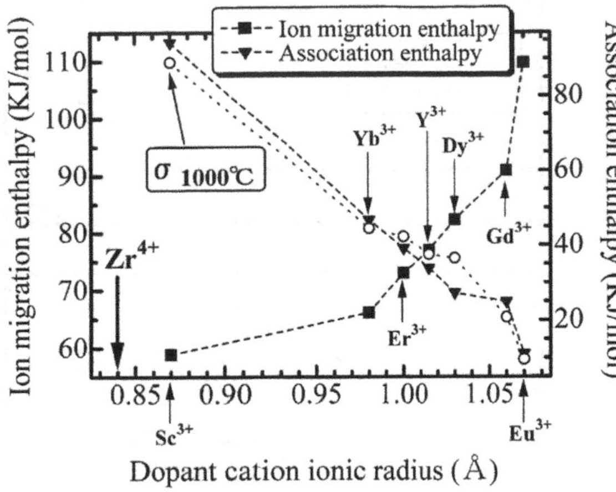

FIGURE 10.1 Ionic conductivity and diffusion enthalpies as a function of ionic radius of dopant cations to ZrO_2.[31]

provided the dopant remains cost-effective. The current higher cost and known aging phenomena of Sc-doped ZrO_2 make it less attractive in commercializing SOFC. Figure 10.1 illustrates the ionic conductivity and diffusion enthalpies as a function of ionic radius of dopant cations to ZrO_2.[31]

10.2.2 CeO_2-Based Materials

Doped CeO_2 materials are considered intermediate-temperature (500 to 700°C) solid electrolytes for SOFCs.[32–35] This is due to their higher oxygen ion conductivity ($Ce_{0.9}Gd_{0.1}O_{1.95}$: 0.025 $\Omega^{-1}\cdot cm^{-1}$ at 600°C) compared to zirconia-based electrolyte materials (<0.005 $\Omega^{-1}\cdot cm^{-1}$). The principal problem with CeO_2 is its structural instability under reducing exposure conditions. Reduction of ceria also results in significant increase in electronic conductivity and dimensional change due to the formation of oxygen vacancies and the associated reduction of Ce^{4+} to Ce^{3+}. Operation of cells containing ceria electrolyte at intermediate temperatures eliminates these disadvantages. The highest oxygen ion conductivity has been observed in Gd- or Sm-doped CeO_2. The ionic radii are 1.053, 1.079, and 0.97 Å for Gd^{3+}, Sm^{3+}, and Ce^{4+}, respectively, when the coordination number is 8.[36]

A doped CeO_2 electrolyte containing SOFCs has been considered very attractive for low-temperature operation (<600°C) because of higher ionic conductivity (than YSZ) and reduced electrical losses in the electrolyte. In general, for SOFCs operating at about 0.7 V and at current densities in the range of 0.5 to 2 amps/cm², the aim is to achieve a cell electrolyte ASR of < 0.2 Ω cm². Figure 10.2 illustrates the ASR of the electrolyte as a function of electrolyte thickness at 500 and 600°C. In the IT regime, the electrode overpotential, particularly cathode overpotential, is of the order of the ASR of the electrolyte. Therefore, a maximum allowed electrolyte thickness is directly determined by the cell design.

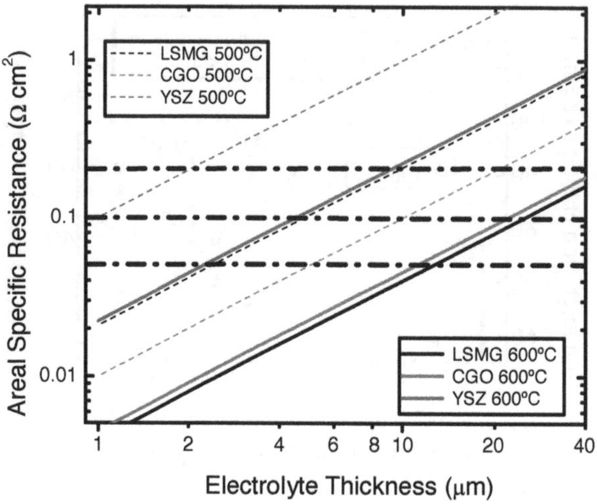

FIGURE 10.2 Area-specific resistance vs. electrolyte thickness for YSZ, CGO, and LSMG at 500 and 600°C.

10.2.3 Bi_2O_3-Based Materials

Stabilized Bi_2O_3 has a fluorite structure with a highly deficient oxygen sublattice that attributes to its higher ionic conductivity compared to ZrO_2- or CeO_2-based materials.[9,37–42] On the other hand, the highly defective structure results in several disadvantages of Bi_2O_3-based electrolytes, particularly the thermodynamic instability, which significantly limits its application in fuel cells. Bi_2O_3-based materials exhibit a complex array of structures and properties, depending upon the dopant concentrations, temperature, and exposure atmosphere. Degradation of conductivity at 500°C as a function of time was observed by Jiang and Wachsman,[43] who employed several 25 mol% lanthanide-doped Bi_2O_3 (figure 10.3). The level of dopant concentrations was chosen, as it corresponds to a one-to-one array between the number of dopant cations and oxide vacancies. The degradation rate is apparently a function of the ionic radius of the dopant cation. The larger the dopant cations are, the slower the conductivity degradation is. Hence, doping with Yb (0.985 Å) results in the fastest degradation, whereas Dy (1.027 Å) dopants yield the slowest conductivity decay.[43] All of these dopants are smaller than Bi^{3+} (1.17 Å).[36]

There exists an ordering/disordering of oxygen vacancies in Bi_2O_3-based oxides, which plays a substantially important role in the ionic conductivity. The order–disorder transition takes place at ~600°C in the doped Bi_2O_3.[38,44] The sublattice for oxygen (and oxygen vacancies) is ordered when the annealing temperature is less than the transition temperature. As a result, oxygen ion conductivity undergoes degradation with time, as shown in figure 10.3. Transmission electron microscopy and neutron diffraction[45] results indicated that the vacancy ordering takes place along <111>, as shown in figure 10.4, which is similar to the YSZ system.[44,46] It is therefore postulated that the ordering of oxygen vacancies in <111> may also be common in fluorite oxides at high vacancy concentrations.

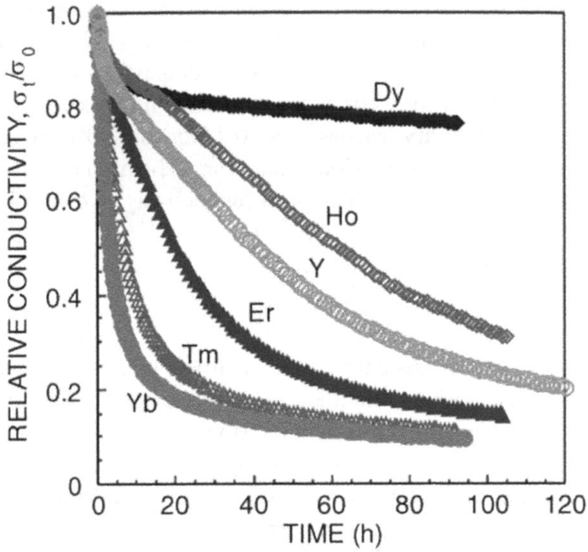

FIGURE 10.3 Conductivity as a function of time for doped Bi_2O_3, measured at 500°C.

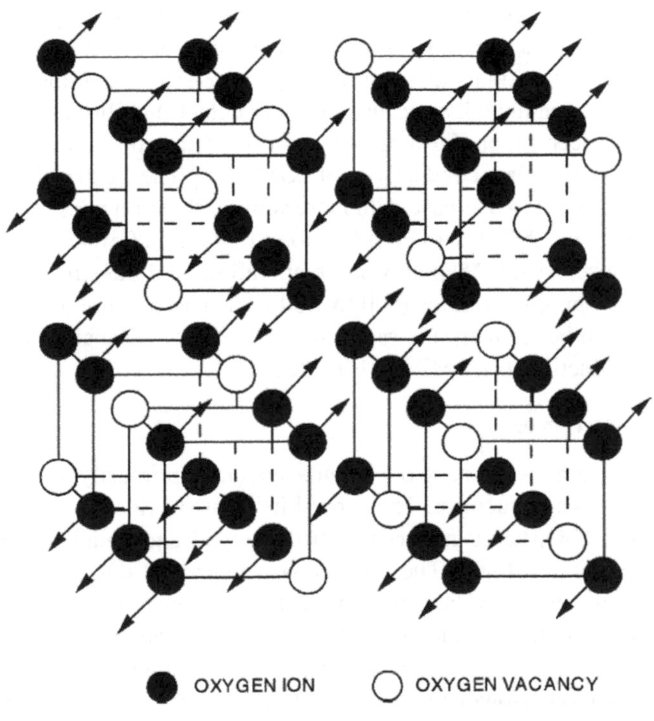

FIGURE 10.4 Model of ordered structure showing both occupancy ordering (filled vs. vacant) and positional ordering.[38]

The stability of Bi_2O_3 was reported to improve by forming solid solution with vanadium, which can be partially substituted with Co, Ni, or Cu. A so-called BIMEVOX (bismuth metal–ion vanadium oxide) consists of oxygen vacancies in a perovskite-like V-O layer, which possesses extremely high oxygen ion conductivity. For instance, ionic conductivity reaches ~3×10^{-3} S/cm at 300°C, which is nearly two orders higher than any other oxygen ion conductors. The high reactivity and instability, however, hinder its further application in fuel cells as an electrolyte.

10.3 PEROVSKITES

10.3.1 $LaGaO_3$

The perovskite structure is basically cubic with the general formula of ABO_3, in which A, the large cation site, is an alkali, alkaline earth, or rare earth ion, and B, the small cation site, is a transition metal cation. The large cations are in 12-fold coordination with oxygen, while the small cations fit into octahedral positions. The occupancy of these sites by different cations is determined primarily by ionic radius rather than the valence. This opens the door for the materials scientists to substitute selectively for either the A or B ion by introducing isovalent or aliovalent cations. An oxygen ion conductor can be tailored because of the geometrical and chemical flexibility of the perovskite structure. This is borne out by $(La,Sr)(Mg,Ga)O_3$ (LSMG),[47–51] which has attracted great attention since its discovery.[52] There exist, however, two drawbacks for LSMG electrolytes: (1) the uncertainty in the cost of Ga sources and (2) the chemical and mechanical stability of LSMG. It is apparent that ordering occurs (sometimes at specific temperatures) that significantly decreases the oxygen ionic conductivity because of lower defect mobility and reduced effective vacancy concentration. Stevenson et al. studied the role of microstructure and nonstoichiometry on ionic conductivity of LSMG.[53–55] The electrical conductivity of sintered LSGM tends to decrease with increasing A/B cation nonstoichiometry. The flexural strength of LSGM with an A/B cation ratio of 1.00 was also measured and found to be closer to 150 MPa at room temperature, and the strength decreased to 100 MPa at higher temperatures (600 to 1,000°C). The fracture toughness, as measured by notched beam analysis, was closer to 2.0 to 2.2 MPa at room temperature, with similar reduction to 1.0 MPa at 1,000°C.

10.3.2 Other Phases

There are many other solid-state oxide ion conductors, primarily derived from either fluorite or perovskite structures. The perovskite-related oxide ion conductors include (1) $Ln(Al,In,Sc,Y)O_3$-based materials, (2) the doped and undoped brownmillerite $Ba_2In_2O_5$, and (3) $La_2Mo_2O_9$. The transference number of doped $La_2Mo_2O_9$ can be higher than 0.99 in an oxidant environment. The drawbacks of $La_2Mo_2O_9$-based materials are instability in reducing conditions, a relatively large thermal expansion coefficient (>16 ppm/K for $La_{1.7}Bi_{0.3}Mo_2O_9$), and the order of the anion sublattice. Doped $LaAlO_3$ has reasonable ionic conductivity (~0.006 S/cm at 800°C) and excellent thermal expansion coefficient (TEC) match with other components (~11 ppm/K); however, it is rather challenging to sinter $LaAlO_3$-based oxides and to prevent the formation of highly insulating grain boundary phases (Al_2O_3).

10.4 DISCUSSION

10.4.1 Solid Solution of ZrO_2-CeO_2

Bi-layers of Y-stabilized ZrO_2 and doped CeO_2 have been used to further improve either the electrolyte stability or the electrode performance. For example, a thin layer of YSZ (<0.5 μm) has been used between the doped CeO_2 electrolyte and the anode to prevent the reduction of doped CeO_2.[56] A thin CeO_2 layer placed between the YSZ electrolyte and the cathode prevents interaction between them.[57] Also, mixtures of the above two phases have been used as electrolyte layers. The objective for building these structures has been to utilize the best properties for each component. As has been reported in the literature, however, considerable interdiffusion between ZrO_2- and CeO_2-based materials occurs at elevated temperatures (>1,400°C).[58–61] Solid solutions between ZrO_2 and CeO_2 are formed during high-temperature sintering, which can result in several problems. For example, the dimensional stability and the electrical conductivity may be altered, which can affect the successful operation of SOFCs. It has been shown that the overall electrical conductivity exhibited a minimum when the fraction of Gd-doped CeO_2 (CGO) was about 50% in the system of CGO_xYSZ_{1-x}.[60]

10.4.1.1 Reaction between CGO Film and YSZ[62]

CGO film has been used as the protective layer to reduce the reaction between YSZ and the La- and Sr-containing cathodes, which can form $La_2Zr_2O_7$ and $SrZrO_3$. These compounds exhibit a much higher resistance than YSZ. Figure 10.5 illustrates a plot of d spacing of the YSZ phase and CGO phase as a function of annealing temperature. Over the composition ranges used in this study, all of the sintered compositions

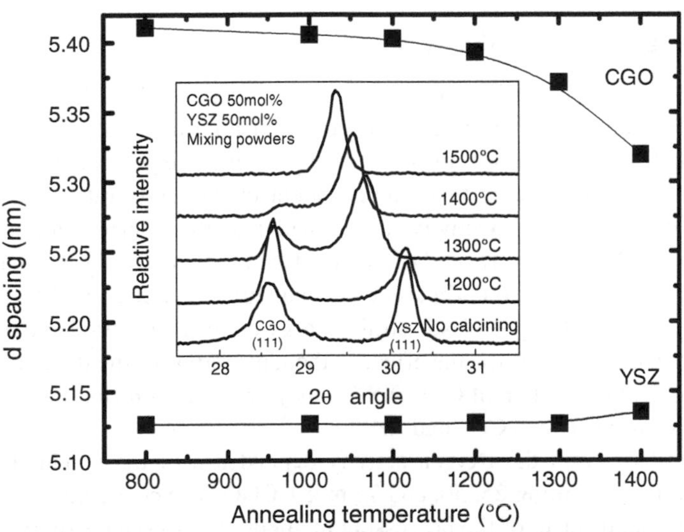

FIGURE 10.5 A plot of d spacing for CGO films on YSZ substrates as a function of annealing temperature. The inset illustrates the XRD of CGO and YSZ powders annealed at various temperatures.[62]

were single phase with a fluorite structure. The lattice parameter followed Vegard's rule for solid solution.[58,62] The lattice parameter for YSZ was nearly constant when the annealing temperature (T_a) was less than 1,300°C, whereas peaks for the CGO phase started to shift to higher angles when $T_a > 1{,}000$°C. This indicates a decreasing lattice parameter for CGO, as shown in figure 10.5. Diffusion of Y and Zr into the CGO lattice was attributed to a decreasing lattice parameter for CGO because Zr^{4+} is smaller than Ce^{4+}. Since the YSZ substrates in this study were dense tapes with an average grain size of ~5 μm, it can be concluded that CGO thin films possessed a higher reactivity than YSZ substrates at $T_a < 1{,}300$°C, which allowed Y and Zr to diffuse into the thin film.

10.4.1.2 Reaction between CGO and YSZ Powders

The mixing of CGO and YSZ powders with a similar average particle size enabled the study of interdiffusion by simply annealing the mixture at an elevated temperature, then studying powder diffraction results. Figure 10.5 illustrates X-ray diffraction (XRD) patterns for a mixture of CGO and YSZ powders annealed from 1,200 to 1,500°C. The (111) peak for YSZ continuously shifted to a lower angle as the annealing temperature increased, whereas the (111) peak for CGO was nearly constant. Therefore, one can conclude that CGO diffuses into YSZ in powder mixtures at high temperatures, which is contrary to what occurred for the system of CGO films on a YSZ substrate. Hence, an increase of lattice parameter for the composition of YSZ can be attributed to the diffusion of Ce and Gd. On the other hand, if Ce and Gd did diffuse into YSZ, the size of the CGO particles would become smaller during this reaction process, which was confirmed by a line broadening of the (111) diffraction peak of CGO.

10.4.1.3 Electrical Conductivity

Solid solutions of CGO-YSZ exhibited lower conductivity than that of either CGO or YSZ. The activation energy (E_a) and preexponential factor (σ_0) can be obtained by plotting ln(σT) versus 1/T for CGO_xYSZ_{1-x} at high oxygen activity; the preexponential factor, $\sigma_{\_}$, appears to be independent of the CGO ratio. Therefore, the observed decrease in conductivity, in particular at the intermediate-temperature regime (~600°C), might be related to an increase in E_a, which indicates a decrease in the mobility of the oxide ion. Theoretical calculations by Butler et al.[63] showed that the E_a was dependent on the elastic strain energy on the association enthalpy for the defect pair, which in turn was the effect of the ionic radius of the dopant.[11,64] Hence, changes of E_a in the system of CGO_xYSZ_{1-x} may be due to contributions by the ionic radius difference between Ce^{4+} and Zr^{4+}.

Figure 10.6 shows the oxygen activity dependence of the conductivity of the solid solutions containing 25, 50, and 75 mol% CGO. The conductivity values were found to be independent of the processing methods. The oxygen activity–dependent electronic conductivity behavior for the solid solution CGO_xYSZ_{1-x} is somewhat surprising in that this system can still be considered an acceptor- (Y and Gd) doped conductor. In the CeO_2 system, the electronic conduction is well known to be due to

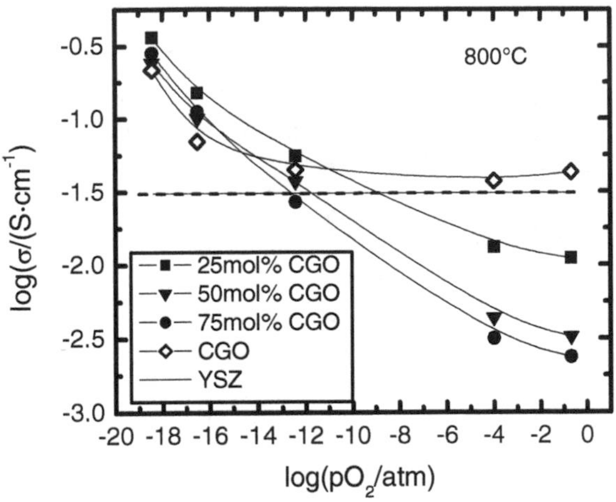

FIGURE 10.6 A plot of log(σ) vs. log(pO_2) for CGO_xYSZ_{1-x} (x = 1, 0.75, 0.5, and 0.25) solid solution measured at 800°C.[62]

the redox reaction between Ce^{4+} and Ce^{3+} oxidation states, which are oxygen pressure (activity) dependent. This reaction is expressed as

$$2Ce_{Ce}^{\times} + O_O^{\times} \rightarrow 1/2\,O_2 + V_O^{\bullet\bullet} + 2Ce_{Ce}' \quad \text{(Kröger–Vink notation)}^{65} \qquad (10.2)$$

The reduction energy has been calculated and found to be reduced by the introduction of zirconia because of the formation of defect clusters, such as $Ce_{Zr}' - V_O^{\bullet\bullet} - Ce_{Zr}'$.[66] This type of defect cluster increases the electronic conductivity in two ways: (1) the oxygen ion mobility decreases because of the trapping of oxygen vacancies, and (2) the electronic conductivity increases since the carrier concentration, n = $[Ce'_{Ce}]$ increases due to the decrease in the reduction energy.

10.4.2 Size Effect on Ionic Conduction in YSZ

A large discrepancy exists in the electrical conductivity of nanocrystalline-doped ZrO_2 thin films. On one hand, enhanced electrical conductivity[67–70] and oxygen diffusivity[71] are reported in the literature, whereas on the other hand, conductivity in agreement with microcrystalline specimens,[72,73] or even a decreasing conductivity, is reported in YSZ thin films.[74]

Kosacki et al.[67–69,75] studied the electrical conductivity of yttria- and scandia-doped zirconia thin films deposited onto either single-crystal alumina or magnesia substrates. Their study showed that the electrical conductivity of YSZ can be enhanced significantly at thickness <60 nm[68] (figure 10.7). The highly textured YSZ films were deposited by pulsed laser ablation. After annealing, the films were well crystallized. Epitaxial growth of the YSZ films on the MgO substrates was obtained, which was not the case for Guo et al.'s films.[74] When the film thickness is reduced from 2,000 nm to 60 nm, and both the current path length and film width are kept the same, the resistance of the films is inversely proportional to film thickness.

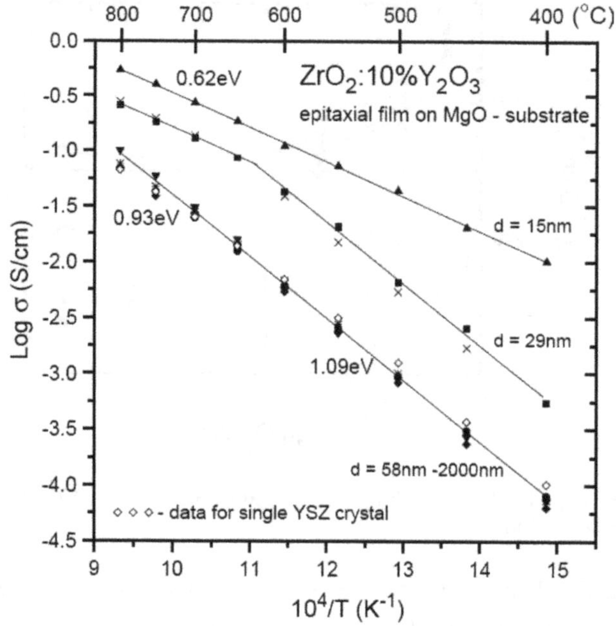

FIGURE 10.7 Electronic conductivity as a function of temperature for nanocrystalline YSZ films.[68]

This scale effect indicates that the electrical conductivity is nearly constant for the films varying in thickness from 60 to 2,000 nm. The measured resistance, however, decreased when the film thickness further reduced. Both DC and AC conductivity measurements indicated that there was an enhanced conductivity for film thickness of <60 nm. They further proposed three orders of magnitude larger conductivity in 1.6-nm-thick films than lattice conductivity. Since the grain size was not provided,[68] it is unknown whether only the grain size plays a role when a film's thickness is less than 60 nm. Guo et al.[74] deposited YSZ thin films by pulsed laser deposition on MgO substrates with thicknesses of 12 and 25 nm. The electrical conductivity was measured in both dry and humid O_2. The electrical conductivity in thin films, however, was found to be four times lower than ionic conductivity in microcrystalline specimens, as shown in figure 10.8. Furthermore, they found that there is not any remarkable proton conduction in the nanostructured films when annealed in water vapor.

10.4.3 Grain Size and Grain Boundary Thickness

Mondal and Hahn[72] used XRD to determine grain size. XRD line broadening provides a characteristic size of a material, as x-ray coherence length can be considered the size for a single crystal in a specimen. In addition to obtaining the accurate XRD line-broadening analysis, it is also considered that the x-ray coherence length may be different from the grain size. Unfortunately, electron microscopy was not employed to examine the microstructure. Moreover, the DC conductivity measurements were not provided to compare with AC impedance results. It is therefore difficult to com-

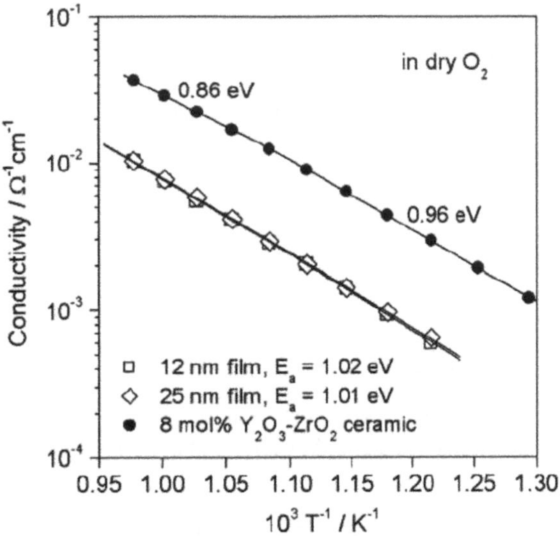

FIGURE 10.8 Conductivity as a function of 1/T for nanocrystalline YSZ film.[74]

pare their results with conductivities in thin films. The activation energies for bulk transport in $Y_{0.06}Zr_{0.94}O_{1.97}$, however, are ~0.1 to 0.2 eV smaller than those from the literature, indicating a possible enhanced conductivity in nanocrystalline materials.

Guo et al.[74] proposed that there exists a "de-doping" effect in nanometer-thick YSZ films, which results in a lower bulk conductivity in nanocrystalline YSZ (grain size ~ 80 nm, thickness = 12 and 25 nm) than in the microcrystalline specimen. They predicted that the conductivity of nanostructured YSZ (e.g., ≤5 nm) will be even smaller, analyzing from a space charge model. Because XRD results were not provided, neither the crystallinity nor the existence of the second phase is known in YSZ films grown by Pulsed Laser Deposition (PLD). However, electrical measurements were carefully carried out in both dry and wet O_2, and the overall conductivity in their YSZ films is lower than that of bulk YSZ (grain size > 15 μm) by a factor of 4 (figure 10.8).

10.4.4 Stability of CeO_2 for Low-Temperature Operation

It is known that Gd-doped CeO_2 can be reduced at very low oxygen partial pressures. Table 10.2 is a list of P^*O_2 (atm) for various dopant levels, y, in $Ce_{1-y}Gd_yO_{1-y/2}$ at different temperatures.[76] P^*O_2 represents the the oxygen partial pressure at which the ionic transference number of the solid solution becomes 0.5. Because the electrons have a much higher mobility (~3 orders higher) than oxygen ions, a transfer number of 0.5 only indicates that the electron concentration (Ce^{3+}) is about 0.1% of the oxygen vacancy concentration. However, further reduction will result in lattice expansion due to the formation of a large number of Ce^{3+}, as shown by Yasuda and Hishinuma[77] (figure 10.9).

TABLE 10.2
P^*O_2 (atm) for Various Dopant Levels, y, in $Ce_{1-y}Gd_yO_{1-y/2}$ at Different Temperatures

y	Temperature (°C)			
	700	750	800	850
0.1	1.3×10^{-17}	2.53×10^{-16}	2.43×10^{-15}	3.56×10^{-14}
0.2	1.24×10^{-19}	3.92×10^{-18}	9.2×10^{-17}	1.73×10^{-15}
0.3	3.16×10^{-19}	5.6×10^{-18}	8.5×10^{-17}	9.0×10^{-16}
0.4	9.3×10^{-19}	8.8×10^{-18}	6.3×10^{-17}	3.85×10^{-16}
0.5	1.46×10^{-16}	1.38×10^{-15}	1.12×10^{-14}	6.5×10^{-14}

10.4.5 Grain Boundary Effects

Lower-temperature operation does pose a problem due to the higher activation energy of the grain boundary resistivity, ρ_{gb}. A high ρ_{gb} can be due to many factors, includ-

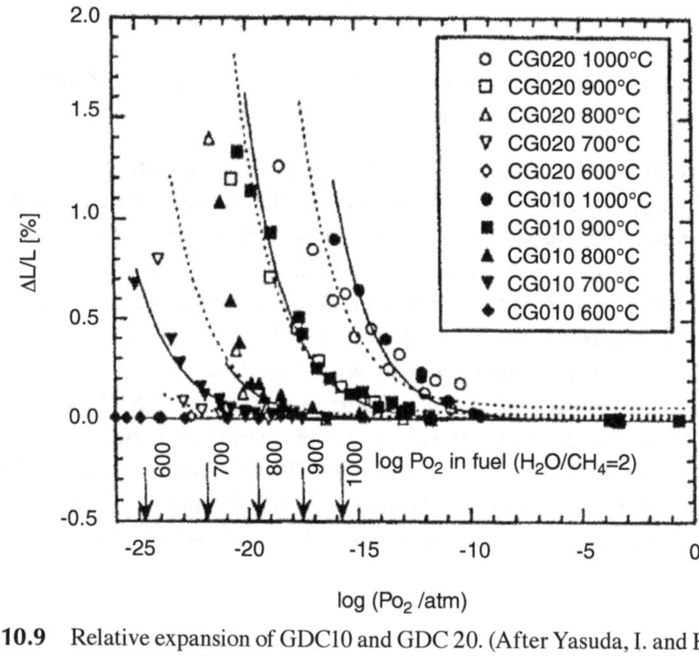

FIGURE 10.9 Relative expansion of GDC10 and GDC 20. (After Yasuda, I. and Hishinuma, M., *Electrochem. Soc. Proc.*, 97, 178, 1997.)

ing (1) amorphous phases, (2) dopant segregation, (3) an altered local defect chemistry due to space charge effects, and (4) intergranular porosity (small effect). These effects are all strongly related to grain size and the associated grain boundary area. Among these, the first factor is typically predominant, as impurities such as silicon form insulating phases that tend to wet the grains, and hence effectively block the ionic current. An example is that the lattice conductivity in $Ce_{0.90}Gd_{0.10}O_{1.95}$ (CGO10) is higher than that in $Ce_{0.80}Gd_{0.20}O_{1.90}$ (CGO20); however, CGO20 often has higher

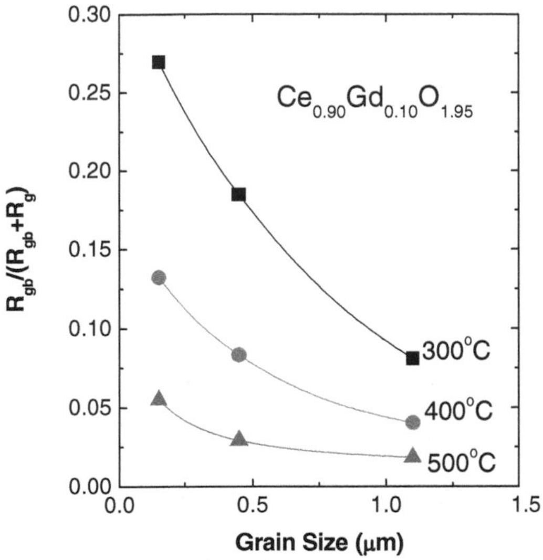

FIGURE 10.10 A plot of R_{gb}/R_t as a function of grain size for doped CeO_2.[78]

total conductivity, likely due to the greater grain boundary conductivity in CGO20. Moreover, in reducing environments, the electronic contribution to the overall conductivity in CGO10 is larger than that in CGO20, indicating that the cerium ions in CGO10 can be reduced more easily than in CGO20.

The grain boundary resistance, shown in Figure 10.10, is a plot of R_{gb}/R_t as a function of grain size for $Ce_{0.90}Gd_{0.10}O_{1.95}$ measured between 300 and 500°C. A reduction of the R_{gb}/R_t ratio is observed as the grain size increases. The raw materials used for conductivity measurements were of high purity (99.95%), indicating low levels of impurities (particularly Si). Hence, the insulating glassy phase is considered negligible over the measuring temperature range. This behavior is explained by a trapping phenomenon, which has been observed in the transport of oxide ions in oxide conductors.[33] This phenomenon has been modeled for several oxygen ion conductors at the temperature where acceptor dopant–oxygen vacancy complexes dissociate. It is worth mentioning that the trend shown in Figure 10.10 is contrary to undoped CeO_2,[78] in which the grain boundary contribution to the total resistance increases with increasing grain size.

10.5 CONCLUSIONS

In SOFCs, the difference in chemical potential or activity of oxygen across the electrolyte surfaces provides the electromotive force, and thus the electrical potential. Extensive research over the past decades has resulted in the development of cost-effective processes for the fabrication of thin and dense electrolyte layers. YSZ has been considered one of the best choices for high-temperature applications (>650°C) because of its feasibility of fabrication of a thin membrane, reasonable ionic conductivity, large ionic domain, and, most importantly, chemical and mechanical stability in an oxidizing and reducing environment. Doped CeO_2, on the other hand, is a leading candidate for the fuel cells operating at temperatures below 600°C, during which the chemically induced expansion is negligible and its high ionic conductivity is fully taken advantage of.

REFERENCES

1. U.S. DOE (Department of Energy), *Fuel Cells: Power for the 21st Century*, U.S. DOE, Washingtn, DC. 2004.
2. Singh, P., Minh, N.Q., Solid oxide fuel cells: technology status, *International Journal of Applied Ceramic Technology*, 2004, 1, 5–15.
3. Steele, B.C.H., Material science and engineering: the enabling technology for the commercialisation of fuel cell systems, *Journal of Materials Science*, 2001, 36, 1053–1068.
4. Steele, B.C.H. and Heinzel, A., Materials for fuel-cell technologies, *Nature*, 2001, 414, 345–352.
5. Badwal, S.P.S. and Foger, K., Solid oxide electrolyte fuel cell review, *Ceramics International*, 1996, 22, 257–265.
6. Stoukides, M., Solid-electrolyte membrane reactors: current experience and future outlook, *Catalysis Reviews: Science and Engineering*, 2000, 42, 1–70.
7. Jiang, S.P., A review of wet impregnation: an alternative method for the fabrication of high performance and nano-structured electrodes of solid oxide fuel cells, *Materials Science and Engineering A: Structural Materials Properties Microstructure and Processing*, 2006, 418, 199–210.
8. Kharton, V.V., Marques, F.M.B., and Atkinson, A., Transport properties of solid oxide electrolyte ceramics: a brief review, *Solid State Ionics*, 2004, 174, 135–149.
9. Sammes, N.M., Tompsett, G.A., Nafe, H., and Aldinger, F., Bismuth based oxide electrolytes: structure and ionic conductivity, *Journal of the European Ceramic Society*, 1999, 19, 1801–1826.
10. Shuk, P., Wiemhofer, H.D., Guth, U., Gopel, W., and Greenblatt, M., Oxide ion conducting solid electrolytes based on Bi_2O_3, *Solid State Ionics*, 1996, 89, 179–196.
11. Kilner, J.A. and Brook, R.J., A study of oxygen ion conductivity in doped nonstoichiometric oxides, *Solid State Ionics*, 1982, 6, 237–252.
12. Alcaide, F., Cabot, P.L., and Brillas, E., Fuel cells for chemicals and energy cogeneration, *Journal of Power Sources*, 2006, 153, 47–60.
13. Paydar, M.H., Hadian, A.M., and Fafilek, G., A new look at oxygen pumping characteristics of BICUVOX.1 solid electrolyte, *Journal of Materials Science*, 2006, 41, 1953–1957.
14. Kharton, V.V., Naumovich, E.N., Yaremchenko, A.A., and Marques, F.M.B., Research on the electrochemistry of oxygen ion conductors in the former Soviet Union. IV. Bismuth oxide-based ceramics, *Journal of Solid State Electrochemistry*, 2001, 5, 160–187.
15. Simner, S.P., SuarezSandoval, D., Mackenzie, J.D., and Dunn, B., Synthesis, densification, and conductivity characteristics of BICUVOX oxygen-ion-conducting ceramics, *Journal of the American Ceramic Society*, 1997, 80, 2563–2568.
16. Iharada, T., Hammouche, A., Fouletier, J., Kleitz, M., Boivin, J.C., and Mairesse, G., Electrochemical characterization of BIMEVOX oxide-ion conductors, *Solid State Ionics*, 1991, 48, 257–265.
17. Corbel, G. and Lacorre, P., Compatibility evaluation between $La_2Mo_2O_9$ fast oxide-ion conductor and Ni-based materials, *Journal of Solid State Chemistry*, 2006, 179, 1339–1344.

18. Yang, J.H., Wen, Z.Y., Gu, Z.H., and Yan, D.S., Ionic conductivity and micro structure of solid electrolyte $La_2Mo_2O_9$ prepared by spark-plasma sintering, *Journal of the European Ceramic Society*, 2005, 25, 3315–3321.
19. Lacorre, P., Goutenoire, F., Bohnke, O., Retoux, R., and Laligant, Y., Designing fast oxide-ion conductors based on $La_2Mo_2O_9$, *Nature*, 2000, 404, 856–858.
20. Shlyakhtina, A.V., Abrantes, J.C.C., Levchenko, A.V., Stefanovich, S.Y., Knot'ko, A.V., Larina, L.L., and Shcherbakova, L.G., New oxide-ion conductors Ln(2 + x)Ti(2 – x)O(7 – X/2) (Ln = Dy-Lu; x = 0.096), in *Advanced Materials Forum III*, Parts 1 and 2, Vols. 514–516, Trans Tech Publications Ltd., Zurich-Uetikon, Switzerland, 2006, pp. 422–426.
21. Bae, J.M. and Steele, B.C.H., Properties of pyrochlore ruthenate cathodes for intermediate temperature solid oxide fuel cells, *Journal of Electroceramics*, 1999, 3, 37–46.
22. Shimura, T., Komori, M., and Iwahara, H., Ionic conduction in pyrochlore-type oxides containing rare earth elements at high temperature, *Solid State Ionics*, 1996, 86–88, 685–689.
23. Takamura, H. and Tuller, H.L., Ionic conductivity of Gd_2GaSbO_7-Gd_2Zr_2O7 solid solutions with structural disorder, *Solid State Ionics*, 2000, 134, 67–73.
24. Yu, T.H. and Tuller, H.L., Electrical conduction and disorder in the pyrochlore system $(Gd_{1-x}Ca_x)(2)Sn_2O7$, *Journal of Electroceramics*, 1998, 2, 49–55.
25. Yu, T.H. and Tuller, H.L., Ionic conduction and disorder in the $Gd_2Sn_2O_7$ pyrochlore system, *Solid State Ionics*, 1996, 86–88, 177–182.
26. Kramers, S.A. and Tuller, H.L., A novel titanate-based oxygen-ion conductor: $Gd_2Ti_2O_7$, *Solid State Ionics*, 1995, 82, 15–23.
27. Kramer, S., Spears, M., and Tuller, H.L., Conduction in titanate pyrochlores: role of dopants, *Solid State Ionics*, 1994, 72, 59–66.
28. Tuller, H.L., Mixed ionic electronic conduction in a number of fluorite and pyrochlore compounds, *Solid State Ionics*, 1992, 52, 135–146.
29. Sansom, J.E.H., Najib, A., and Slater, P.R., Oxide ion conductivity in mixed Si/Ge-based apatite-type systems, *Solid State Ionics*, 2004, 175, 353–355.
30. Yaremchenko, A.A., Shaula, A.L., Kharton, V.V., Waerenborgh, J.C., Rojas, D.P., Patrakeev, M.V., and Marques, F.M.B., Ionic and electronic conductivity of $La_{9.83}$-$xPrxSi_{4.5}Fe_{1.5}O_{26}$ +/–delta apatites, *Solid State Ionics*, 2004, 171, 51–59.
31. Arachi, Y., Sakai, H., Yamamoto, O., Takeda, Y., and Imanishai, N., Electrical conductivity of the ZrO_2-Ln(2)O(3) (Ln = lanthanides) system, *Solid State Ionics*, 1999, 121, 133–139.
32. Steele, B.C.H., Materials for IT-SOFC stacks 35 years R&D: the inevitability of gradualness? *Solid State Ionics*, 2000, 134, 3–20.
33. Steele, B.C.H., Appraisal of $Ce_{1-y}Gd_yO_{2-y/2}$ electrolytes for IT-SOFC operation at 500 degrees C, *Solid State Ionics*, 2000, 129, 95–110.
34. Steele, B.C.H., Oxygen-transport and exchange in oxide ceramics. *Journal of Power Sources*, 1994, 49, 1–14.
35. Inaba, H. and Tagawa, H., Ceria-based solid electrolytes: review, *Solid State Ionics*, 1996, 83, 1–16.
36. Shannon, R.D., Revised effective ionic-radii and systematic studies of interatomic distances in halides and chalcogenides, *Acta Crystallographica Section A*, 1976, 32, 751–767.
37. Azad, A.M., Larose, S., and Akbar, S.A., Bismuth oxide-based solid electrolytes for fuel-cells, *Journal of Materials Science*, 1994, 29, 4135–4151.
38. Wachsman, E.D., Effect of oxygen sublattice order on conductivity in highly defective fluorite oxides, *Journal of the European Ceramic Society*, 2004, 24, 1281–1285.
39. Wachsman, E.D., Functionally gradient bilayer oxide membranes and electrolytes, *Solid State Ionics*, 2002, 152, 657–662.

40. Jiang, N.X., Wachsman, E.D., and Jung, S.H., A higher conductivity Bi_2O_3-based electrolyte, *Solid State Ionics*, 2002, 150, 347–353.
41. Wachsman, E.D., Boyapati, S., Kaufman, M.J., and Jiang, N.X., Modeling of ordered structures of phase-stabilized cubic bismuth oxides, *Journal of the American Ceramic Society*, 2000, 83, 1964–1968.
42. Wachsman, E.D., Ball, G.R., Jiang, N., and Stevenson, D.A., Structural and defect studies in solid oxide electrolytes, *Solid State Ionics*, 1992, 52, 213–218.
43. Jiang, N.X. and Wachsman, E.D., Structural stability and conductivity of phase-stabilized cubic bismuth oxides, *Journal of the American Ceramic Society*, 1999, 82, 3057–3064.
44. Bogicevic, A., Wolverton, C., Crosbie, G.M., and Stechel, E.B., Defect ordering in aliovalently doped cubic zirconia from first principles, *Physical Review B*, 2001, 64, 014106.
45. Boyapati, S., Wachsman, E.D., and Chakoumakos, B.C., Neutron diffraction study of occupancy and positional order of oxygen ions in phase stabilized cubic bismuth oxides, *Solid State Ionics*, 2001, 138, 293–304.
46. Goff, J.P., Hayes, W., Hull, S., Hutchings, M.T., and Clausen, K.N., Defect structure of yttria-stabilized zirconia and its influence on the ionic conductivity at elevated temperatures, *Physical Review B*, 1999, 59, 14202.
47. Huang, K.Q., Tichy, R.S., and Goodenough, J.B., Superior perovskite oxide-ion conductor; strontium- and magnesium-doped $LaGaO_3$. I. Phase relationships and electrical properties, *Journal of the American Ceramic Society*, 1998, 81, 2565–2575.
48. Huang, K.Q., Tichy, R., and Goodenough, J.B., Superior perovskite oxide-ion conductor; strontium- and magnesium-doped $LaGaO_3$. III. Performance tests of single ceramic fuel cells, *Journal of the American Ceramic Society*, 1998, 81, 2581–2585.
49. Huang, K.Q., Feng, M., Goodenough, J.B., and Milliken, C., Electrode performance test on single ceramic fuel cells using as electrolyte Sr- and Mg-doped $LaGaO_3$, *Journal of the Electrochemical Society*, 1997, 144, 3620–3624.
50. Yan, J.W., Lu, Z.G., Jiang, Y., Dong, Y.L., Yu, C.Y., and Li, W.Z., Fabrication and testing of a doped lanthanum gallate electrolyte thin-film solid oxide fuel cell, *Journal of the Electrochemical Society*, 2002, 149, A1132–A1135.
51. Lerch, M., Boysen, H., and Hansen, T., High-temperature neutron scattering investigation of pure and doped lanthanum gallate, *Journal of Physics and Chemistry of Solids*, 2001, 62, 445–455.
52. Ishihara, T., Matsuda, H., and Takita, Y., Doped $LaGaO_3$ perovskite-type oxide as a new oxide ionic conductor, *Journal of the American Chemical Society*, 1994, 116, 3801–3803.
53. Stevenson, J.W., Hasinska, K., Canfield, N.L., and Armstrong, T.R., Influence of cobalt and iron additions on the electrical and thermal properties of $(La,Sr)(Ga,Mg)O_3$-delta, *Journal of the Electrochemical Society*, 2000, 147, 3213–3218.
54. Baskaran, S., Lewinsohn, C.A., Chou, Y.S., Qian, M., Stevenson, J.W., and Armstrong, T.R., Mechanical properties of alkaline earth-doped lanthanum gallate, *Journal of Materials Science*, 1999, 34, 3913–3922.
55. Stevenson, J.W., Armstrong, T.R., Pederson, L.R., Li, J., Lewinsohn, C.A., and Baskaran, S., Effect of A-site cation nonstoichiometry on the properties of doped lanthanum gallate, *Solid State Ionics*, 1998, 115, 571–583.
56. Kim, S.-G., Yoon, S.P., Nam, S.W., Hyun, S.-H., and Hong, S.-A., Fabrication and characterization of a YSZ/YDC composite electrolyte by a sol-gel coating method, *Journal of Power Sources*, 2002, 110, 222.
57. Simner, S.P., Bonnett, J.F., Canfield, N.L., Meinhardt, K.D., Sprenkle, V.L., and Stevenson, J.W., Optimized lanthanum ferrite-based cathodes for anode-supported SOFCs, *Electrochemical Solid-State Letters*, 2002, 5, A173.

58. Eguchi, K., Akasaka, N., Mitsuyasu, H., and Nonaka, Y., Process of solid state reaction between doped ceria and zirconia, *Solid State Ionics*, 2000, 135, 589–594.
59. Lee, C.H. and Choi, G.M., Electrical conductivity of CeO_2-doped YSZ, *Solid State Ionics*, 2000, 135, 653–661.
60. Tsoga, A., Naoumidis, A., and Stover, D., Total electrical conductivity and defect structure of ZrO_2-CeO_2-Y_2O_3-Gd_2O_3 solid solutions, *Solid State Ionics*, 2000, 135, 403–409.
61. Xiong, Y.P., Yamaji, K., Sakai, N., Negishi, H., Horita, T., and Yokokawa, H., Electronic conductivity of ZrO_2-CeO_2-$YO_{1.5}$ solid solutions, *Journal of the Electrochemical Society*, 2001, 148, E489–E492.
62. Zhou, X.D., Scarfino, B., and Anderson, H.U., Electrical conductivity and stability of Gd-doped ceria/Y-doped zirconia ceramics and thin films, *Solid State Ionics*, 2004, 175 19–22.
63. Butler, V., Catlow, C.R.A., Fender, B.E.F., and Harding, J.H., Dopant ion radius and ionic-conductivity in cerium dioxide, *Solid State Ionics*, 1983, 8, 109–113.
64. Kilner, J.A., Fast oxygen transport in acceptor doped oxides, *Solid State Ionics*, 2000, 129, 13–23.
65. Kroger, F.A. and Vink, H.J., Relationships between the concentration of imperfections in crystalline solids, in *Solid State Physics: Advances in Research and Applications*, Vol. 3, Seitz, F., Turnbull, T., Eds. Academic Press, New York, 1957, p. 307.
66. Balducci, G., Kaspar, J., Fornasiero, P., Graziani, M., Islam, M.S., and Gale, J.D., Computer simulation studies of bulk reduction and oxygen migration in CeO_2-ZrO_2 solid solutions, *Journal of Physical Chemistry B*, 1997, 101, 1750–1753.
67. Kosacki, I., Anderson, H.U., Mizutani, Y., and Ukai, K., Nonstoichiometry and electrical transport in Sc-doped zirconia, *Solid State Ionics*, 2002, 152, 431–438.
68. Kosacki, I., Rouleau, C.M., Becher, P.F., Bentley, J., and Lowndes, D.H., Nanoscale effects on the ionic conductivity in highly textured YSZ thin films, *Solid State Ionics*, 2005, 176, 1319–1326.
69. Kosacki, I., Suzuki, T., Petrovsky, V., and Anderson, H.U., Electrical conductivity of nanocrystalline ceria and zirconia thin films, *Solid State Ionics*, 2000, 136, 1225–1233.
70. Zhang, Y.W., Jin, S., Yang, Y., Li, G.B., Tian, S.J., Jia, J.T., Liao, C.S., and Yan, C.H., Electrical conductivity enhancement in nanocrystalline $(RE_2O_3)(0.08)(ZrO_2)(0.92)$ (RE = Sc, Y) thin films, *Applied Physics Letters*, 2000, 77, 3409–3411.
71. Knoner, G., Reimann, K., Rower, R., Sodervall, U., and Schaefer, H.E., Enhanced oxygen diffusivity in interfaces of nanocrystalline ZrO_2 center dot Y_2O_3, *Proceedings of the National Academy of Sciences of the United States of America*, 2003, 100, 3870–3873.
72. Mondal, P. and Hahn, H., Investigation of the complex conductivity of nanocrystalline Y_2O_3-stabilized zirconia, *Berichte Der Bunsen-Gesellschaft-Physical Chemistry Chemical Physics*, 1997, 101, 1765–1768.
73. Jiang, S.S., Schulze, W.A., Amarakoon, V.R.W., and Stangle, G.C., Electrical properties of ultrafine-grained yttria-stabilized zirconia ceramics, *Journal of Materials Research*, 1997, 12, 2374–2380.
74. Guo, X., Vasco, E., Mi, S.B., Szot, K., Wachsman, E., and Waser, R., Ionic conduction in zirconia films of nanometer thickness, *Acta Materialia*, 2005, 53, 5161–5166.
75. Kosacki, I., Petrovsky, V., and Anderson, H.U., Band gap energy in nanocrystalline ZrO_2: 16%Y thin films, *Applied Physics Letters*, 1999, 74, 341–343.
76. Mogensen, M., Sammes, N.M., and Tompsett, G.A., Physical, chemical and electrochemical properties of pure and doped ceria, *Solid State Ionics*, 2000, 129, 63–94.

77. Yasuda, I. and Hishinuma, M., Electrical conductivity, dimensional instability and internal stresses of CeO_2-Gd_2O_3 solid solutions, *Electrochemical Society Proceedings*, 1997, 97, 178–187.
78. Zhou, X.D., Huebner, W., and Anderson, H.U., Size effect on the electronic properties of doped and undoped ceria, in *Defects and Diffusion in Ceramics: An Annual Retrospective VII*, Vols. 242–244, Trans Tech Publications Ltd., Zurich-Uetikon, Switzerland, 2005, pp. 277–289.

11 Corrosion and Protection of Metallic Interconnects in Solid-Oxide Fuel Cells

Zhenguo Yang, Jeffry W. Stevenson, and Prabhakar Singh

CONTENTS

11.1 INTRODUCTION

Energy security and increased concern over environmental protection have spurred a dramatic worldwide growth in research and development of fuel cells, which electrochemically convert incoming fuel into electricity with no or low pollution. Fuel cell technology has become increasingly attractive to a number of sectors, including utility, automotive, and defense industries. Among the various types of fuel cells, solid-oxide fuel cells (SOFCs) operate at high temperature (typically 650 to 1,000°C) and have advantages in terms of high conversion efficiency and the flexibility of using hydrocarbon fuels, in addition to hydrogen.[1–5] The high-temperature operation, however, can lead to increased mass transport and interactions between the surrounding environment and components that are required to be stable during a lifetime of thousands of hours and up to hundreds of thermal cycles. For stacks with relatively low operating temperatures (<800°C), the interconnects that are used to

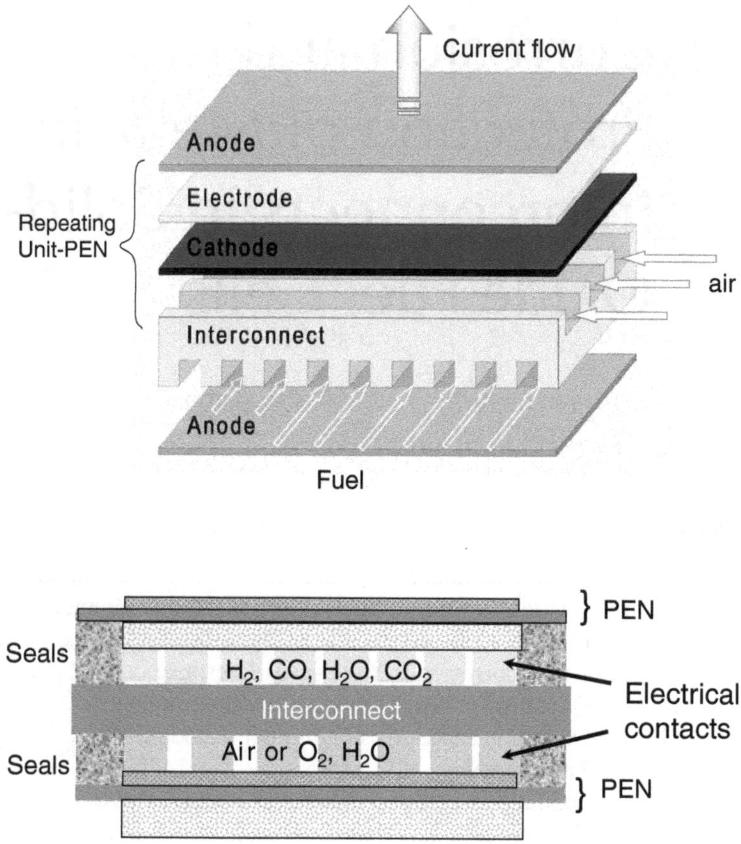

FIGURE 11.1 Planar design of solid-oxide fuel cell: (a) a stack repeat unit and (b) details of a possible design.

electrically connect a number of cells in series (see figure 11.1) are typically made from cost-effective metals or alloys. The metallic interconnects must demonstrate excellent stability in a very challenging environment during SOFC operation, as they are simultaneously exposed to both an oxidizing (air) environment on the cathode side and a reducing environment (hydrogen or a reformed hydrocarbon fuel) on the anode side. Other challenges include the fact that water vapor is likely to be present in both of these environments, and the fuel is likely to contain impurities, such as sulfides. Since the fuel is usually a reformed hydrocarbon fuel, such as natural gas, coal gas, biogas, gasoline, etc., the interconnect is exposed to a wet carbonaceous environment at the anode side. Finally, the interconnect must be stable toward any adjacent components, such as electrodes, seals, and electrical contact materials, with which it is in physical contact.

Until recently, the leading candidate material for the interconnect was doped lanthanum chromite, $La_xSr_{1-x}CrO_3$, a ceramic that could easily withstand traditional 900 to 1,000°C operating temperatures. However, several issues remain, including the high cost of raw materials and fabrication, difficulties in obtaining high-density chromite parts at reasonable sintering temperatures,[1,6–9] and the tendency of the chromite interconnect to partially reduce at the fuel gas–interconnect interface, causing the component to warp and the peripheral seal to break.[1,10] The recent trend in developing lower-temperature (650 to 800°C), more cost-effective cells that utilize anode-supported, thin electrolytes[11,12] or new electrolytes with improved conductivity[13,14] makes it feasible for lanthanum chromite to be supplanted by cost-effective metals or alloys as the interconnect materials.[15–18] The metals are typically those alloys that contain "active" constituents, mainly Cr, Al, or Si, which are preferentially oxidized at the surface to form an oxide scale that minimizes further environmental attack during high-temperature exposure.[19–21] Since alumina (Al_2O_3) and silica (SiO_2) are electrically insulating,[21,22] alloys that form a semiconductive chromia scale (with a conductivity of ~1.0×10^{-2} S-cm^{-1}at 800°C in air[21,23–25]) are the preferred candidates. (However, if an insulating scale can be excluded from the electrical current path, alumina or silica formers, which usually demonstrate higher oxidation resistance than chromia-forming alloys, can be considered for interconnect applications.[16]) Since thermal expansion matching is important, alloys such as Fe-Cr-base ferritic stainless steels that have a coefficient of thermal expansion (CTE) similar to that of the ceramic cells are usually selected instead of other groups of alloys, including Ni(-Fe)-Cr base austenitic compositions.

Different groups of alloys that have been considered potential candidates are schematically represented in figure 11.2, and their suitability is summarized in table 11.1. It should be noted that traditional chromia-forming oxidation-resistant alloys were designed with an emphasis on surface and structural stability, but not the scale electrical conductivity, which is equally important for SOFC interconnect applications. Thus, alloying practices used in the past to enhance surface and structural stability may not be compatible with the desired high scale electrical conductivity. For example, Si, often a residual element in alloy substrates, may help improve alloy oxidation resistance through formation of a silica subscale along the metal–scale interface. (Silica is immiscible with chromia.) However, that subscale is electrically insulating

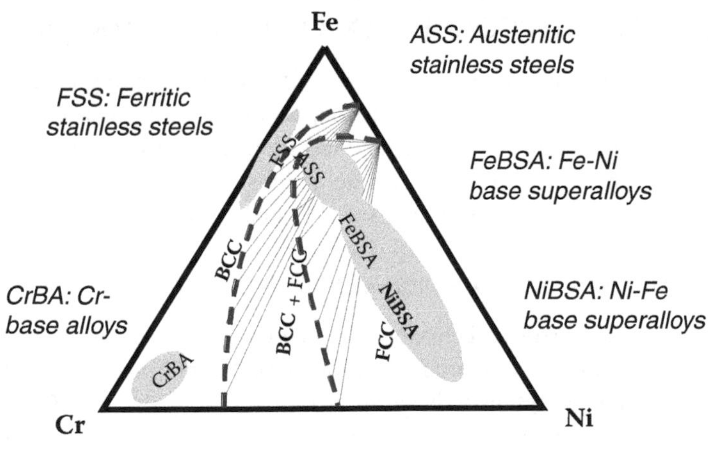

FIGURE 11.2 Schematic of alloy options for SOFC applications.

TABLE 11.1
Comparison of Key Properties of Different Alloy Groups for SOFC Applications

Alloys	Matrix Structure	TEC (10–6·K–1	Oxidation Resistance	Mechanical Strengths	Manufacturability	Cost
CrBA	BCC	11.0–12.5(RT 800°C)	Good	High	Difficult	Very expensive
FSS	BCC	11.5–14.0(RT 800°C)	Good	Low	Fairly easy	Cheap
ASS	FCC	18.0–20.0(RT 800°C)	Good	Fairly high	Easy	Cheap
FeBSA	FCC	15.0–20.0(RT 800°C)	Good	High	Easy	Fairly expensive
NiBSA	FCC	14.0–19.0(RT 800°C)	Good	High	Easy	Expensive

and therefore increases electrical resistance, especially if the subscale is continuous. Several new alloys such as Crofer 22 APU[26-37] and ZMG232[38,39] have been recently developed specifically for SOFC interconnect applications, with an improved scale electrical conductivity. One remaining issue, however, is the long-term surface stability that obviously needs further improvement. This is often accomplished by application of protective layers or coatings on metallic interconnects.

This chapter will provide an overview of oxidation and corrosion behavior of candidate oxidation-resistant alloys under SOFC operating conditions and discuss surface modifications for improved stability and performance of metallic interconnects.

11.2 CORROSION OF OXIDATION-RESISTANT ALLOYS UNDER SOFC INTERCONNECT EXPOSURE CONDITIONS

During SOFC operation, interconnects interact with surrounding gaseous environments on both the cathode and anode side, as well as with adjacent components such as sealing materials, electrodes, and electrical contact layers inserted between interconnects and electrodes. These interactions potentially cause corrosion of metallic interconnects and affect their stability and performance.

11.2.1 Oxidation and Corrosion at Metal–Gas Interfaces

The oxidation and corrosion behavior of oxidation-resistant alloys has been widely investigated in a range of environments for myriad applications.[19,21,40–44] Recently, oxidation-resistant alloys have been studied particularly for SOFC interconnect applications. These studies were often carried out using single atmosphere exposure conditions, either air (or moist air) representing the cathode-side environment[45–50] or a reducing atmosphere simulating the anode-side environment.[29,32,38,48,54–72] Lately, studies have also been performed to determine the oxidation/corrosion behavior of metal and alloys under dual-atmosphere exposure conditions that closely simulate the actual interconnect exposure conditions during SOFC operation.[31,64–69] The alloys studied include both Fe-Cr-base ferritic stainless steels[29,32,38,52,53,56,59–62] and Ni or Ni-Cr-base heat-resistant alloys,[46,47,49,50,54,59–61] as well as Cr or Cr base alloys.[71,72]

11.2.1.1 Oxidation in Air, Cathode-Side Environment

During high-temperature exposure in air, active elements, e.g., Cr, in alloy substrates are preferentially oxidized, forming an oxide scale on the alloy surface. The scale growth is often modeled with a parabolic relationship with time t:[40]

$$(\Delta w)^2 = k_p.t \tag{11.1}$$

where Δw is the weight gain and k_p is the parabolic rate constant.

However, due to the complexity of the scale growth, which involves factors such as scale vaporization and grain boundary diffusion, deviation from the parabolic law is often observed.[15,73,74] Nevertheless, oxidation behavior can frequently be approximated by the parabolic relationship, and the parabolic growth rate is thus often used to report oxidation resistance of different alloys. In general, alloys with higher Cr% possess higher oxidation resistance, and therefore exhibit a lower scale growth rate. To remain capable of self-healing and prevent breakaway oxidation, alloy substrates must maintain a Cr reservoir with Cr% above a critical value that marks the transition from external to internal oxidation. For example, Fe-Cr-base alloys require a critical Cr% of 17 to 20%, depending on temperature, minor alloy additions, impurities (S, P, C), etc.[15,21] A recent study also examined the effect of thickness on the stability of Fe-Cr substrates, indicating potential occurrence of breakaway oxidation when the Fe-Cr substrate becomes too thin.[75–77] Also, it has been established that reactive elements such as rare earth metals in a trace amount (a few hundredths or tenths of a percent) can significantly increase alloy oxidation resistance by improving scale adherence and suppressing scale growth by Cr outward diffusion.[15,78–81] Most (if not all) newly developed alloys for the interconnect application contain reactive element additions, e.g., La in Crofer 22 APU.[15]

11.2.1.2 Oxidation and Corrosion in Fuel, Anode-Side Environment

In comparison to the oxidizing, cathode-side environment, the reducing, anode-side environment is more complex, particularly when a hydrocarbon fuel is used. The presence of high water vapor partial pressure, carbon activity, and residual components such as sulfur makes metallic interconnects susceptible to varied forms of corrosion. Though the oxygen partial pressure (10^{-18} ~ 10^{-12} bar) at the anode side is much lower than that at the cathode side, formation of oxides such as chromia and $(Mn,Cr)_3O_4$ is nevertheless still thermodynamically favored. In moist hydrogen, Fe-Cr-base stainless steels exhibit a scale growth rate that is comparable to that in air.[32,38,57,61,62] The scale grown in the hydrogen fuel is usually comprised of the same major phases as found in air, although their morphology and minor components can be different. X-ray diffraction analysis on scales grown on Crofer 22 APU in either moist hydrogen or moist air indicated that both scales were comprised of Cr_2O_3 and $(Mn,Cr)_3O_4$ phases.[31,32,65] Additionally, amorphous phases such as silica and MnO that could not be detected by x-ray diffraction formed in the fuel environment, as revealed by transmission electron microscopy analyses.[31] The spinel phase tended to grow with a large aspect ratio in moist hydrogen, in comparison to well-defined, equiaxed crystallites in air. For Ni-Cr-base alloys, NiO formation is suppressed in the scales since oxidation of nickel is thermodynamically unfavorable in low oxygen partial pressure fuels. Overall, it appears Ni-Cr base alloys exhibit higher oxidation resistance in moist hydrogen than in air, and even those with a relatively low Cr% appear to be promising for interconnect applications in terms of oxidation resistance.[48,58]

Further differences in oxidation resistance between Fe-Cr base and Ni(-Fe)-Cr-base alloys are also observed in high water vapor environments. For Fe-Cr-base alloys, exposure to high water vapor environments potentially leads to breakaway oxidation through formation of Fe or Fe-rich oxides.[82–85] For example, Shen et al.[85] reported that during oxidation in an oxygen stream containing 2 to 10% H_2O at 900 to 1,000°C, Fe-Cr alloys (with 15 and 20% Cr) exhibited an initial protective behavior due to formation of a Cr-rich scale, which was followed by nonprotective breakaway oxidation due to the formation of a scale containing Fe_2O_3 and $(Fe,Cr)_3O_4$ phases. The breakaway oxidation was sensitive to the water vapor content in the atmosphere and %Cr in the Fe-Cr alloys: the higher the water vapor content or the lower the Cr%, the earlier the breakaway oxidation took place. In contrast, Ni(-Fe)-Cr-base alloys exhibited enhanced scale adherence and suppressed formation of NiO in the scale after exposure to high vapor environments.[86,87] Similarly, Cr or Cr base alloys exhibited enhanced chromia scale adherence to the metal or alloy substrate in $Ar/H_2O/H_2$.[71,72]

When a hydrocarbon fuel is fed, either directly or after reformation, into the anode chamber, metallic interconnects are exposed to an environment with a carbon activity, and therefore potentially could suffer carbon-induced corrosion. It has been found that in carbon-bearing gas environments, alloys, including Fe(-Ni)-Cr- and Ni-Fe-Cr-base alloys, are susceptible to metal dusting at temperatures in the 400 ~850°C range, which leads to alloy degradation by disintegration of the metal matrix into tiny particles.[88–90] For Fe base alloys, the proposed mechanism involves saturation of the alloy substrate with carbon and the subsequent formation of Fe_3C that decomposes into metal particles and carbon when the carbon activity approaches unity.[91,92] A different mechanism was proposed for Ni-base alloys that also starts with carbon saturation, but in which the saturated substrate directly decomposes into metal particles and graphite when the internal carbon activity approaches unity.[93,94] Recently, several publications reported and discussed the danger of encountering carbon-induced corrosion for oxidation-resistant alloys under SOFC interconnect exposure conditions at the anode side.[59,60] Overall, it appears that metal dusting is likely to occur in a hydrocarbon fuel with a carbon activity of ≥1. Toh et al.[60] reported metal dusting of some selected oxidation-resistant alloys tested in CO–26% H_2–6% H_2O (vol%), corresponding to a carbon activity of 2.9, at 650°C under thermal cycling. The resistance to metal dusting depended on alloy composition. For example, with electro-polished surfaces, chromia-forming austenitic alloys, including Alloy 800, Inconel 601, 690, and 693, and Alloy 602CA, suffered a rapid metal dusting, which did not occur until after about 50 one-hour cycles for the ferritic steel Fe–27 Cr–0.0001 Y. The alloy with the best performance was Inconel 625, which was still protected by its Cr_2O_3 scale after 500 cycles. Zeng and Natesan[59] evaluated a number of high-temperature oxidation-resistant alloys at 593°C in carbonaceous gases of high carbon activity (>7.0) and found pit formation and metal dusting in tested alloys. In comparison, Ni-base alloys, which developed scales with less spinel content, performed better than the Fe-base alloys, which formed scales with more spinel content. When exposed to a carbonaceous environment with a carbon activity less than unity, oxidation-resistant alloys are much less likely to suffer metal dusting or carbon-induced corrosion. Horita et al.[61] examined both Fe-Cr- and Ni-Cr-base alloys at 800°C in a CH_4-H_2O atmosphere with an equilibrium carbon activity of 0.8 and found no carbide formation after nearly 300 h.

In addition to bulk alloys, Jian et al.[63] investigated a clad structure that had Ni layers at both sides of a ferritic stainless steel substrate. After testing at 750°C for 1,000 h in 53.1% N_2–25.2% H_2–18.3% CO–3.3% CO_2–0.17% CH_4, the clad structure exhibited severe structural degradation due to carbon penetration through the Ni layers and formation of carbides in the ferritic stainless steel substrate along grain boundaries.

11.2.1.3 Oxidation/Corrosion under Air–Fuel Dual-Exposure Conditions

During SOFC operation, interconnects are simultaneously exposed to air at the cathode side and fuel at the anode side, and therefore experience a hydrogen partial pressure gradient from the fuel side to the air side. Previous investigations of Fe-Cr-base stainless steels, Ni-Cr-base alloys, and elemental metals, under hydrogen–air dual-exposure conditions found that the oxidation/corrosion behavior of the metals or alloys under the dual exposures can be very different from that under single-exposure conditions.[31,48,55,64–68,70,95] In particular, the composition and microstructure of the scale grown on the air side differed significantly from the behavior when exposed to air on both sides, while the oxidation/corrosion behavior at the hydrogen fuel side was comparable to that when exposed to the hydrogen fuel at both sides.

Fe-Cr-base ferritic stainless steels, in particular those with a relatively low Cr%, were reported to be susceptible to hematite (Fe_2O_3) phase nodular growth in the scale grown on the air side of the air–hydrogen sample. For example, Yang et al.[64] found that AISI 430, with 17% Cr, formed hematite nodules (see figure 11.3a and b) on the air side of the air–hydrogen sample during isothermal heating at 800°C after 300 h, potentially resulting in localized attack. In comparison, there was no hematite phase formation on the air–air sample, on the hydrogen side of the air–hydrogen sample, or on the hydrogen–hydrogen sample. The oxidation behavior on the hydrogen side of the air–hydrogen sample was similar to that on the hydrogen–hydrogen sample. The potential detrimental effects of the dual exposures appeared to be dependent on the alloy composition, in particular Cr% in the Fe-Cr substrate. For Crofer 22 APU, with 22 ~ 23% Cr, no hematite phase formation or nodule growth was observed under the same test conditions as for AISI 430. Instead, it was found that the spinel top layer of the scale on the air side of the hydrogen–air sample was enriched in

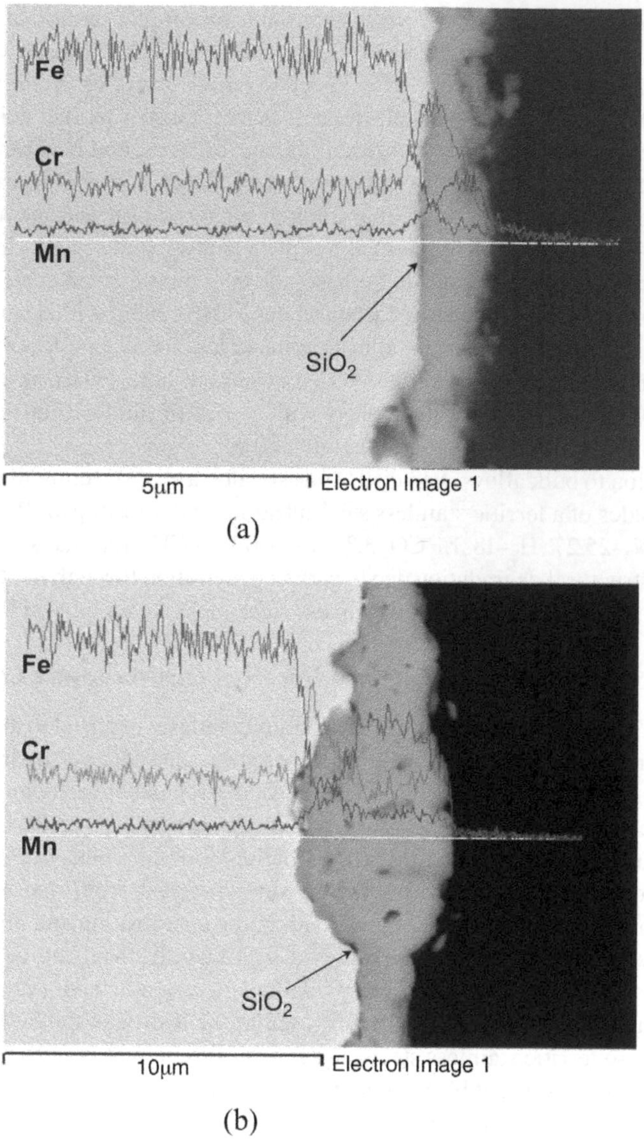

FIGURE 11.3 SEM cross-sections of AISI 430 coupons after 300 h of oxidation at 800°C in air under different exposure conditions: (a) both sides exposed to air and (b) on the air side of the air–(H_2 + 3% H_2O) exposure.[64]

iron and grew into a different morphology from that on the air–air sample under the same conditions.[31,65] However, when ambient air (~1% H_2O) was replaced by moist air (3% H_2O), the hematite phase was observed in the scale on the air side of the air–hydrogen sample.[67] This anomalous oxidation thus appears to be a result of combined effects from both the hydrogen flux from the fuel side to the air side and increased water vapor partial pressure on the air side. E-brite, with 27% Cr, appeared to be more resistant to formation of hematite nodules at 800°C in the scale grown on the air side of the air–hydrogen sample, though the surface microstructure of the scale was different from the air-only sample. At higher temperatures (900°C), Meier et al.[96] observed iron oxide formation in the scale grown on the air side of E-brite during air–hydrogen dual exposures. Similar anomalous oxidation behavior was also observed by Ziomek-Moroz et al.[68] and Holcomb et al.[69] not only on ferritic stainless steels, but also austenitic stainless steels as reported earlier by Singh et al.[97] In addition to the hydrogen–air dual exposures, an early publication by Nakagawa et al.[95,98] examined the oxidation behavior of ferritic stainless steels in (argon + hydrogen)–air dual environments as well as a steam–air dual environment that simulated the boiler tube exposure conditions in steam turbines. Anomalous oxidation was found on the air side of the dual-exposure sample, which exhibited a significantly increased oxidation rate on the air side due to formation of the hematite phase, which was attributed to hydrogen permeation from the (argon + hydrogen) side or the steam side to the air side of the stainless steels. The anomalous oxidation/corrosion behavior of oxidation-resistant alloys observed under the dual-exposure conditions appears to be similar to that found in a high partial pressure water vapor single environment.[40,67,82–85,87,99]

In addition to the Fe-Cr-base stainless steels, dual-exposure effects were also reported on Ni-Cr-base alloys.[48,100] For Ni-Cr-base alloys, the dual exposures also resulted in different oxidation/corrosion behavior from that in a single-atmosphere exposure. But unlike the ferritic chromia-forming alloys, nickel and Ni-Cr-base alloys formed a uniform, well-adherent scale on the air side of the air–hydrogen sample that was free from any nodule growth. Also, the dual exposures tended to eliminate the porosity that was often observed along the scale–metal interface in a single-air exposure, likely resulting in an improved scale adherence. The absence of detrimental effects of the air–hydrogen dual exposures on the scale stability on the Ni-Cr-base alloys in comparison with the Fe-Cr-base alloys is consistent with reported results in water vapor by Pint.[86]

In an effort to gain mechanistic understanding, elemental metals were also studied. Singh et al.[66] reported the destructive effects of air–hydrogen dual exposures on silver at elevated temperatures, as shown in figure 11.4. As found in this study, simultaneous exposures to fuel and oxidant environments lead to extensive porosity development in bulk silver at elevated temperatures. The porosity formation takes place predominantly along the grain boundary. The formation of the water vapor phase is attributed to the nucleation and growth of high-pressure steam bubbles that connect to form the porosity and fissures in the solid silver. The water phase that evolves into a high-pressure steam is related to higher solubility and fast diffusivity of H and O in the bulk metal. Thermodynamic modeling indicates that the reaction $2[H]_{Ag} + [O]_{Ag} = H_2O(g)$ is highly favorable. In contrast, Yang et al.[31] found minimal effects of air–hydrogen dual exposures on the scale growth on Ni metal. Similar observations were also reported by Meier.[96] The anomalous oxidation behavior of the metals or alloys on the air side of the air–hydrogen samples is currently attributed to the transport of hydrogen through the metal substrate from the fuel side to the air side, and its subsequent presence at the oxide scale–metal interface and in the scale.

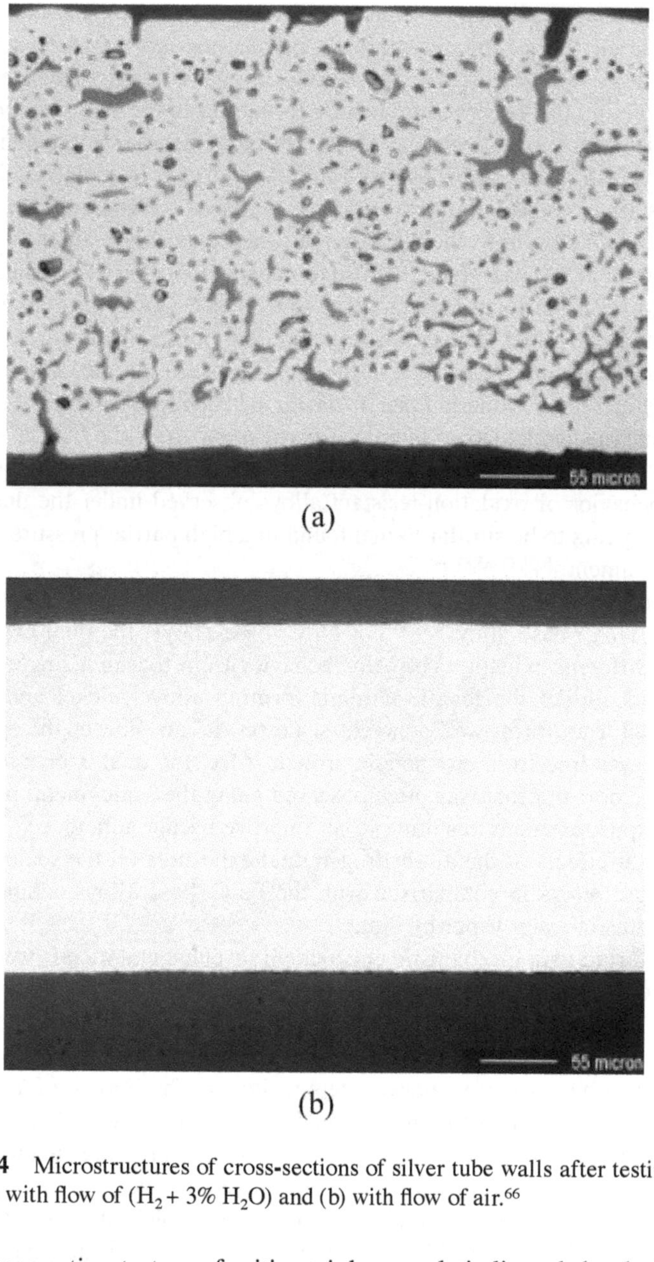

FIGURE 11.4 Microstructures of cross-sections of silver tube walls after testing at 700°C for 100 h: (a) with flow of (H_2 + 3% H_2O) and (b) with flow of air.[66]

Hydrogen permeation tests on ferritic stainless steels indicated that hydrogen can diffuse through the alloys, though the permeation was drastically decreased by formation of chromia scale on the alloys.[95,101] The mechanisms by which the presence of hydrogen or protons at the air side affects the oxide scale structure and growth are not clearly understood at this time. Several mechanisms have been proposed to tentatively explain the observed anomalous oxidation behavior.[65,69]

The anomalous oxidation/corrosion on the air side of the air–fuel sample appears to be sensitive to surrounding environments, alloy composition, preexisting conditions, and other factors. Kurokawa et al.[101] tested AISI 430 by flowing 80% Ar–20% H_2 through a water saturator at room temperature on one side and air on the other, and observed minimal effect of the hydrogen potential gradient on the stainless steel oxidation behavior after up to 300 h at 800°C. It should be noted that the hydrogen gradient was lower and the sample was thicker (2.0 mm) than those in the aforementioned studies that found anomalous oxidation. Preoxidation in air to form a scale on ferritic stainless steel was found to mitigate or prevent the hematite phase growth in the additional scale grown on Crofer 22 APU during air–hydrogen dual exposures. Overall, it appears that further systematic work is required to gain clearer insight into the oxidation/corrosion behavior in the dual environment and its effects on metallic interconnect long-term stability.

11.2.2 Corrosion at Interfaces with Adjacent Components

A typical example of interactions between metallic interconnects and adjacent components involves rigid glass–ceramic seals, including those made from barium–calcium–aluminosilicate (BCAS) base glasses.[102–106] Previous work[107–109] found that ferritic stainless steel interconnect candidates reacted extensively with the BCAS sealing glass–ceramic, resulting in an interface that was more prone to defects. For traditional chromia-forming stainless steels, the extent and nature of their interactions with the glass–ceramic depends on the exposure conditions and proximity of the interface of the sealing glass and ferritic stainless steel to the ambient air. At or near the edges, where oxygen from the air is accessible, the chromia scale grown on the steel and Cr-containing vapor species reacted with BaO in the glass–ceramic, forming $BaCrO_4$, presumably via the following reactions:

$$2\,Cr_2O_3\,(s) + 4\,BaO(s) + 3\,O_2(g) = 4\,BaCrO_4(s) \tag{11.2}$$

$$CrO_2(OH)_2(g) + BaO(s) = BaCrO_4(s) + H_2O(g) \tag{11.3}$$

Due to the large thermal expansion mismatch between barium chromate and the sealing glass or ferritic stainless steel (e.g., AISI 446),[110] the extensive formation of barium chromate resulted in crack initiation and growth between the sealing glass and alloy coupons, as shown in figure 11.5.

In the interior seal regions, where access of oxygen from the air was blocked, chromium or chromia dissolved into the BCAS sealing glass to form chromium-rich solid solutions. The stainless steel also reacted with residual species in the sealing glass–ceramic to generate porosity in the glass–ceramic along the interface in the interior regions. Recently Haanappel et al.[111,112] further investigated the compatibility of ferritic stainless steels and sealing glasses under air–hydrogen dual exposures. It was found that the corrosion at the interface of the sealing glass and the chromia-forming steel was substantially different from that when exposed to hydrogen or air only. At the air side, iron oxide nodules formed on the ferritic stainless steel near or at the triple-phase boundary of air–glass–metal, causing short-circuiting of the glass–ceramic seals. In contrast, no iron oxide formation was found at the interface

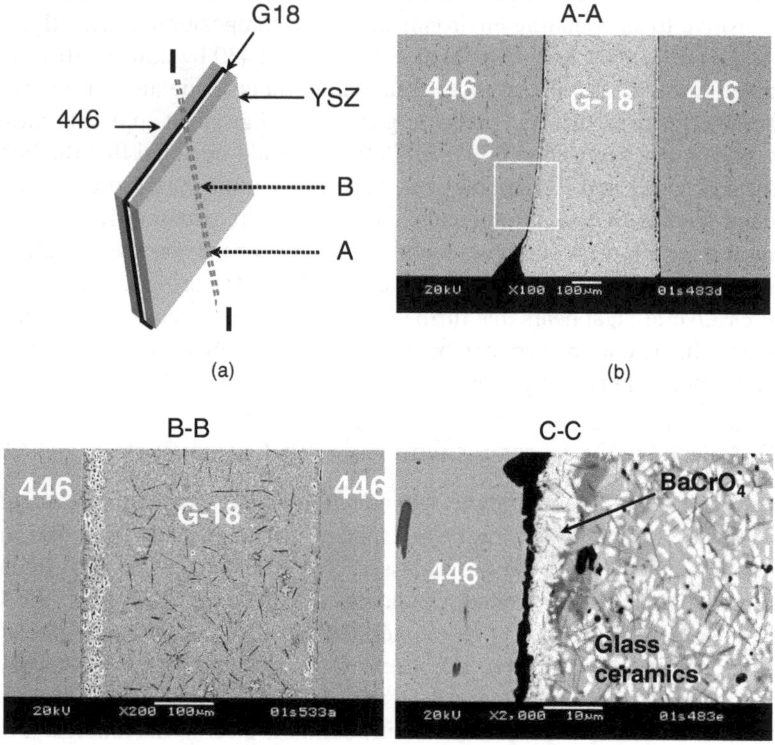

FIGURE 11.5 Interfacial reactions between G18 sealing glass and 446 stainless steel. (a) A schematic of the joined couple (446/G18/446) and SEM images of the interfacial cross-section (b) at the edge area A, (c) at the interior region, and (d) from the region marked C in (b). The 446 coupons (12.7 × 12.7 × 0.5 mm) were joined to the G18 through heat treatment at 850°C for 1 h, followed by 750°C for 4 h in air.[107]

between the glass–ceramics and the ferritic steel on the hydrogen side. However, it is not clear how the dual exposures led to the iron oxide formation and the subsequent seal degradation. Besides the BCAS sealing glass, Haanapel et al.[113] and Batfalsky et al.[114] also investigated compatibility of PbO-containing glass–ceramics and ferritic stainless steels, observing extensive internal and external oxidation of ferritic stainless steels.

Besides the glass seal interfaces, interactions were also reported at the interfaces of the metallic interconnect with electrical contact layers, which are inserted between the cathode and the interconnect to minimize interfacial electrical resistance and facilitate stack assembly.[115] For example, perovskites that are typically used for cathodes and considered potential contact materials have been reported to react with interconnect alloys. Reaction between manganites and chromia-forming alloys led to formation of a manganese-containing spinel interlayer that appeared to help minimize the contact ASR.[115–117] Sr in the perovskite conductive oxides can react with the chromia scale on alloys to form $SrCrO_4$.[115,118] Alternatively, Tietz et al.[119,120] examined chemical compatibility of

Sr-containing cuprate superconductors, as potential contact materials, with ferritic stainless steel X 10CrAl18 and found formation of $SrCrO_4$ and extensive reaction between the alloy and the conductive oxides.

11.3 SURFACE MODIFICATION FOR IMPROVED STABILITY

For satisfactory long-term stability against oxidation and corrosion, metallic interconnects are often coated with a protection layer that helps minimize electrical contact resistance and mitigate or prevent potential cell degradation due to chromia scale evaporation. The discussion in this chapter mainly focuses on the protection layer on the cathode side, considering the oxidizing environment and the susceptibility of SOFC cathodes to chromium poisoning. Functionally, the protection layer is intended first to serve as a mass barrier to both chromium cation outward and oxygen inward transport (via solid-state diffusion if the barrier contains no open porosity), as shown in figure 11.6. The difference in chromium chemical potential across the protection layer drives chromium cations (e.g., Cr^{3+}) to possibly diffuse into and through the protection layer. If this occurs, chromium can volatilize from the protection layer surface. Therefore, the material for the protection layer ought to either exhibit no solubility to Cr or possess very low chromium cation diffusivity. Or alternatively, the protection layer can react with the alloy or the scale grown on the alloy to form a reaction product layer that can function as a Cr barrier. Opposite to the chromium outward diffusion, the oxygen anions (O^{2-}) potentially diffuse inward, driven by the oxygen chemical potential gradient across the protection layer. This oxygen flux leads to selective oxidation of the substrate alloy, and therefore subsequent growth of an oxide scale between the protection layer and the bulk alloy. Extensive growth of the oxide interlayer plays a negative role, increasing the electrical resistance and likelihood of spallation, especially during thermal cycling, and degrading the mechanical integrity of the interconnect. Thus, the protection layer should have an oxygen ionic conductivity that is as low as possible. In addition to the mass transport properties, thermomechanical and chemical stability of the

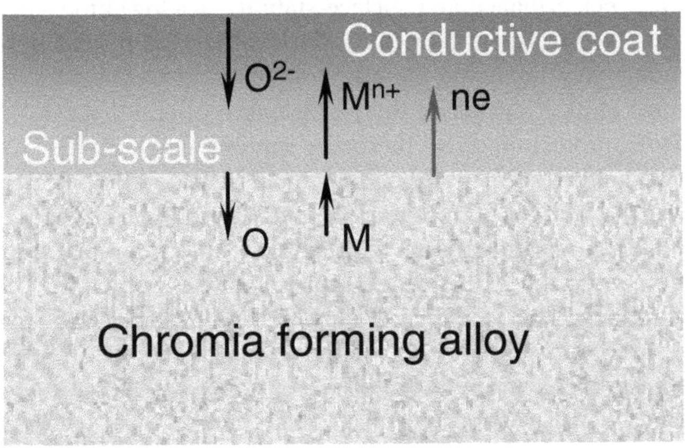

FIGURE 11.6 Schematic of mass transport in a conductive oxide protection layer on a chromia-forming alloy.

protection layer is essential to maintain its structural integrity during SOFC operation. Accordingly, candidate materials for the protection layer are also required to have a good thermal expansion match to the substrate alloy and be thermal-chemically stable and compatible with adjacent stack components in SOFC stacks.

The aforementioned requirements on surface stability are typical for all exposed areas of the metallic interconnect, as well as other metallic components in a SOFC stack (e.g., some designs use metallic frames to support the ceramic cell). In addition, the protection layer for the interconnect, or in particular the active areas that interface with electrodes and are in the path of electric current, must be electrically conductive. This conductivity requirement differentiates the interconnect protection layer from many traditional surface modifications as well as nonactive areas of interconnects and other components in SOFC stacks, where only surface stability is emphasized. While the electrical conductivity is usually dominated by their electronic conductivity, conductive oxides for protection layer applications often demonstrate a nonnegligible oxygen ion conductivity as well, which leads to scale growth beneath the protection layer. With this in mind, a high electrical conductivity is always desirable for the protection layers, along with low chromium cation and oxygen anion diffusivity.

As noted, for nonactive areas, such as the edge sealing area or areas without direct contact with cathodes, their protection layers are not required to be electrically conductive. For example, the channel walls in a corrugated design, as shown in figure 11.1a, can be aluminized for improved surface stability and reduced Cr volatility. Similarly, nonconductive coatings (e.g., applied by aluminizing[121]) can be applied onto edge areas of metallic interconnects for improved sealing and interfacial compatibility. Surface modification of the active areas of metallic interconnects is mainly discussed in this chapter.

Early reported examples of protection layers include overlay coatings of the conductive perovskite compositions that are often used as cathode and interconnect materials in SOFCs. For example, Linderoth[122] and Sakai et al.[123] reported the effectiveness of a $(La,Sr)CrO_3$ protection layer on Ducralloy $Cr5FeY_2O_3$ in improving its electrical performance and surface stability. Kadowaki et al.[124] found that $(La,Sr)CoO_3$ protection layers fabricated via low-pressure plasma spray on Ni-Cr base alloys effectively improved the alloy interconnect electrical conductivity. In contrast, Batawi et al.[125] evaluated the performance of $(La,Sr)CrO_3$, $(La,Sr)CoO_3$, and $(La,Sr)MnO_3$ protection layers thermally sprayed onto both $Cr5FeY_2O_3$ and Ni-Cr base alloys, indicating that all coatings increased the alloy oxidation rate. As pointed out by the authors, the $(La,Sr)CoO_3$ coatings were ineffective because of rapid diffusion of chromium through the coatings and formation of thick interfacial reaction layers, while the $(La,Sr)MnO_3$ protection layers on $Cr5FeY_2O_3$ exhibited the best performance due to the sluggish kinetics of interlayer growth and slow diffusion of chromium through the coatings. Quadakkers et al.[117] also observed significant transport of chromium into plasma-sprayed $(La,Sr)CoO_3$ coatings on $Cr5FeY_2O_3$ alloy. Recent work by Fujita et al.[126] found that $(La,Sr)CoO_3$ protection layers spin coated onto ferritic stainless steels AISI 430 and ZMG 232 helped improve the alloy interconnect surface stability and cell performance by reducing chromium poisoning. Overall, it appears that the chromites, which exhibit a lower oxygen ionic conductivity than many other perovskites, such as cobaltites, provided better protection to the metal substrates by more effectively inhibiting the scale growth beneath the perovskite protection layer. Besides, the chromites offer a better CTE match to alloy substrates than cobaltites if a BCC substrate alloy such as a ferritic stainless steel is used, thus providing better thermomechanical stability. However, one potential concern is that the chromites will release Cr via vaporization, albeit at a relatively low rate,[127,128] which may still lead to an unacceptable degradation in cell performance. In comparison, non-Cr-containing perovskites such as cobaltites with higher electrical conductivity and higher oxygen ion conductivity than the chromites offer more effective improvement in surface conductivity. However, the higher ionic conductivity often leads to a higher growth rate of the scale beneath the protection layer, thus negatively affecting the surface stability of the coated metallic interconnect. Furthermore, there is a concern regarding potential Cr outward diffusion through the nonchromium perovskite layers during high-temperature exposure, which can eventually lead to the presence of Cr at the surface of the protection layer and subsequent migration into cells and cell poisoning. Recent work by Yang et al.[129] found Cr diffusion through the $(La,Sr)FeO_3$ layers after 300 h of exposure at 800°C in air. The work also clearly demonstrated that the growth rate of the scale beneath the protection layers increased with ionic conductivity of the conductive oxides. The scale growth approximately followed a parabolic relationship with time during high-temperature exposure.

In addition to the perovskite compositions, late work extended to include the use of spinel compositions. In particular, non-Cr-containing spinels become more favorable, due to the fact that Cr-containing ones, such as the $(Mn,Cr)_3O_4$ spinel that is thermally grown as a top layer on Crofer 22 APU, help improve surface and electrical stability, but still release Cr.[37,130] Previous work of Larring and Norby[131] on Ducrolloy $Cr5FeY_2O_3$ indicated that a $(Mn,Co)_3O_4$ spinel contact layer could be a promising barrier to chromium migration. Recently, Yang et al.[116,132–134] investigated thermal growth of $(Mn,Co)_3O_4$ spinel layers, with a nominal composition of $Mn_{1.5}Co_{1.5}O_4$, onto a number of ferritic stainless steels, including AISI 430 and Crofer 22 APU (see figure 11.7) for interconnect applications in intermediate-temperature SOFCs. Chen et al.[135] reported $MnCo_2O_4$ coatings onto ferritic stainless steel AISI 430 via slurry coating followed by mechanical compaction and air heating. The authors found that the coated spinel significantly improved oxidation resistance of the 430 substrate. The Mn-Co spinel protection layer appeared to be an effective mass barrier to both chromium outward and oxygen inward transport, as indicated by long-term evaluations.[136] It also demonstrated an excellent thermomechanical stability against spallation, mainly due to the excellent thermal expansion matching to the ferritic stainless steel substrate. With electrical conductivities (60 $S{\cdot}cm^{-1}$ at 800°C) that are 3 ~ 4 and 1 ~ 2 orders of magnitude higher than those of chromia and $MnCr_2O_4$, respectively,[137–139] the spinel protection layers drastically improved interfacial contact resistance. Electrochemical evaluation by Simner et al.[140] found that the $Mn_{1.5}Co_{1.5}O_4$ protection layers of ferritic stainless steels were effective in blocking Cr migration, which resulted in long-term stability of the $(La,Sr)MnO_3$ cathode. Overall, it appears that the $(Mn,Co)_3O_4$ spinels are promising coating materials to improve the surface

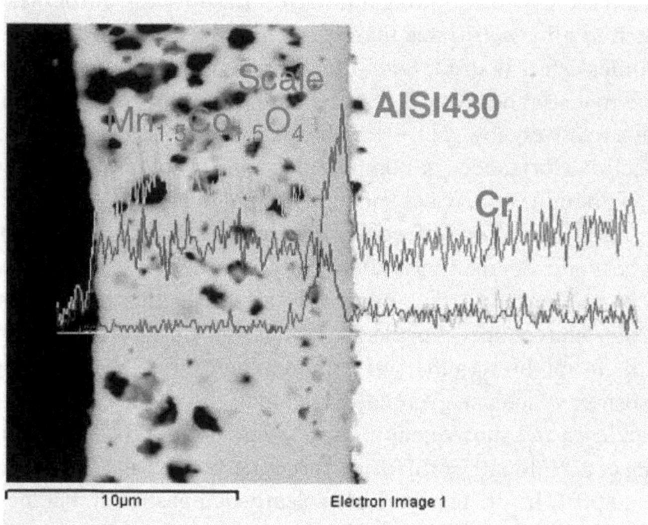

(a)

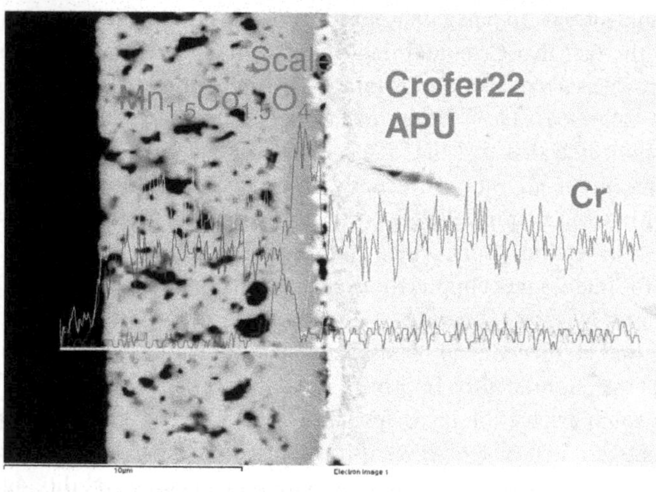

(b)

FIGURE 11.7 SEM cross-sections and elemental distributions of (a) AISI430 and (b) Crofer 22 APU coated with $Mn_{1.5}Co_{1.5}O_4$. The slurry samples were first heat treated in Ar + 2.75 H_2 + 3% H_2O at 800°C for 4 hours and then oxidized in air for 100 h.[134]

stability of ferritic stainless steel interconnects, minimize contact resistance, and seal off chromium in the metal substrates.

11.4 SUMMARY

During SOFC operation, the interconnects that electrically connect neighboring cells and act as a gas separator in stacks must perform in a very challenging environment, as they are simultaneously exposed to air, an oxidizing environment, on the cathode side, and fuel, a reducing environment, on the anode side. Driven by cost reduction and encouraged by the advances in materials technology, high-temperature oxidation-resistant alloys have found increased use as interconnect materials. They nevertheless have to demonstrate satisfactory stability against oxidation and corrosion under the challenging SOFC operating conditions. The candidate alloys are oxidized by reaction with oxygen in air on the cathode side, and with oxygen in the fuel on the anode side. When a hydrocarbon fuel is used or fuel impurity is present, corrosion by carburization, for example, potentially occurs on the anode side of metallic interconnects. Additionally, recent studies found that metallic interconnects potentially suffer anomalous oxidation or corrosion on the air side under air–fuel dual exposures. The anomalous oxidation is likely a result of hydrogen transport from the fuel side to the air side, driven by the difference in the hydrogen partial pressures between these two sides. For Fe-Cr-base stainless steels, the anomalous oxidation is detrimental to the surface and structural stability of the metallic interconnects. In addition to the oxidation and corrosion at the metal–gas interface, metallic interconnects may also be degraded by interactions at the interfaces between interconnects and adjacent components. As an example, the sealing glass can react with the oxidation-resistant alloys, affecting the stability of the metallic interconnects, as well as that of the glass seals.

Substantial progress has been achieved in the past few years in improving the surface stability of bulk metallic interconnects. However, it appears that the use of high-temperature oxidation-resistant alloys, in particular ferritic stainless steels that demonstrate a CTE match to the ceramic cells, remains challenging in terms of satisfactory lifetime surface stability at the temperatures that are allowed by current SOFC technologies. An alternative approach to bulk alloy development, surface modification of metallic interconnects via application of protection layers, has proved to be viable for improvement of their stability and mitigation of adverse interactions with cell and stack components. In particular, protection layers fabricated from non-Cr-containing conductive oxides, e.g., $(Mn,Co)_3O_4$ spinels, are among the most promising candidates. Further progress in terms of materials understanding and optimization will be necessary to achieve satisfactory cost-effective, long-term interconnect performance.

REFERENCES

1. Minh, N. Q., *J. Am. Ceram. Soc.*, 76, 563, 1992.
2. Yamamoto, O., *Electrochimica Acta*, 45, 2423, 2000.
3. Steele, B. C. H. and Heinzel, A., *Nature*, 414, 345, 2001.
4. Singhal, S. C., *Solid State Ionics*, 152-153, 405, 2002.

5. William, M. C., Strakey, J. P., and Surdoval, W. A., *Int. J. App. Ceram. Tech.*, 2, 295, 2005.
6. Paulick, S. W., Baskaran, S., and Armstrong, T. R., *J. Mater. Sci.*, 33, 2397, 1998.
7. Peck, D-H., Miller, M., and Hilpert, K., *Solid State Ionics*, 143, 391, 2001.
8. Chick, L.A., Liu, J., Stevenson, J. W., Armstrong, T. R., McCready, D.E., Maupin, G. D., Coffey, G. W., and Coyle, C. A., *J. Am. Ceram. Soc.*, 80, 2109, 1997.
9. Fergus, J.W. *Solid State Ionics*, 171, 1, 2004.
10. Anderson, H.C., and Tietz, F., in *High Temperature Solid Oxide Fuel Cells: Fundamentals, Design and Applications*, Eds., S.S. Singhal and K. Kendall, 173, Oxford: Elsevier. 2003.
11. de Souza, S., Visco, S.J., and De Jonghe, L.C., *Solid State Ionics*, 98, 57, 1997.
12. de Souza, S., Visco, S.J., and De Jonghe, L.C., *J. Electrochem. Soc.*, 144, L35, 1997.
13. Ishihara, T., Matsuda, H., and Takita, Y., *J. Am. Chem. Soc.*, 116, 3801, 1994.
14. Huang, Q., Tichy, R., and Goodenough, J. B., *J. Am. Ceram. Soc.*, 81, 2565, 1998.
15. Quadakkers, W.J., Piron-Abellan, J., Shemet, V., and Singheiser, L., *Mater. High Temp.*, 20, 115, 2003.
16. Yang, Z., Weil, K.S., Paxton, D.M., and Stevenson, J.W., *J. Electrochem. Soc.*, 150, A1188, 2003.
17. Zhu, W.Z. and Deevi, S.C., *Mater. Sci. & Eng.*, A348, 227, 2003.
18. Fergus, J.W., *Mater. Sci. & Eng.*, A397, 271, 2005.
19. Sims, C.T., Stoloff, N.S., and Hagel, W.C., *Superalloys II*, New York: John Wiley & Sons, 1987.
20. Davis, J.R. *ASM Specialty Handbook: Stainless Steels*, Materials Park, OH: ASM International®, 1994.
21. Wasielewski, G.E. and Robb, R.A., *High Temperature Oxidation in The Superalloys*, Eds. C.S. Sims and W. Hagel, 287, New York: John Wiley & Sons, Inc., 1972.
22. Kofstad, P. *Nonstoichometry, Diffusion and Electrical Conductivity in Binary Metal Oxides*, Malabar, FL: Robert E. Krieger, 1983.
23. Kofstad, P. and Bredesen, R., *Solid State Ionics*, 52, 69, 1992.
24. Holt, A., and Kofstad, P., *Solid State Ionics*, 69, 137, 1994.
25. Holt, A. and Kofstad, P., *Solid State Ionics*, 69, 127, 1994.
26. Abeilan, J.P., Shemet, V., Tietz, F., Singheiser, L., and Quadakkers, W.J., in *SOFC VII-Electrochemical Society Proceedings PV-2001-16*, Eds., S.C. Singhal and M. Dokiya, 811, The Electrochemical Society, Pennington, NJ, 2001.
27. Teller, O., Meulenberg, W.A., Tietz, F., Wessel, E., and Quadakkers, W.J., in *SOFC VII- Electrochemical Society Proceedings PV2001-16*, Eds., S.C. Singhal and M. Dokiya, 895, The Electrochemical Society, Pennington, NJ, 2001.
28. Quadakkers, W.J., Shemet, V., and Singheiser, L. U.S. Patent. 2003059335, 2003.
29. Quadakkers, W.J., Malkow, T., Piron-Abellan, J., Flesch, U., Shemet, V., and Singheiser, L., in *Proceedings of the 4th European SOFC Forum*, Vol. 2, 827, European Fuel Cell Forum, Switzerland, 2000.
30. Buchkremer, H.P., Diekmann, U., de Haart, L.G. J., Kabs, H., Stover, D., and Vinke, I.C., in *Proceedings of the 3rd European SOFC Forum*, Ed., P. Stevens, Vol. 1, 143, European Fuel Cell Forum, Switzerland, 1998.
31. Yang, Z., Xia, G.-G., Walker, M.P., Wang, C.-M., Stevenson, J.W., and Singh, P., *Inter. J. Hydrogen Energy*, 2006, in press.
32. Yang, Z., Hardy, J.S., Walker, M.S., Xia, G., Simner, S.P., and Stevenson, J.W., *J. Electrochem. Soc.*, 151, A1825, 2004.
33. Park, J.H. and Natesan, K. *Oxid. Met.*, 33, 31, 1990.
34. Sasamoto, T., Sumi, N., Shimaji, A., Yamamoto, O., and Abe, Y., *J. Mater. Sci. Soc.*, 33, 32, 1996.

35. Fava, F.F., Barraille, I., Lichanot, A., Larrieu, C., and Dovesi, R., *J. Phys. Condense Mater.*, 9, 10715, 1997.
36. Lu, Z., Zhu, J., Payzant, E.A., and Paranthaman, M.P., *J. Am. Ceram. Soc.*, 88, 1050, 2005.
37. Virkar, A.V. and England, D.M., U.S. Patent 6054231, 2000.
38. Horita, T., Xiong, Y., Yamaji, K., and Sakai, N., *J. Electrochem. Soc.*, 150, A243, 2003.
39. Uehara, T. Ohno, T., and Toji, A., in *Proceedings of the 5th European SOFC Forum*, ed., J. Huijsmans, 281, European Fuel Cell Forum, Switzerland, 2002.
40. Quadakkers, W.J., Greiner, H., Kock, W., Buchkremer, H.P., Hilpert, K., and Stover, D., in *Proceedings of the 2nd European SOFC Forum*, Ed., B. Thorstensen, 297, European Fuel Cell Forum, Switzerland, 1996.
41. Kofstad, P. *High Temperature Corrosion*, 2nd ed., London: Elsevier Applied Science, 1988.
42. Birks, N., Meier, G.H., and Pettit, F.S., *Introduction to the High Temperature Oxidation of Metals*, 2nd ed., London: Cambridge University Press, 2006.
43. Young, D.J. and Watson, S., *Oxid. Met.*, 44, 239, 1995.
44. Gesmundo, F. and Gleeson, B., *Oxid. Met.*, 44, 211 1995.
45. F.H. Stott, G.C. Wood, and J. Stringer, Oxid. Met., 1995, 44, 113.
46. Linderoth, S., Hendriksen, P.V., Mogensen, M., and Langvad, N., *J. Mater. Sci.*, 31, 5077, 1996.
47. England, D.M. and Virkar, A.V., *J. Electrochem. Soc.*, 146, 3196, 1999.
48. Yang, Z., Xia, G.-G., Singh, P., and Stevenson, J.W., *J. Power Sources*, 160, 1104, 2006.
49. Geng, S.J., Zhu, J.H., and Lu, Z.G., *Solid State Ionics*, 177, 559, 2006.
50. Geng, S.J., Zhu, J.H., and Lu Z.G., *Script. Mater*, 55, 239, 2006.
51. Alman, D.E. and Jablonski, P.D. In *Fuel Cell Seminar,* Washington, DC: Courtesy Associate, 2004.
52. Kofstad, P. and Bredesen, R., *Solid State Ionics*, 52, 69, 1992.
53. Huang, K., Hou, P.Y., and Goodenough, J.B., *Solid State Ionics*, 129, 237, 2000.
54. England, D.M. and Virkar, A.V., *J. Electrochem. Soc.*, 148, A330, 2001.
55. Holcomb, G.R. and Alman, D.E., *J. Mater. Eng. & Perform.*, 2006, 15, 394.
56. Meulenberg, W.A., Uhlenbruck, S., Wessel, E., Buchkremer, H.P., and Stover, D., *J. Mater. Sci.,* 2003, 38, 507.
57. Brylewski, T., Nanko, M., Maruyama, T., and Przybylski, K., *Solid State Ionics*, 143, 131, 2001.
58. Geng, S.J., Zhu, J.H. and Lu, Z.G., *Electrochem. & Solid-State Lett.*, 9, A211 2006.
59. Zeng, Z. and Natesan, K., *Solid State Ionics*, 167, 9, 2004.
60. Toh, C.H., Munroe, P.R., Young, D.J., and Foger, K., *Mater. High Temp.*, 20, 129, 2003.
61. Horita, T., Xiong, Y., Kishimoto, H., Yamaji, K., Sakai, N., Brito, M.E., and Yokokawa, H., *J. Electrochem. Soc.*, 152, A2193, 2005.
62. Horita, T., Xiong, Y., Kishimoto, H., Yamaji, K. Sakai, N., and Yokokawa, H., *Surf. Interface Anal.*, 36, 973, 2004.
63. Jian, L., Huezo, J., and Ivey, D.G., *J. Power Sources*, 123, 151, 2003.
64. Yang, Z., Walker, M.S., Singh, P., and Stevenson, J.W., *Electrochem. & Solid State Lett.*, 6, B35, 2003.
65. Yang, Z., Walker, M.S., Singh, P., Stevenson, J.W., and Norby, T., *J. Electrochem. Soc.,* 151, B669, 2004.
66. Sing, P., Yang, Z., Viswanathan, V, and Stevenson, J.W., *J. Mater. Perform. Eng.*, 13, 287, 2004.

67. Yang, Z., Xia, G-G., Singh, P., and Stevenson, J.W., *Solid State Ionics*, 176, 1495, 2005.
68. Ziomek-Moroz, M., Cramer, S.D., Holcomb, G.R., Covino, B.S., Jr, Bullard, S.J., and Singh, P., in *Corrosion*, NACE International, Houston, TX, 2005, paper 10.
69. Holcomb, G.R., Ziomek-Moroz, M., Cramer, S.D., Covino, B.S., Jr., and Bullard, S.J., *J. Mater. Eng. & Perform.*, 15, 404, 2006.
70. Kurokawa, H., Kawamura, K., and Maruyama, T., *Solid State Ionics*, 168, 13, 2004.
71. Quadakkers, W.J., Hansel, M., and Rieck, T., *Mater. & Corro.*, 49, 252, 1999.
72. Larring, Y., Hangsrud, R., and Norby, T., *J. Electrochem. Soc.*, 150, B374, 2003.
73. Bongartz, K., Quadakkers, W.J., Pfeifer, J.P., and Becker, J.S., *Surf. Sci.*, 292, 196, 1993.
74. Rapp, R.A., *Metall. Trans. A*, 15A, 765, 1984.
75. Huczkowski, P., Shemet, V., Piron-Abellan, J., Singheiser, L., Quadakkers, W. J., and Christiansen, N., *Mater. & Corrosion*, 55, 825, 2004.
76. Huczkowski, P., Ertl, S., Piron-Abellan, J., Christiansena, N., Ho, T., Shemet, V., Singheiser, L., and Quadakkers, W.J., *Mater. High Temp.*, 22, 253, 2005.
77. Huczkowski, P., Christiansen, N., Shemet, V., Piron-Abellan, J., Singheiser, L., and Quadakkers, W. J., *J. Fuel Cell Sci. & Tech.*, 1, 30, 2004.
78. Hou, P. and Stringer, J., *Oxid. Met.*, 38, 323, 1992.
79. Golightly, F., Stott, H., and Wood, G., *Oxid. Met.*, 10, 163, 1976.
80. Pint, B., *Oxid. Met.*, 45, 1, 1996.
81. Pieraggi, B. and Rapp, R.A., *J. Electrochem. Soc.*, 140, 2844, 1993.
82. Kofstad, P., *Oxid. Met.*, 44, 3, 1995.
83. Douglass, D.L., Kofstad, P., Rahmel, A., and Wood, G.C., *Oxid. Met.*, 45, 529, 1996.
84. Kvernes, I., Oliveira, M., and Kofstad, P., *Corrosion Sci.*, 17, 237, 1977.
85. Shen, J., Zhou, L., and Li, T., *Oxid. Met.*, 38, 347, 1997.
86. Pint, B.A., *J. Eng. Gas Turbine and Power*, 128, 370, 2006.
87. Fujii, C.T. and Meussner, R.A., *Corro. Iron & Steel*, 111, 1215, 1964.
88. Lefrancois, P.A. and Hoyt, W.B., *Corrosion*, 19, 360, 1963.
89. Grabke, H.J., Muller-Lorenz, E.M., Eltester, B., and Lucas, M., *Mater. High Temp.*, 17, 339, 2000.
90. Toh, C.H., Munroe, P.R., and Young, D.J., *Mater. High Temp.*, 20, 527, 2003.
91. Hochman, R.F., in *Proceedings of the 4th International Congress Metal Corrosion*, Ed., N.E. Hammer, NACE, 258, 1972.
92. Grabke, H.J., Bracho-Troconis, C.B., and Muller-Lorenz, E.M., *Werkstoffe und Korrosion*, 45, 215, 1994.
93. Zeng, Z. and Natesan, K., *Chem. Mater.*, 15, 872, 2003.
94. Schneider, R., Pippel, E., Woltersdorf, J., Strauss, S., and Grabke, H.J., *Steel Research*, 68, 326, 1997.
95. Nakagawa, K., Matsunaga, Y., and Yanagisawa, T., *Mater. High Temp.*, 20, 67, 2003.
96. Meier, G. H., *2005 Proceedings of the U.S. DOE SECA Core Technology Peer Review*, National Energy Technology Laboratory, http://www.netl.doe.gov/publications/ proceedings/05/SECA_PeerReview/SECAPeerReview05.html.
97. Singh, P., Paetsch, L., and Maru, H.C., in *Corrosion 86/87*, NACE, Houston, TX, 2006, paper 86.
98. Nakagawa, K., Matsunaga, Y., and Yanagisawa, T., *Mater. High Temp.*, 18, 51, 2001.
99. Wood, G.C., Wright, I.G., Hodgkiess, T., and Whittle, D.P., *Werkst Korros*, 21, 900, 1970.
100. Yang, Z., Xia, G.-G., Singh, P., and Stevenson, J.W., *J. Electrochem. Soc.*, 153, A1873, 2006.
101. Kurokawa, H., Oyama, Y., Kawamura, K., and Maruyama, T., *J. Electrochem. Soc.*, 151, A1264, 2004.

102. Eichler, K., Solow, G., Otschik, P., and Schafferath, W., *J. Europ. Ceram. Soc.,* 19, 1101, 1999.
103. Meinhardt, K.D., Vienna, J.D., Armstrong, T.R., and Peterson, L.R., U.S. patent 6430966, 2001.
104. Sohn, S.B., Choi, S.Y., Kim, G. H., Song, H.S., and Kim, G.D., *J. Non-Cryst. Solids*, 297, 103, 2002.
105. Schwickert, T., Geasee, P., Janke, A., Diekmann, U., and Conradt, R., in *Proceedings of the International Brazing and Soldering Conference*, 116, Albuquerque, NM, 2000.
106. Sohn, S.-B., Choi, S.-Y., Kim, G.-H., Song, H.-S., and Kim, G.-D., *J. Amer. Ceram. Soc.*, 87, 254, 2004.
107. Yang, Z., Meinhardt, K.D., and Stevenson, J.W., *J. Electrochem. Soc.*, 150, A1095, 2003.
108. Yang, Z., Stevenson, J.W., and Meinhardt, K.D., *Solid State Ionics*, 160, 213, 2003.
109. Yang, Z., Xia, G.-G., Meihardt, K.D., Weil, K.S., and Stevenson, J.W., *J. Mater. Eng. & Perform.,* 13, 327, 2004.
110. Pistorius, C.W.F.T. and Pistorius, M.C., *Z. Krist*, 117, 259, 1962.
111. Haanapel, V.A.C., Shemet, V., Vinke, I.C., and Quadakkers, W.J., *J. Power Sources*, 141, 102, 2005.
112. Haanapel, V.A.C., Shemet, V., Vinke, I.C., Gross, S.M., Koppitz, Th., Menzler, N.H., Zahid, M., and Quadakkers, W.J., *J. Mater. Sci.,* 2005, 40, 1583.
113. Haanapel, V.A.C., Shemet, I.C., Gross, S.M., Koppitz, Th., Menzler, N.H., Zahid, M., and Quadakkers, W.J., *J. Power Sources*, 2005, 150, 86.
114. Batfalsky, P., Haanapel, V.A.C., Malzbender, J., Menzler, N.H., Shemet, V., Vinke, I.C., and Steinbrech, R.W., *J. Power Sources,* 2005, 155, 128.
115. Yang, Z., Xia, G.-G., Singh, P., and Stevensoin, J.W., *J. Power Sources*, 155, 246, 2006.
116. Badwal, S. P. S., Deller, R., Foger, K., Ramprakash, Y., and Zhang, J. P., *Solid State Ionics,* 99, 297, 1997.
117. Quadakkers, W.J., Greiner, H., Hansel, M., Pattanaik, A., Khanna, A.S. and Mallener, W., *Solid State Ionics*, 91, 55, 1996.
118. Maruyama, T., Inoue, T., and Nagata, K., in *SOFC VII-Electrochemical Society Proceedings PV-2001-16*, Ed., S.C. Singhal and M. Dokiya, Editors, 889, The Electrochemical Society, ennington, NJ, 2001.
119. Arul Raj, I., Tietz, F., Gupta, A., Jungen, W., and Stover, D., *Acta Mater.*, 49, 1987, 2001.
120. Tietz, F., Arul Raj, I., Jungen, W., and Stover, D. *Acta Mater.*, 49, 803, 2001.
121. Yang, Z., Coyle, C.A., Baskaran, S., and Chick, L.A., U.S. Patent 6843406, 2005.
122. Linderoth, S., *Surf. & Coating Tech.*, 80, 185, 1996.
123. Sakai, N., Yamaji, K., Horita, T., Lshikawa, M., Yokokawa, H. and Dokiya, M., in *Proceedings of the 3rd European SOFC Forum*, Ed., P. Stevens, 333, European Fuel Cell Forum, Switzerland, 1993.
124. Kadowaki, T., Shiomitsu, T., Matsuda, E., Nakagawa, H., and Tsuneisumi, H., *Solid State Ionics*, 67, 65, 1993.
125. Batawi, E., Honegger, K., Diethelm, D., and Wettstein, M. in *Proceedings of the 2nd European SOFC Forum*, Ed., B. Tharstensen, 307, European Fuel Cell Forum, Switzerland, 1996.
126. Fujita, K., Ogasawara, K., Matsuzaki, Y., and Sakurai, T., *J. Power Sources*, 131, 261, 2004.
127. Hilpert, K. Das, D., Miller, M., Peck, D.P., and Weib, R., *J. Electrochem. Soc.*, 143, 3642, 1996.
128. Jacobson, N., Myers, D., Opila, E., and Copland, E., *J. Phys. Chem. Sol.*, 66, 471, 2005.

129. Yang, Z., Xia, G.-G., Maupin, G.D., and Stevenson, J.W., *J. Electrochem. Soc.*, 153, A1852, 2006.
130. Qu, W., Jian, L., Hill, J.M., and Ivey, D.G., *J. Power Sources*, 153, 114, 2006.
131. Larring, Y. and Norby, T., *J. Electrochem., Soc.*, 147, 3251, 2000.
132. Yang, Z., Xia, G., and Stevenson, J.W., *Electrochem. & Solid State Lett.*, 8 A168, 2005.
133. Yang, Z., Xia, G.-G., Simner, S.P., and Stevenson, J.W., *J. Electrochem. Soc.*, 152, A1896, 2005.
134. Yang, Z., Xia, G.-G., Li, X.-H., and Stevenson, J.W., *Int. J. Hydrogen Energy*, 2006 in press.
135. Chen, X., Hou, P.Y., Jacobson, C.P., Visco, S.J., and De Jonghe, L.C., *Solid State Ionics*, 176, 425, 2005.
136. Yang, Z., Xia, G-G., Maupin, G., Simner, S., Li, X., Stevenson, J., and Singh, P. In Fuel Cell Seminar, paper 253, 2006, Courtesy Associate, Washington.
137. Yokoyama, T., Abe, Y., Meguro, T., Komeya, K., Kondo, K. Kaneko, S., and Sasamoto, T., *Japan J. Appl. Phys.*, 35, 5775, 1996.
138. Yang, Z., Li, X.-H., Maupin, G.D., Singh, P., Simner, S.P., Stevenson, J.W., Xia, G.-G., and Zhao, X.-D., *Ceram. Sci. & Eng. Proc.*, 27, 231, 2006.
139. Ling, H. and Petric, A., in *SOFC IX-Electrochem. Soc. Proc. PV2005-07*, ed., S.C. Singhal and J. Mizusaki, 1866, The Electrochemical Society, Pennington, NJ, 2005.
140. Simner, S.P., Anderson, M.D., Xia, G.-G., Yang, Z., and Stevenson, J.W., *Ceram. Eng. & Sci. Proc.*, 26, 83, 2005.

12 Materials for Proton Exchange Membrane Fuel Cells

Bin Du, Qunhui Guo, Zhigang Qi, Leng Mao, Richard Pollard, and John F. Elter

CONTENTS

12.1 INTRODUCTION

Proton exchange membrane (PEM) fuel cell technology is a promising alternative for a secure and clean energy source in portable, stationary, and automotive applications. However, it has to compete in cost, reliability, and energy efficiency with established energy sources such as batteries and internal combustion engines. Many of the major challenges in PEM fuel cell commercialization are closely related to three critical materials considerations: cost, durability, and performance. The challenge is to find a combination of materials that will give an acceptable result for the three criteria combined. For example, Hamilton Standard (a subsidiary of United Technologies Corporation) demonstrated individual cell lifetimes of over 87,600 run hours on at least three individual test cells operated continuously at 0.54 A/cm^2 using a thick membrane (Nafion® 120, 250 µm thick) and Pt black electrodes (>10 mg Pt/cm^2).[1,2] They also achieved stable voltage (decay rate ~ 1 µV/h) for 40,000 h on a four-cell stack operated continuously at low current density (CD) (~0.13 A/cm^2). These lifetime performances met or exceeded the Department of Energy (DOE) target (40,000 h) for stationary applications.[3] However, the cost of these systems is prohibitively high for commercial applications (DOE targets: $30/kW for transportation applications using neat H_2 and $750/kW for stationary power applications using natural gas reformate). On the other hand, state-of-the-art PEM fuel cells, using thinner membranes (<40 µm) and Pt/C electrodes (<1 mg Pt/cm^2) for cost reduction, are less expensive (but still higher than DOE cost targets) but only have a demonstrated lifetime of less than 15,000 h operating on reformate.[3–5] There are numerous reviews on general PEM fuel cell technology,[5–11] fuel cell components,[12–15] electrode catalysts,[16–24] membrane electrolytes,[25–32] bipolar plates,[33,34] and system reliability and compatibility.[4,35,36] This chapter summarizes the current status of materials-related aspects of PEM fuel cell research and development, including basic functional requirements, state-of-the-art materials, and technical challenges for each individual component. Hydrogen production, distribution, and storage are covered in sections 12.1 to 12.3.

The idea of using an ion-conductive polymeric membrane as a gas–electron barrier in a fuel cell was first conceived by William T. Grubb, Jr. (General Electric Company) in 1955.[37,38] In his classic patent,[37] Grubb described the use of Amberplex C-1, a cation exchange polymer membrane from Rohm and Haas, to build a prototype H_2–air PEM fuel cell (known in those days as a solid-polymer electrolyte fuel cell). Today, the most widely used membrane electrolyte is DuPont's Nafion due to its good chemical and mechanical stability in the challenging PEM fuel cell environment. A perfluorinated polymer with pendant sulfonated side chains, Nafion was initially developed in 1968 by Walther G. Grot of DuPont for the chlor-alkali cell project of the National Aeronautics and Space Administration (NASA) space program.[39] Several manufacturers provide other perfluorinated polymers, composite polymers, and hydrocarbon polymers as membrane electrolytes.[25–32]

Figure 12.1 is a schematic view of a typical PEM fuel cell. A membrane electrode assembly (MEA) usually refers to a five-layer structure that includes an anode gas diffusion layer (GDL), an anode electrode layer, a membrane electrolyte, a cathode electrode layer, and a cathode GDL. Most recently, several MEA manufacturers started to include a set of membrane subgaskets as a part of their MEA packages. This is often referred to as a seven-layer MEA. In addition to acting as a gas and

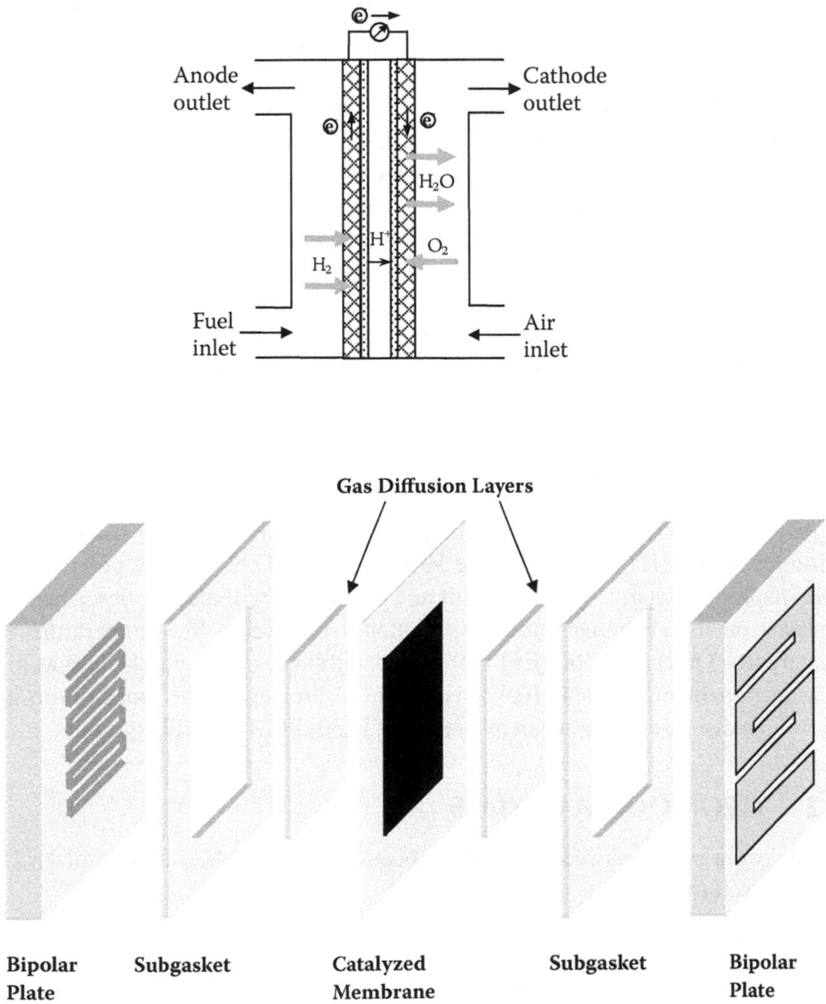

FIGURE 12.1 Schematic views of a PEM fuel cell and a seven-layered MEA.

electron barrier, a membrane electrolyte transports protons (H^+) from the anode, where H_2 is oxidized to produce H^+ ions and electrons, to the cathode, where H^+ ions and electrons recombine with O_2 to produce H_2O. Small organic molecules, such as CH_3OH and HCOOH, can also be used as the anode fuel in place of H_2, but they pose special challenges for various MEA components, especially the catalysts (poisoning) and the membrane (swelling and fuel crossover). For economic reasons, air is usually used as the cathode feed rather than pure O_2. Electrons are carried from the anode to the cathode through the external electric circuit. The anode and cathode electrode layers are typically made of Pt or Pt alloys dispersed on a carbon support for maximum catalyst utilization. Ionomers and polytetrafluoroethylene (PTFE) resins can be added to the electrode layers. The former extends the proton transport path beyond the electrode–membrane interfaces; the latter facilitates liquid water removal from the electrode layers. Both can also help bind together various electrode components. GDLs are made of porous media such as carbon paper or carbon cloth to facilitate the transport of gaseous reactants to the electrode layers, as well as the transport of electrons and water away from the electrode layers. An MEA is sandwiched between two bipolar plates to form a single fuel cell. The word *bipolar* refers to a plate's bipolar nature in a series of single cells (known as a stack) in which a plate (or a set of half plates) is anodic on one side and cathodic on the other side. Bipolar (half) plates often have gas channels on the side facing an MEA and channels for temperature control on the other side and, together with the GDLs, they provide structural support for the MEAs in addition to serving as transport media for reactants/products, electricity, and heat.

In the following sections, a brief overview of the basic electrochemical processes in a H_2/O_2 PEM fuel cell is given, followed by information on individual fuel cell components: anode, cathode, catalyst support, membrane, GDLs, and bipolar plates. The focus is on the specific functionalities and material requirements for each individual component. The subgaskets of a seven-layer MEA will be discussed in conjunction with materials compatibility in a separate section, which also covers the materials selection criteria for coolant, hoses, and other system components. A high-temperature (HT) version of the H_2/O_2 PEM fuel cell using a polybenzimidazole–phosphoric acid (PBI-PA) membrane electrolyte will also be described, with emphasis on its advantages and disadvantages relative to low-temperature (LT) counterparts. Other types of PEM fuel cells using small organic molecules as direct fuels, such as direct methanol fuel cells (DMFCs), are beyond the scope of this book and will be discussed only when relevant to a H_2/O_2 PEM fuel cell system.

12.2 ELECTRODE MATERIALS

The hydrogen oxidation reaction (HOR) occurs at the anode electrode of an H/O_2 PEM fuel cell (reaction 1):

$$H_2 \leftrightarrow 2\,H^+ + 2\,e^- \qquad E^0 = 0\text{ V} \qquad (1)$$

This is a thermodynamically reversible process that often serves as a standard reference electrode, known as the reversible hydrogen electrode (RHE), for all other electrochemical processes.

At the cathode electrode, the thermodynamically irreversible four-electron oxygen reduction reaction (ORR) is the dominant electrochemical process (reaction 2):

$$O_2 + 4\,H^+ + 4\,e^- \leftrightarrow 2\,H_2O \qquad E^0 = 1.229\text{ V} \tag{2}$$

When connected through an external circuit, the net result of these two half-cell reactions is the production of H_2O and electricity from H_2 and O_2. Heat is also generated in the process. In the absence of a proper catalyst, however, neither of these two half reactions takes place at meaningful rates under PEM fuel cell operating conditions (50 to 80°C, 1 to 5 atm). Despite decades of effort in search of cheaper alternatives, platinum is still the catalyst of choice for both the HOR and ORR.

In a real fuel cell, the apparent cell voltage is significantly lower than 1.229 V, the standard potential difference between the two half reactions. The difference between the ideal and apparent cell voltage is known as the overpotential, which includes catalyst activation loss, mass transport loss, and ohmic loss (figure 12.2). Most of the activation losses originate from a sluggish ORR kinetics,[40] as the overpotential for the HOR on a Pt anode is generally negligible except at a very high CD or in the presence of certain catalyst-poisoning species (such as CO). These overpotentials are responsible for the reduced efficiency of an electrochemical cell. For an HT system, some of the energy lost may be recuperated through a heat recovery process for internal or external usage, but the quality of the heat from an LT system may be too low for this to be worthwhile.

Hydrogen peroxide is also formed as the two-electron ORR-byproduct (reaction 3):

$$O_2 + 2\,H^+ + 2\,e^- \leftrightarrow H_2O_2 \qquad E^0 = 0.695\text{ V} \tag{3}$$

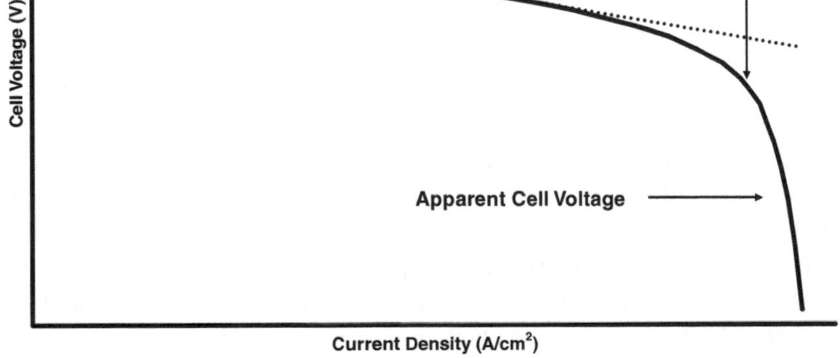

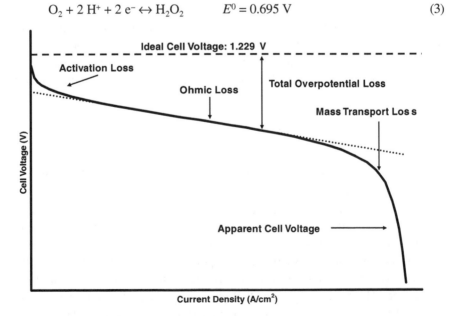

FIGURE 12.2 Schematic view of various overpotential losses: ideal and apparent fuel cell voltage–current characteristics.

The free radicals (·OH, ·OOH, ...) from H_2O_2 decomposition are a primary cause of membrane and ionomer chemical degradation.[25] The H_2O_2-related membrane degradation mechanism will be discussed in more detail in section 12.3.1. The remainder of section 12.2 is divided into four subtopics: anode, cathode, catalyst support, and engineered nanostructured electrodes.

12.2.1 Anode Catalyst Materials

As mentioned above, an anode serves as the HOR site in a H_2/O_2 fuel cell. As such, it must fulfill the following basic functional requirements: (1) transport H_2 to the catalyst sites, (2) catalyze the HOR process, (3) carry protons away from the reaction sites to the membrane electrolyte, (4) remove electrons from the anode, and (5) transfer heat in or out of the reaction zone. Water management is also an important consideration, as it is for all PEM fuel cell components.

The HOR process is believed to proceed through the following steps (Reactions 4 to 6; M = metal catalyst):[41,42]

$$H_2 + 2\,M \rightarrow 2\,MH_{ads} \qquad \text{Tafel reaction} \tag{4}$$

or

$$H_2 + H_2O + M \rightarrow MH_{ads} + H_3O^+ + e^- \qquad \text{Heyrovsky reaction} \tag{5}$$

and

$$MH_{ads} + H_2O \rightarrow H_3O^+ + e^- + M \qquad \text{Volmer reaction} \tag{6}$$

The rate-determining step varies depending on the specific catalysts and the reaction conditions. For a PEM fuel cell with a Pt anode, the HOR process involves only the Tafel and Volmer reactions, with the Tafel reaction being the rate-determining step.[41] The rate of the overall HOR process can be expressed in the Butler–Volmer form (equation 12.1):

$$j = j_0\left[e^{(1-\beta_1)F\eta_s/RT} - e^{\beta_1 F\eta_s/RT}\right] \tag{12.1}$$

The exchange current density (j_0) depends on the nature of the catalyst morphology, the catalyst–electrolyte interface, the properties of the reaction media (pH, electrolyte, temperature, concentration, etc.), and the levels of contaminants such as CO, Cl^-, and sulfur species. For the HOR, the measurement of j_0 is further complicated by the lowered H_2 gas diffusivity in a strong electrolyte solution known as the "salting out" effect.[42] As a result, the reported value ranges from 10^{-5} to 10^{-2} A/cm^2 for different Pt electrodes in various acidic media.[42] However, the rate-limiting process of a H_2/O_2 fuel cell is the ORR on the cathode electrode because j_0 for the ORR is $10^{-6} \sim 10^{-11}$ A/cm^2.[40] Anode materials research has been centered mostly on Pt-loading reduction, CO-tolerant catalysts for DMFCs and systems operating with CO-contaminated fuels (such as reformate), and low-cost Pt alternatives.

12.2.1.1 Pt-Loading Reduction

Over the last 15 years or so, reduction of Pt loading has been accomplished primarily through the transition from Pt black catalysts (~10 mg Pt/cm^2 used in the NASA space program) to high surface area carbon-supported Pt catalysts (nominally 0.8 mg Pt/cm^2 for representative commercial MEAs).[6,7,43,44] It has been estimated that at j_0 = 27 mA/cm^2, an anode with 0.05 mg Pt/cm^2 loading would be sufficient to support the HOR up to 1 A/cm^2 with less than 10 mV overpotential loss (figure 12.3).[44] Johnson Matthey reported the half-cell test results of an anode catalyst with as low as 0.025 mg Pt/cm^2 (20% Pt on Vulcan® XC72R) and observed less than a 5-mV increase in anode overpotential.[23] An MEA with a $PtRu_{20}$/C anode catalyst provided by Adzic's group at Brookhaven National Laboratory (anode, 0.022 mg Pt/cm^2; cathode, commercial Pt/C catalyst-coated GDL with 0.4 mg Pt/cm^2) demonstrated a 10- to 15-mV improvement over a commercial anode (1:1 alloy, 0.6 mg Pt/Ru/cm^2) when 100% H_2 was used as the fuel.[45–47] The $PtRu_{20}$/C anode catalyst was prepared by depositing a monolayer of Pt over approximately 1/8 of the surface of carbon-supported Ru nanoparticles through the spontaneous deposition of Pt on Ru.[48–50] The metallic Ru surface undergoes facile oxidation without dissolution, which ensures selective Pt deposition on Ru but not at the carbon support. Long-term (>1,000 h) performance evaluation of this anode catalyst, by both Plug Power and Los Alamos National Laboratory (~0.017 mg Pt/cm^2), demonstrated excellent cell voltage stability using neat H_2 as the fuel.[45–47] However, this catalyst appeared to be highly susceptible to Ru oxidation and quickly lost its CO tolerance when operating with a CO-containing reformate.[45] Zeis et al. prepared Pt-plated nanoporous gold leafs as low Pt loading (10 to 100 μg Pt/cm^2) and carbon-free electrodes.[51] The long-term stability of low Pt

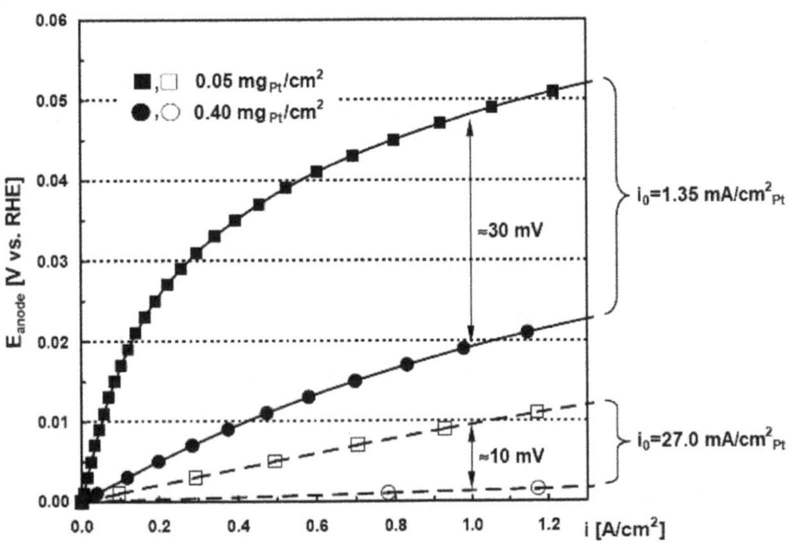

FIGURE 12.3 Calculated anode overpotential as a function of current density and Pt loading. (From Gasteiger, H. A. et al., *J. Power Sources*, 127, 162, 2004. With permission.)

loading electrodes is still not settled, and more experiments are needed using either neat H_2 or reformate.[23,44–47]

12.2.1.2 Non-Pt Anode Catalysts

As mentioned earlier, the rate-determining step in the HOR process for a H_2/O_2 PEM fuel cell is the Tafel reaction.[41] It involves the dissociative chemisorption of H_2 on a catalyst surface to form MH adatom species. Figure 12.4 is a volcano diagram depicting the hydrogen evolution reaction (HER, a thermodynamically reversible process of the HOR) exchange current density over different metal surfaces as a function of the calculated hydrogen chemisorption energy.[52] There is a clear correlation between hydrogen chemisorption energy and exchange current density. Pt is a better HOR catalyst than other metals because the Tafel reaction is energetically neutral on Pt at the equilibrium potential.[52] A few other metals, such as Pd and Re, also possess HOR exchange current densities that are comparable to that of a Pt electrode. However, chemical instability limits the application of these elements as viable anode catalysts under PEM fuel cell operating conditions. Furthermore, the cost of Pd or Re, although considerably lower than that of Pt, would still be high for a commercial fuel cell.

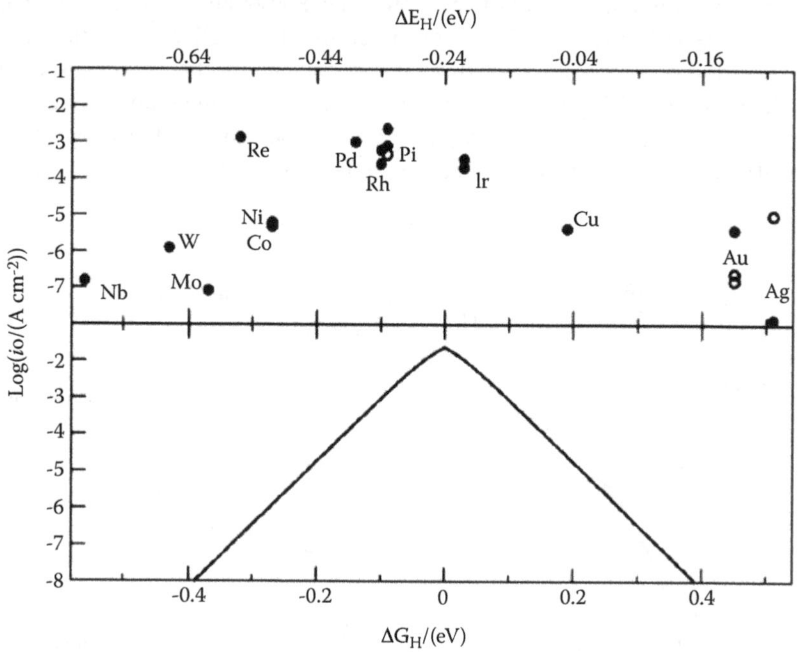

FIGURE 12.4 Top: Experimentally measured exchange current, $\log(i_0)$, for the HER over different metal surfaces plotted as a function of the calculated hydrogen chemisorption energy per atom, ΔE_H (top axis). Single crystal data are indicated by open symbols. Bottom: The result of the simple kinetic model plotted as a function of the free energy for hydrogen adsorption, $\Delta G_{H*} = \Delta E_H + 0.24$ eV. (From Nørskov, J. K. et al., *J. Electrochem. Soc.*, 152, J23, 2005. With permission.)

Nonprecious metal alloys, carbides, or oxides may hold the solution to the cost and chemical instability problems. WC_x and WO_x are the most studied systems as low-cost alternatives to a Pt anode because of their excellent stability in acidic media. They also have high CO tolerance because CO does not readily adsorb onto their surfaces.[53–56] WC_x is particularly attractive because its electron density states near the Fermi level are similar to those of Pt.[57,58] Yang and Wang obtained a high CD (0.9 A/cm^2) from a H_2–air PEM fuel cell with a WC anode (0.48 mg WC/cm^2).[59] The CD was limited by the WC anode in contrast to a typical PEM fuel cell with a Pt anode, which is cathode (ORR) limited and can reach over 2 A/cm^2 under the same test conditions. Nevertheless, this represents a CD increase of two orders of magnitude over the previously reported WC anode catalysts.[60,61]

Limoges et al. looked at the HOR catalytic activities of a series of heteropolyacids (HPAs) containing Mo and V.[62] The CD is too low (a few mA/cm^2) for them to be used as stand-alone anode catalysts, although it should be pointed out that the HPA loading of the anode used in this study was one to two orders of magnitude lower on a molar basis than that of a typical Pt anode. However, HPAs have been shown to be promising proton-conductive membrane/ionomer fillers and effective catalysts for H_2O_2 decomposition.[63,64] On this basis, they may eventually become a part of fuel cell electrodes.

12.2.1.3 Carbon Monoxide–Tolerant Anode Catalysts

The current lack of a national hydrogen infrastructure dictates that on-site hydrogen generation will be the choice of many H_2–air PEM fuel cell applications in the foreseeable future. In many respects, water hydrolysis using electricity generated through renewable solar or wind energy would be ideal for on-site H_2 generation. However, the most technically and economically viable on-site H_2 generation technology today is still reforming of natural gas or other readily available hydrocarbon fuels. The ubiquitous CO in a reformate fuel poses a significant challenge to anode materials because even a few ppm of CO can induce a considerable cell voltage loss (figure 12.5).[65] This is because the Pt-CO adlayer formation is far more exothermic than the energetically neutral Pt-H adatom formation.[50,66,67] The Pt-CO adlayer coverage can reach over 98% even when just a few ppm of CO is present in a reformate.[23] In this situation, the HOR can occur at only a few bare Pt sites in a compact CO monolayer.[68] The result is an elevated anode overpotential even at a CD as low as 0.1 A/cm^2 (~0.8 V for an E-TEK 20% Pt/Vulcan® anode) and, in turn, a lower cell voltage.[69]

Development of CO-tolerant anode materials is also driven by DMFC applications in which CO is one of the methanol oxidation products.[6,7] CO poisoning at the DMFC anode leads to low power density, a critical parameter for portable applications. Pt/Ru alloys are the state-of-the-art materials for CO-tolerant anodes. The optimal Pt/Ru molar ratio is generally found to be 1:1, but it varies depending on the exact nature of a Pt/Ru alloy and its fabrication process.[23,70] Pt alloys of other metals (such as W, Sn, and Mo) and non-Pt alloys have also been examined as CO-tolerant anode materials.[14,23,71]

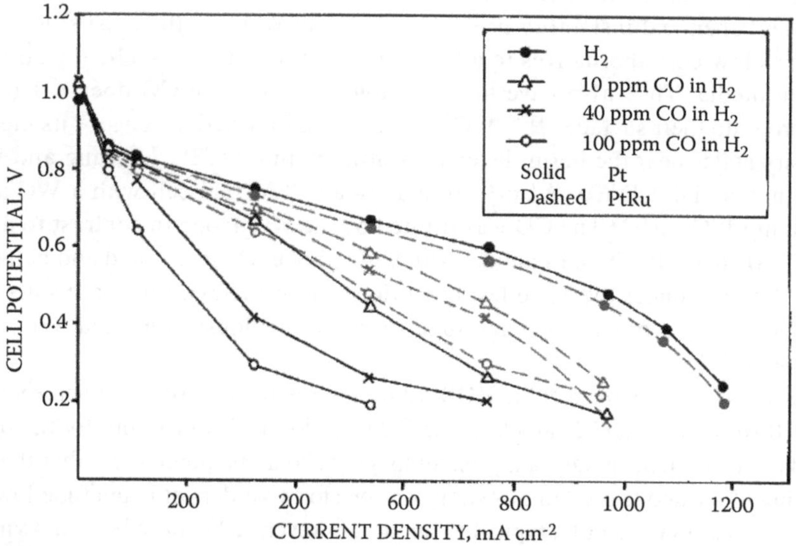

FIGURE 12.5 Progressive poisoning from 10, 40, and 100 ppm CO on pure Pt and $Pt_{0.5}Ru_{0.5}$ alloy anodes. Increased CO tolerance is shown by the $Pt_{0.5}Ru_{0.5}$ alloy anodes. The MEAs are based on catalyzed substrates bonded to Nafion 115. The single cell is operated at 80°C, 308/308 kPa, 1.3/2 stoichiometry with full internal membrane humidification. (From Ralph, T. R. and Hogarth, M. P., *Platinum Metal Rev.*, 46, 117, 2002. With permission.)

The generally accepted bifunctional mechanism for Pt/Ru-catalyzed CO oxidation involves the formation of either a Ru-activated H_2O molecule or a Ru(OH) surface complex adjacent to Pt-CO sites (reactions 7 to 10):[72–74]

$$Ru + H_2O \rightarrow Ru\text{-}H_2O \quad (7)$$

$$Ru\text{-}H_2O + Pt\text{-}CO \rightarrow CO_2 + 2\ H^+ + 2\ e^- \quad (8)$$

or

$$Ru + H_2O \rightarrow Ru(OH) + H^+ + e^- \quad (9)$$

$$Ru(OH) + Pt\text{-}CO \rightarrow CO_2 + H^+ + e^- \quad (10)$$

There appears to be a close link between the actual CO oxidation mechanism and the nature of Pt-CO bonds: the route involving Ru-H_2O is associated with a CO molecule bridge bonded to two adjacent Pt sites,[72] whereas the route involving Ru(OH) is linked to a CO molecule linearly bonded to a single Pt site.[73]

For reformate with a high CO concentration (>10 ppm), a PEM fuel cell with a Pt/Ru alloy anode still suffers from a substantial cell voltage loss, especially in the high-CD region, because the maximum CO oxidation current occurs from 0.39 to 0.6 V.[75] At its onset potential (<0.1 V), the CO oxidation current density of a Pt/Ru anode is capable of oxidizing only a few ppm CO. The ignition potential, defined as the potential at which the CD increases by approximately two orders of magnitude within a narrow potential range, was found to be as high as 0.45 V in some cases.[76] A Pt/Ru anode may experience an overpotential from 0.39 V for a state-of-the-art Pt/Ru catalyst to as high as 0.6 V for an average Pt/Ru catalyst when the Pt surface is saturated with CO.[68–76] Gottesfeld and Pafford discovered that CO could be chemically oxidized to CO_2 when a small amount of air was bled into the anode of a PEM fuel cell.[77,78] This CO oxidation process is electrochemically promoted and is catalyzed by Pt.[79,80] Unlike the electrochemical CO oxidation catalyzed by Pt/Ru, the air bleed (AB) process prefers a hydrophobic environment for facile O_2 diffusion. Excess O_2, usually at $O_2/CO > 100$ (or 200 in stoichiometry), is required for effective CO removal.[75] The excess O_2 reacts with H_2. In addition to the consumption of valuable H_2, it also leads to the formation of a significant amount of H_2O_2 at the anode.[81–84] Free radicals (HO·, HOO·) generated by decomposition of H_2O_2 have been identified as the primary cause for ionomer/membrane chemical degradation.[29,85,86] We have developed a pulsed air bleed (PAB) technology to minimize H_2O_2 formation, and hence to reduce the rate of membrane/ionomer degradation.[75] For a reformate with 10 ppm CO, PAB reduced the amount of air needed by more than 80% relative to a continuous AB under otherwise the same operating conditions. This led to a reduction in fluoride release rate (FRR) of >70%, and it improved cell performance in single-cell and short-stack endurance tests.[75,87] PAB is also a simple yet effective CO mitigation strategy for systems with variable CO concentrations or transient high CO concentrations because it automatically adjusts its pulsing frequency in response to changes in CO concentration.

An anode configuration closely related to the AB approach is the so-called reconfigured anode, in which a thin layer of metal (such as Pt/C) or metal oxide (such as FeO_x) is added to the outside of the anode GDL facing the flow field.[79,88] Unlike a normal anode electrode layer that is impregnated with ionomers for facile proton transport, this ionomer-free CO oxidation layer is hydrophobic for improved gas diffusion to help maximize the interaction between CO and O_2.

With almost endless possible combinations, finding the right alloy catalysts for the HOR and CO-tolerant anode, and for that matter good ORR catalysts, requires rational design strategies with a set of sound guidelines. Strasser et al. employed a density functional theory (DFT) calculation to map out detailed adsorption energies and activation barriers for a variety of model ternary PtRuM alloys as potential CO-tolerant catalysts.[89] They found a similar trend for electrocatalytic activity as a function of the alloy composition as observed experimentally. Greeley and Mavrikakis suggested the use of a plot of CO binding energy vs. surface segregation energy to assist the selection of near-surface alloys (NSAs) as candidates for further screening.[90] NSAs are alloys where a solute metal is present near the surface of a host metal in concentrations different from the bulk. A minute amount of solute metal in the surface region can drastically change the catalytic properties of the corresponding pure metals.[90–92] At the NSA dilution limit, it is expected that defect sites on or near the surface can catalyze the HOR by providing a local environment that resists poisoning.[90] High-throughput sample preparation and fast screening technology are essential for a successful implementation of such a vast undertaking.[89,93] One attractive fast screening method was demonstrated by Stevens et al., who devised a 64-electrode PEM fuel cell to study the effect of composition for a series of $(Pt_{1-x}Ru_x)_{1-y}Mo_y$ alloys in a single experimental run under realistic fuel cell operating conditions.[94]

The equilibrium CO coverage on Pt decreases as the temperature increases because CO adsorption on Pt is an exothermic process. It has been demonstrated in phosphoric acid fuel cells that at temperatures above 180°C, one can operate with a reformate containing 1% CO or higher.[6,7] The ability to operate with high CO concentrations can greatly simplify the reforming subsystem for reduced cost and improved system reliability. Plug Power has been working on a PBI-based PEM fuel cell system with an operating temperature range from 160 to 180°C.[95] Compared to its LT counterpart, this system eliminates the need for LT shift and preferential oxidation (PROX) reactors in the fuel processing system. It also does not require water management components for its fuel cell system. Furthermore, it enables a combined heat and power (CHP) system design that provides high-quality heating and improved system efficiency. A Pt/C anode is used in place of a Pt/Ru alloy because of the enhanced CO tolerance and the good chemical stability of Pt at the elevated temperature.

12.2.2 Cathode Catalyst Materials

A cathode serves as the site for the ORR in a H_2/O_2 fuel cell. It should fulfill the following basic functional requirements: (1) transport O_2 to the catalyst sites, (2) carry protons from the membrane electrolyte to the catalyst sites, (3) move electrons to the reaction sites, (4) catalyze the ORR, (5) remove product water, and (6) transfer heat to or from the reaction zone.

The exact ORR mechanism is still a topic of much debate.[40] Two representative mechanisms, namely, a dissociative mechanism (reactions 11, 15, and 16) and an associative mechanism (reactions 12 to 16), are illustrated here:

$$1/2\ O_2 + M \rightarrow M\text{-}O_{ads} \tag{11}$$

or

$$O_2 + M \rightarrow M\text{-}O_2 \tag{12}$$

$$M\text{-}O_2 + H^+ + e^- \rightarrow M\text{-}O_2H \tag{13}$$

$$M\text{-}O_2H + H^+ + e^- \rightarrow H_2O + M\text{-}O_{ads} \tag{14}$$

The final two steps are the same for both mechanisms:

$$M\text{-}O_{ads} + H^+ + e^- \rightarrow M\text{-}OH \tag{15}$$

$$M\text{-}OH + H^+ + e^- \rightarrow H_2O + M \tag{16}$$

Recent studies pointed to the formation of a peroxy intermediate on the Pt surface, suggesting that the more complex associative mechanism is at work on a Pt electrode.[96–98] However, the DFT calculations by Nørskov et al. suggested that the associative mechanism was only the dominant pathway at ORR overpotentials greater than 0.8 V.[99] At realistic ORR overpotentials (<0.8 V), the two pathways run parallel to each other. Regardless of the actual ORR mechanism, one thing is clear: the ORR is the rate-limiting process in a H_2/O_2 fuel cell. The slow ORR kinetics is responsible for the steep slope in the activation polarization region (figure 12.2). A small but noticeable H_2 crossover current at the open circuit (OC) is largely to blame for the open-circuit voltage (OCV) loss even for a state-of-the-art membrane electrolyte. Much of the cathode research has been directed at finding a cathode catalyst to improve the slow ORR kinetics and to find a cheap replacement for Pt.

12.2.2.1 Pt and Pt Alloy Cathode Catalysts

Pt and Pt alloys are the most active catalysts for the ORR.[100] The DFT calculations by Nørskov et al. indicated that the ORR activity is a function of both the O and OH binding energy.[99] They generated two ORR volcano plots (figure 12.6) for the ORR activities of various metals: one based on the O binding energy (dissociative mechanism) and the other on both the O and OH binding energies (associative mechanism). These plots explain why Pt is the best elemental catalyst material and why certain Pt alloys display a better ORR activity than elemental Pt, i.e., metals such as Ni, Co, Fe, and Cr have smaller O binding energies than Pt.[99] Markovic and Ross observed that Pt alloys of Ni, Co, or Fe with a Pt monolayer on their surfaces displayed higher ORR activities than the corresponding Pt alloys (skin effect).[96] Calculations by Kitchin et al. indicated that the oxygen dissociative adsorption energy on the surface Pt was weakened as its *d*-band was broadened and lowered in energy by interactions with the underlayer 3*d* metals.[101] This explains the enhanced ORR activities of these "skin" alloys. Studies like these have provided guidelines in the search for an effective ORR catalyst. Adzic et al. synthesized a series of PtM (M = Ni, Co, Cr, Pd, Au, Ru, Ir, Rh) alloy cathode catalysts.[46,48–50,102] These alloys exhibited better ORR activity than pure Pt, with the highest half-wave potential increase (45 mV) obtained on a Pt/PtCo skin alloy.[102] Stamenkovic et al. demonstrated that the $Pt_3Ni(111)$ surface is 10 times more active for the ORR than the corresponding Pt(111) surface, and 90 times more active than the current state-of-the-art Pt/C catalysts for PEM fuel cells.[103] Xu et al. studied the skin effect of Pt-Co and Pt-Fe alloys.[104] Teliska et al. found that the OH chemisorption decreased in the direction of Pt > Pt-Ni > Pt-Co > Pt-Fe > Pt-Cr, which correlated directly with the observed fuel cell performance.[105] Balbuena et al. developed a thermodynamic design guideline for bimetallic Pt alloy ORR catalysts.[106,107] Tamizhmani and Capuano showed that $Pt\text{-}Cr\text{-}Cu_{1-x}(CuO)_{x>0.3}$ was six times more active than pure Pt, and that Pt-Cr and Pt-Cr-Cu alloys were about twice as active as Pt.[108] Mukerjee et al. studied the effect of alloy preparation conditions on electronic and structural properties, and ORR electrocatalytic activities.[109] Pt alloys can be made by either depositing a base metal onto pre-made Pt particles or depositing Pt and base metals simultaneously. A sintering step at about 600°C or higher temperatures is often needed for the formation of a true alloy, which may inadvertently cause the sintering and coalescence of alloy particles. Pt alloys of other precious metals have also been shown to display higher ORR activities than Pt alone. For example, Ioroi and Yasuda showed that Pt alloys with 5 to 20 wt% Ir enhanced the ORR activity by a factor of more than 1.5 at 0.8 V.[110]

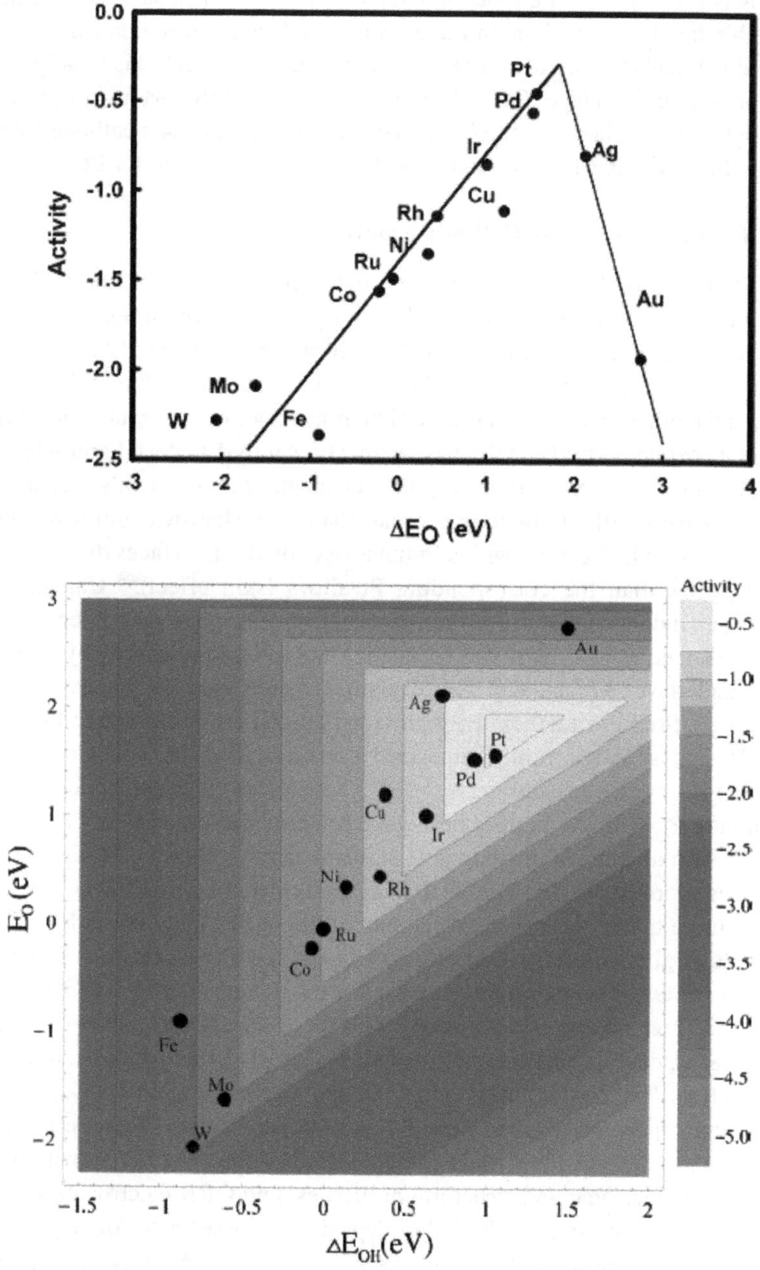

FIGURE 12.6 Trends in ORR activity as a function of (a) the O binding energy and (b) both the O and OH binding energy. (From Norskov, J. K. et al., *J. Phys. Chem. B*, 108, 17886, 2004. With permission.)

The stability and durability of Pt alloys, especially those involving a 3*d* transition metal, are the major hurdles preventing them from commercial fuel cell applications.[111,112] The transition metals in these alloys are not thermodynamically stable and may leach out in the acidic PEM fuel cell environment. Transition metal atoms at the surface of the alloy particles leach out faster than those under the surface of Pt atom layers.[113] The metal cations of the leaching products can replace the protons of ionomers in the membrane and lead to reduced ionic conductivity, which in turn increases the resistance loss and activation overpotential loss.[16] Gasteiger et al. showed that preleached Pt alloys displayed improved chemical stability and reduced ORR overpotential loss (in the mass transport region), but their long-term stability has not been demonstrated.[16,114] These alloys experienced rapid activity loss after a few hundred hours of fuel cell tests, which was attributed to changes in their surface composition and structure.[114]

Bouwman et al. demonstrated that Pt can be used in the ionic form (Pt^{2+} and Pt^{4+}) by dispersing it in a matrix of hydrous iron phosphate (FePO) via a sol-gel process (Pt-FePO).[115] The hydrous FePO possesses micropores of approximately 2 nm. It has ~3 H_2O molecules per Fe atom and is thought to also serve as a proton transport medium. The Pt-FePO catalyst exhibited a higher ORR activity than Pt/C catalysts. This catalyst was also found to be less sensitive to CO poisoning because CO did not adsorb onto the catalyst surface. The ORR catalytic activity was attributed to the adsorption and storage of oxygen on the FePO, presumably as Fe–hydroperoxides. However, these catalysts have poor electrical conductivity. There is no published data on the long-term stability of these catalysts in fuel cell environments.

12.2.2.2 Non-Pt Cathode Catalysts

There are many active programs pursuing nonprecious metal ORR catalysts, but none of them have demonstrated acceptable ORR activity for practical usage.[17] The most promising candidates are a class of Fe– and Co-N_x–C complexes on a carbon support that show reasonably good ORR activity.[17,116,117] Fe/Co-based PEM fuel cell catalysts are often made by pyrolyzing metal porphyrins and other macrocycles. The chelating reagents often contain four N atoms coordinated to the metal center (N_4-M). In 1964, Jasinski discovered that some N_4–Co macrocycles were capable of catalyzing the ORR.[118] Since then, a host of N_4-M (M = Fe, Co) macrocycles have been prepared and studied for PEM fuel cell applications.[119–128] The common chelating reagents are tetraazaannulene, phthalocyanine, and tetraphenylporphyrin. The last two are porphyrin derivatives closely related to the heme unit found in biological systems such as heme oxygenase, a common enzyme catalyzing the ORR process in living organisms.[129] It was proposed that the catalytic site is the N_4-M macrocycle bound to the carbon support through a heat treatment.[130,131] N_4-M was shown to improve not only the activity of these catalysts for the ORR, but also their chemical stability, even in an acidic medium.[116] A second active site was recently identified as Fe-N_2/C based on its typical $FeN_2C_4^+$ ion signature in time-of-flight secondary ion mass spectometry, although its full coordination is not yet known.[131] There is evidence that this Fe-N_2/C site is catalytically more active than Fe-N_4/C for oxygen reduction, converting >95% oxygen to water.[131] Most recently, Bashyam and Zelenay found that Co–polypyrrole–carbon exhibited a good ORR activity and a remarkable stability with a Co loading as low as 0.06 mg/cm^2 in PEM fuel cells.[132]

N_4-M catalysts can also be made from nonmacrocyclic precursors.[133–138] In the presence of a nitrogen source, they can be prepared by pyrolyzing transition metal salts or complexes adsorbed on a carbon support. Examples of salts/complexes include acetate,[127,139] $Fe(OH)_2$ (derived from $FeSO_4$),[134] and phenanthroline complexes.[137] The nitrogen sources may come from an external supply such as NH_3,[139] or from nitrogen surface groups of a N-enriched carbon support.[127,139]

There are, however, many hurdles for transition metal-based catalysts to be used in PEM fuel cells. The two critical ones are the low catalytic activity relative to commercial Pt catalysts and the significant peroxide generation as a side reaction. The low catalytic activity is attributed to a low surface nitrogen content of the carbon support, which is required to anchor metal atoms to the carbon.[127] Even with a N-enriched carbon support, the best catalyst was reported to have a catalyst activity of ~0.1 A/cm^2 at 0.6 V,[127] compared to >0.6 A/cm^2 for commercial Pt catalysts. The significant peroxide generation presents two major problems for PEM fuel cells: (1) Nafion membrane degradation as a result of peroxide free radical attack[29] and (2) Fe/Co dissolution in acidic conditions catalyzed by peroxide.[131] Transition metal dissolution further accelerates the membrane degradation because Fe/Co ions serve as free radical initiators. Fe/Co ions can also replace protons within the Nafion, leading to a lower proton conductivity.

Fernández et al. proposed guidelines based on simple thermodynamic principles for the improved design of noble metal–base metal alloy electrocatalysts for the ORR in acidic media.[140] They assumed a simple mechanism where one metal breaks the O–O bond of molecular O_2 and the other metal acts to reduce the resulting adsorbed atomic oxygen. Analysis of the Gibbs free energies of these two reactions helped select combinations of metals that can produce alloy surfaces with enhanced activity for the ORR relative to the individual constituents. On this basis, they prepared M-Co (M = Pd, Ag, and Au), Pd-Ti, and Pd-Co-M (M = Mo, Au) alloys of various compositions, each as a binary or ternary array on a glassy carbon substrate.[140–142] These arrays were subject to rapid screening using scanning electrochemical microscopy technology. Co was shown to reduce the ORR overpotential of Pd, Ag, and Au, with the Pd-Co alloy displaying an ORR activity similar to that of Pt.[140] Pd-Ti and Pd-Co-M (M = Mo, Au) alloys showed even better ORR activity than Pt, with Pd-Co-Mo also displaying a remarkable stability in acidic media.[141,142] Long-term fuel cell testing is required to assess these catalysts further.

Tantalum oxynitride ($TaO_{0.92}N_{1.05}$) showed some ORR catalytic activity, but it was much lower than that of Pt.[143] ZrO_x showed high stability in an acidic electrolyte and was also found to possess some ORR catalytic activity.[144]

12.2.2.3 Stability of Pt Cathode Catalysts

Cathode lifetime durability presents a special challenge in PEM fuel cells. The major problems associated with the catalysts are Pt agglomeration, sintering, dissolution, and redistribution. The cathode environment is highly oxidative and corrosive due to high voltage (e.g., 0.6 to 1.0 V), low pH, elevated temperature, and the presence of water and oxygen. Paik et al. showed that the Pt surface oxidation increased with cathode potential, O_2 concentration, and exposure time.[145] Although Pt has low solubility at normal cell operating voltages, its solubility increases significantly with cell voltage and reaches the highest dissolution rate at around 1.1 V.[146] Furthermore, freshly formed PtO_x and $Pt(OH)_x$ are less stable in acidic media than materials that have been aged. This makes a PEM fuel cell very susceptible to Pt dissolution when its voltage cycles are between 0.75 and 1.2 V.[147–149] Some dissolved Pt ions (e.g., Pt^{2+}) will migrate into the ionomers or membrane, where they are reduced to Pt by H_2 that has diffused from the anode side.[147] These Pt particles in membrane/ionomers are unlikely to participate in the ORR process because of the lack of electrical continuity.[148] Some dissolved Pt can redeposit onto other Pt particles, which results in the growth of the Pt particles.[147–149] Yasuda et al. found that when a catalyst layer was made of Pt black, the dissolved Pt preferentially deposited onto other Pt particles, but when Pt/C was used instead, the dissolved Pt preferentially migrated into the membrane,[150] and a Pt band was observed within a membrane.[147] It is believed that the exact band location is determined by the H_2 crossover rate. There is no indication of cell shorting due to the formation of such a Pt band because the band is quite narrow in width.

Xie et al. observed that both anode and cathode catalysts migrate toward the membrane.[151] They found that Pt particles migrated more deeply into the membrane than Pt_3Cr particles, indicating that the alloy particles were more stable in terms of both bonding to the carbon support and resistance to oxidation. They also found that Pt agglomeration occurred primarily during the first 500 h of operation, and speculated that the fuel cell activity decay afterward was mainly due to the degradation of the ionomers within the catalyst layer. Ferreira et al. showed that Ostwald ripening of Pt particles and migration of soluble Pt species (then redeposited within the ionomer) each accounted for about 50% of the overall Pt electrochemical active surface area loss.[147]

We demonstrated that a multilayered cathode with a thin Pt black layer near the membrane and a Pt/C layer near the GDL was more stable than a single Pt/C layer.[152] Zhang et al. found that adding gold clusters to Pt/C catalysts prevented Pt dissolution under the oxidizing conditions of the ORR, and with potential cycling between 0.6 and 1.1 V for over 30,000 cycles.[153] They observed only insignificant changes in the activity and the surface area of Au-modified Pt over the course of cycling, compared to rapid losses with the pure Pt catalyst under the same conditions. The increased Pt stability was attributed to the raised Pt oxidation potential by the gold clusters. This finding is sure to draw renewed interest in Pt-Au catalysts, such as the ultra low Pt-loading catalysts that use nanoporous gold foils (<100 nm in thickness) as the catalyst support.[51] This class of catalysts is also attractive because no carbon is present.

12.2.3 Electrode Support Materials

The primary functions of a good catalyst support are to (1) maximize catalyst utilization, (2) transport electrons, and (3) transfer heat. Other desirable attributes include high chemical and electrochemical stability, good mechanical integrity, and, last but not least, low cost. Carbon materials, such as Vulcan-X72 by Cabot Corp., are widely used as PEM fuel cell catalyst supports because of their high surface area, low cost, excellent electric/thermal conductivity, and adequate stability and mechanical properties. The use of carbon-supported Pt in place of Pt black is directly responsible for the reduction of Pt loading in PEM fuel cells.[44,154] However, this also substantially increases the thickness of the electrode layers. Since both the HOR and ORR processes involve gaseous reactants (H_2/O_2), protons, and electrons, the active catalyst sites must have access to these species at the same time. Such reaction sites are often referred to as catalyst–electrolyte–reactant triple-phase boundaries (TPBs).[154] In order to increase catalyst utilization and reduce the activation overpotential loss, it is critical to maximize the TPB regions within an electrode. Proton-conducting materials (such as Nafion ionomers) are usually added to the electrode layer to improve the proton transport beyond the electrode–membrane interface, thus increasing the Pt utilization.[155,156] The ionomer and PTFE resin also serves as a binder to keep the Pt/C particles together and to create hydrophobic/hydrophilic domains for facile gas and water transport.

12.2.3.1 Stability of Carbon Support

In the past few years, the issue of carbon corrosion under various PEM fuel cell operating conditions has come under intensive scrutiny.[157–176] Kangasniemi et al. showed that carbon underwent surface electrochemical oxidation under typical PEM fuel cell operating conditions.[157] Pt catalyzes both the chemical and electrochemical carbon oxidation processes.[158–160] Surface carbon corrosion weakens the Pt–carbon interaction and promotes Pt agglomeration. Rapid carbon corrosion has been linked to adverse fuel cell operating conditions such as repeated start–stop cycling,[161–170] voltage cycling,[171,172] fuel starvation,[23,173–175] and cell flooding.[176]

Despite its low equilibrium potential,[177] the rate of the carbon oxidation reaction (COR) (reaction 17) is negligible at potentials less than 1.8 V because of its very small exchange current density ($j_o = 6 \times 10^{-19}$ A/cm^2).[161–163]

$$C + 2\,H_2O \rightarrow CO_2 + 4\,H^+ + 4\,e^- \quad E^0 = 0.207\text{ V} \tag{17}$$

Substantial carbon corrosion occurs in a PEM fuel cell when a reverse current is imposed on one of its electrodes.[23,161,163] This can happen at the anode during fuel starvation, in which a reverse current is imposed by either an electric load or normal fuel cells adjacent to a starving cell.[23] It can also happen on the cathode during a fuel cell start-up or shutdown, in which a fuel–air front is formed on the

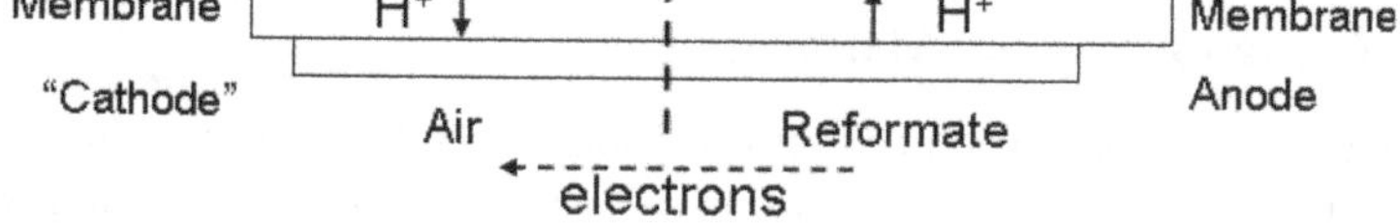

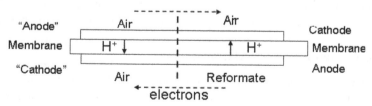

FIGURE 12.7 A schematic view of the fuel–air front formed during a fuel cell start-up or shutdown.

anode side (figure 12.7).[161,163] In this case, a reverse current is generated *in situ* in the fuel–air segment (figure 12.7, right), and it drives reactions on the air–air segment (figure 12.7, left) of the same MEA through an internal circuit.

Under these conditions, the COR current is dictated by the amount of fuel available in the fuel–air segment as well as the extent of a competing water oxidation reaction (WOR) (reaction 18) in the air–air segment:

$$H_2O \rightarrow \tfrac{1}{2}\,O_2 + 2\,H^+ + 2\,e^- \qquad E^0 = 1.229\ \text{V} \tag{18}$$

Both are a function of the ionomer water activity (controlled by its hydration state and relative humidity (RH)) and the CD (controlled by the fuel–air segment).[163]

The exchange current density for the WOR is ~1×10^{-9} A/cm^2. The WOR ignition potential is ~1.4 V. In the presence of a sufficient amount of water, figure 12.8 shows that the WOR oxidation potential will not exceed the COR ignition potential (~1.8 V) below 0.5 A/cm^2. This implies that carbon is protected by virtue of the WOR unless a cell is subjected to a CD that is not sustainable by the WOR alone. In a real PEM fuel cell, there is a finite water supply during a start-up or a shutdown. As the water activity gradually decreases, the WOR overpotential increases and the curve bends toward the COR region (dashed line in figure 12.8). The rate of the COR is therefore the greatest under low RH and high CD conditions.[161–163]

Many start-up/shutdown procedures have aimed to reduce the corrosion current through practices such as anode purging and shunting.[166–169] Others have attempted to use more stable carbon materials such as graphitized carbon[169–171] and carbon nanotubes (CNTs).[178,179] To delay the onset of the COR during fuel starvation, WOR catalysts have been incorporated into the anode electrodes.[23] Increasing anode ionomer content has also been recommended for a fuel starvation-resistant anode because it increases the amount of water available for the WOR.[23,165] In general, the materials approaches are applicable to all kinds of carbon corrosion.

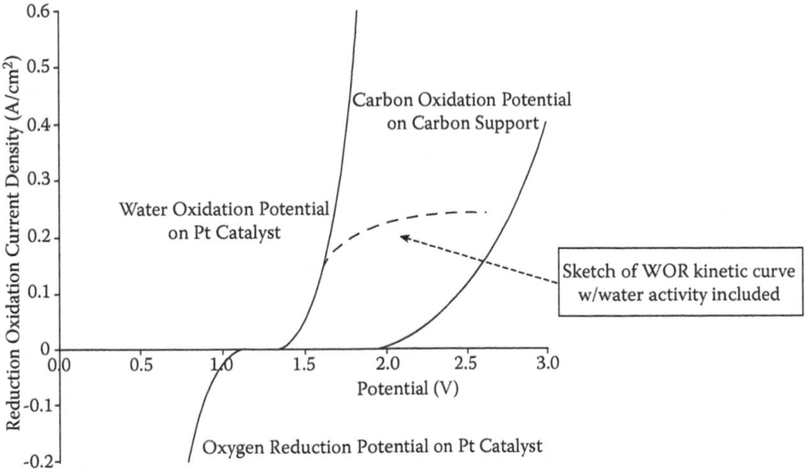

FIGURE 12.8 A schematic view of the WOR and COR overpotentials.

12.2.3.2 Modified Carbon and Noncarbon Support Materials

To increase catalyst utilization, it is desirable to maximize the TPB regions of Pt particles. Several innovative approaches have been pursued to improve the Pt utilization through catalyst support modifications. Qi and Pickup first proposed that a catalyst support can be made to conduct both electrons and protons,[180] which requires only a two-phase (catalyst-reactant) interface instead of a TPB. It can be accomplished through the use of either a functionalized carbon support or a conductive polymer with both proton- and electron-conducting capabilities.[180–185] Qi and Pickup used conducting poly(3,4-ethylenedioxythiophene)/poly(styrene-4-sulfonate) composites as the Pt support and produced currents as high as 0.4 A/cm^2 at 0.5 V.[181] They also showed that carbon surfaces could be oxidized to improve the fuel cell performance.[182] Qi et al. also demonstrated the feasibility of using a sulfonated carbon support as a dual proton–electron conductor.[183–185] For example, by functionalizing an E-TEK Pt/C catalyst with ethanesulfonic acid groups, they were able to achieve a 60% increase in fuel cell peak power and a 50% reduction in ionomer content.[184]

Gullá et al. reported several thin-layer electrodes with superior performance and stability.[186] Using a dual-ion beam-assisted deposition technique, they coated a Pt outer layer (~50 nm thick, 0.08 mg Pt/cm^2) directly onto GDLs with either a Co or Cr inner layer (~50 nm thick). These bilayered electrodes showed a mass-specific Pt activity more than 50% higher at 900 mV than that for a single Pt layer on GDLs. No ionomers were present in the electrodes.

Noncarbon supports, such as metal oxides (i.e., SnO_2 and ZrO_2) and HPAs, have also been investigated. One advantage of these noncarbon supports is that their hydrous surfaces may provide a pathway for proton transport. However, the low electrical conductivity of such supporting materials often requires that they be used with carbon black. An example is the Au/SiO_2/Vulcan catalyst, which is much more active toward ORR than either Au/Vulcan or SiO_2/Vulcan alone. The enhanced catalytic activity was attributed to the Au–SiO_2 interaction.[187]

Polyaniline (PANI) is a class of conductive polymer with good chemical and thermal stability.[188] It contains various benzoid and quinoid fragments in different redox and protonation states linked through N atoms. The ratio of different fragments can be adjusted in a reversible manner by changing the electric potential and acidity for the fine-tuning of its electron and proton conductivities.[188–190] The electron (hole) conductivity of a PANI material can be as high as ~10 S/cm,[188] whereas its proton conductivity can reach up to 10^{-2} S/cm^2.[191–194] Various PANI materials have been used as the catalyst support in PEM fuel cells.[188,193–197] Some of them also showed low but measurable catalytic activity for oxidations of small organic molecules such as HCHO.[194,197] With its tunable electronic properties, there is hope that one might find a suitable PANI material as a Pt-free HOR catalyst, but so far, Pt is still needed in a PANI-based anode.[198] In general, the use of organic conducting polymers is limited by their low chemical and electrochemical stability and durability under PEM fuel cell conditions.[184]

12.2.3.3 Other Components of Electrode Layers

PEM fuel cells developed by GE in the 1960s used an unsupported Pt black catalyst with PTFE as a binder and water-repelling agent. In such an electrode, only the catalyst particles that are in direct contact with the membrane can participate in the reaction because of the limited access for protons. Those Pt black electrodes typically have high Pt loadings and low performance. A carbon-supported Pt catalyst reduces the Pt loading and at the same time improves the fuel cell performance. The addition of an ionomer such as Nafion further improves catalyst performance and utilization by making the entire catalyst layer accessible to protons.[199–201] The ionomer content within the catalyst layer should be controlled to an optimal level in order to provide effective proton transport without adversely affecting water management. The optimal level differs for different types of catalyst supports and for different Pt loadings on the same type of support. Typically, a carbon support with high surface area requires high ionomer content. In addition, a gradient of the ionomer content can be generated from the membrane–catalyst layer interface (higher for optimal proton transport) to the catalyst–GDL interface (lower for facile gas diffusion and water transport).[202] Uchida et al. concluded that the ionomers filled mainly the secondary pores (40 to 1,000 nm) because the volume of these pores decreased linearly with an increase in the ionomer content, whereas the primary pores (20 to 40 nm) remained unchanged.[203] In contrast, Xie et al. found that the ionomers filled both the primary and secondary pores when MEAs were made by a thin film decal method in which the H form of Nafion was first converted to the tetrabutylammonium form during the catalyst ink formulation process.[204] They also noticed that the total pore volume of the catalyst layers increased tremendously after the MEAs had been boiled in an acidic aqueous solution. These results emphasize that the characteristics of an electrode depend on the processing method as well as its constituents since the processing affects the local juxtaposition of phases.

The catalyst ink can be made by directly mixing a Pt/C catalyst with H-form ionomers and a mixture of water and a short-chain alcohol such as isopropanol.[205] Alternatively, the ionomers can be converted into the Na form before mixing with a Pt/C catalyst in order to increase the allowable processing temperature.[199,200] Subsequently, the Na form can be converted back to the H form by an acid treatment. Colloidal ink can be prepared by adding certain organic solvents to the ionomer solution.[203] Other ingredients can also be added into the catalyst layer in order to tailor its properties. For example, Jung et al. added SiO_2 into the catalyst layer to enhance the fuel cell operations at low RH.[206]

There has been a renewed interest in Pt black electrodes because of the corrosion of the carbon support under certain fuel cell operating conditions.[157–176] Carbon supports have also been linked to the formation of H_2O_2, which leads to excessive membrane degradation.[84] We observed that MEAs made of Pt black and a hydrocarbon membrane achieved over 2,000 h of stable performance under voltage cycling at low RH and 85°C, far exceeding published results for hydrocarbon-based MEAs using Pt/C as electrodes, presumably because of a reduction in H_2O_2 formation.[207] We have also succeeded in preparing MEAs with less than 1 mg Pt/cm^2 using Pt black and thin (<50-μm) Nafion membranes. The long-term performance of these MEAs is currently under investigation. It will be interesting to find out if these MEAs can achieve an acceptable lifetime at a reasonable cost.

12.2.4 Engineered Nanostructured Electrodes

In simple terms, the electrodes that are used in PEM fuel cells are porous structures that attempt to maximize the utilization of the catalyst materials while maintaining their other functional requirements. However, the most fundamental functional requirement of the electrode, i.e., creating and maintaining the integrity of TPBs, is still not adequately understood.[154] Accordingly, electrode manufacture is, to a large extent, an art. As such, much of the knowledge of the influence of an electrode's material properties and its structure on performance is held as highly proprietary information by MEA manufacturers.[21]

Attempts have been made to elucidate the influence of electrode structure through the use of models. Macroscopic models that treat the electrode as a random porous structure, while attempting to get at the material property and structural issues underlying charge and material transfer, do so only in a gross sense, and certainly do not fully explain the underlying materials-related mechanisms involved in PEM fuel cell operation. Instead, they indicate the dependence of performance on macrohomogenous parameters such as electrode porosity and effective diffusivity. Nevertheless, a great deal of advancement has been made in our understanding of PEM fuel cell operation and its dependence on macrohomogeneous material properties through the use of such models. Excellent reviews of the models used to describe transport of charge and reactants in fuel cells have been given by Weber and Newman[208] and Wang.[209] More recently, models have been developed using DFT and molecular dynamics (MD) calculations to get a more detailed understanding of the dependence of performance on materials properties at a more fundamental level. An example is the use of MD to study the transport of gases in CNTs.[210,211]

In order to improve electrode performance, researchers are now actively engaged in the development of more highly engineered nanostructured electrodes, with specific attention placed on the control of the catalyst and the nature of its support. For example, researchers are now heavily involved in understanding the performance characteristics of nanostructured materials such as CNTs, carbon nanofibers (CNFs), and carbon nanohorns (CNHs).[8] Early in these studies, results were inconsistent, possibly because of the quality of the materials used. More recently, the results from various research groups are more consistent with enhanced fuel cell performance over the conventional carbon black supports such as Vulcan XC-72R. For example, Yoshitake et al. described the preparation of fine Pt catalysts supported on single-wall CNHs for use in fuel cell applications, citing superior performance over commercial carbon black supports.[212] Li et al. demonstrated the use of CNTs as support for cathode catalysts in DMFCs.[213] Matsumoto et al. claimed higher performance of a Pt/CNT electrode over a commercial Pt/C counterpart.[214] Finally, Xing developed a sonochemical process to create surface functional groups on CNTs for metal nanoparticle deposition.[215] Cyclic voltammetry measurements showed that the Pt nanoparticles on CNTs are more than twice as active as those supported on carbon black. It is clear that there is a growing effort in place to capitalize on the unique properties of CNTs and other nanostructured materials in improving the performance of electrodes.

In some applications, researchers simply replace the carbon black with CNT or CNF supports, while retaining the basic underlying random structure that results from the typical manufacturing processes used with carbon black. Thus, while the CNT does bring some increased structure to an electrode, the full benefit of CNTs might not be achieved until it is further engineered by aligning the CNTs in such a way as to increase their ability to meet all the desired functional properties. For example, Liu et al. prepared vertically aligned CNTs and tested them both as the catalyst support and as the stand-alone ORR catalyst.[216,217] The stand-alone ORR activity is attributed to the –N–M–N– (M = Fe, Co) active centers that were introduced by mixing NH_3 and organometallic Fe– and Co– precursors during the preparation of the aligned CNTs by chemical vapor deposition. An excellent review of the use of CNTs and approaches to nanostructure electrodes, which discusses their electrochemical properties, has been given by Gooding.[218]

It is interesting to note that conventional carbon black supports promote the formation of peroxide, which then decomposes into radicals that attack the membrane.[84] However, the role of graphitized carbon materials (such as CNTs) in peroxide formation is less clear. Smalley suggested that the curvy graphitic structure of CNTs deactivates free radicals by stabilizing them through enhanced delocalization.[219] It would be worthwhile to determine whether the formation and fate of peroxide is any different between the carbon black and the CNT. At any rate, it is well known that the rate of formation of peroxide is greatly reduced by elimination of the carbon black support. Evidence of this is clear from the work we have done on carbonless electrodes (PTFE-bonded Pt black electrodes) and those with a hybrid structure.[152,207]

An excellent example of a highly engineered carbonless electrode is the work done by Debe and co-workers at 3M, in which Pt catalysts are deposited and supported on a nanostructured thin film of oriented crystalline organic "whiskers."[220–222] These nanostructured thin film (NSTF) whiskers have cross-sectional dimensions in the neighborhood of 50 nm, aerial densities of 3 to 5 billion whiskers per cm^2, and controllable aspect ratios (length to width) in the range of 20 to 50. They are made of parylene red, a class of materials used in a variety of applications that demand extreme chemical and thermal stability. The film is formed by vacuum deposition, a dry process amenable to high-volume manufacturing. A typical NSTF-supported electrode is ~0.3 μm in thickness (20 to 30 times thinner than the conventional electrodes made of Pt/C) with a low Pt loading of 0.1 mg Pt/cm^2 and a specific surface area of ~10 m^2/g. Under accelerated aging conditions, the NSTF electrodes showed 75 times less FRR at 120°C than Pt/C electrodes. The results also showed good resistance to corrosion of the support at 1.5 V, negligible mass transport loss at a CD up to 2 A/cm^2, and a maximum of 33% surface area loss after 9,000 cycles. Electrodes made of these crystalline whiskers contain no ionomer but achieved nearly a 100% Pt utilization, challenging the conventional TPB paradigm. The mechanism by which protons can reach all the Pt sites has not been ascertained, although surface transport along the NSTF whiskers or grain boundary transport through the Pt sites are plausible explanations. NSTF-supported Pt ternary alloys such as PtNiMn and

PtCoMn were 5 to 10 times more active than Pt/C and 4.5 times more active than PtCo/C, with 10 to 20 times longer lifetime than Pt/C.[223]

Finally, there is a whole new approach being taken to engineer electrode structures using conventional and nonconventional microfabrication technologies, mainly for miniature fuel cells for portable applications. For example, Shah and co-workers have used vacuum sputtering to deposit electrodes and catalysts directly onto the membrane, as well as other novel techniques for selective deposition using elastomeric shadow masks and direct patterning on membrane electrolytes.[224] It is anticipated that future electrode materials and structures will be developed using nanotechnology with silicon or more resilient alternatives as the underlying building materials. A review of these nanofabrication techniques is outside the scope of the present discussion.

12.3 MEMBRANE ELECTROLYTE MATERIALS

Membranes are a critical and challenging component in PEM fuel cell applications. In order to meet the cost and durability target for residential and automotive applications, a membrane electrolyte must meet several functional requirements: (1) high proton conductivity over a wide RH range; (2) low electrical conductivity; (3) low gas permeability, particularly for H_2 and O_2 (to minimize H_2/O_2 crossover); (4) good mechanical properties under cycles of humidity and temperature; (5) stable chemical properties; (6) quick start-up capability even at subzero temperatures; and (7) low cost.[6,27,28] The lifetime of a membrane is related to its original thickness, its mechanical integrity, and the chemical stability of the constituent ionomers. There are many factors that can affect the membrane lifetime. These include not only the membrane properties (such as its chemical structure, composition, morphology, configuration, and fabrication process), but also several external factors (such as contamination of the membrane, materials compatibility, and the operating conditions of the fuel cell). A large number of possible membrane materials have been the subject of several recent review articles.[25–28,225] Here we focus on the materials aspects of recent research and development activities.

12.3.1 PERFLUOROSULFONIC ACID MEMBRANE MATERIALS

Perfluorinated membranes are still regarded as the best in the class for PEM fuel cell applications.[226–228] These materials are commercially available in various forms from companies such as DuPont, Asahi Glass, Asahi Chemical, 3M, Gore, and Solvay. Perfluorosulfonic acid (PFSA) polymers all consist of a perfluorocarbon backbone that has side chains terminated with sulfonated groups.

The equivalent weight (EW) of these polymers is typically in the range of 700 ~ 1,200 g/mol for PEM fuel cell applications. With a low EW, the density of ionic groups is generally high (hence high proton conductivity), but the dimensional stability and mechanical strength are both low. PFSA membranes with short side chains, such as the Dow polymer with $-OCF_2CF_2SO_3H$ (now available from Solvay), contain more ionic groups than Nafion with the same EW, and consequently, they have a higher proton conductivity. It achieved a threefold improvement in current density at 0.5 V relative to Nafion 117 under similar fuel cell operating conditions.[229] The perfluorocarbon backbone (resembling PTFE, a.k.a. Teflon®) provides the chemical stability and structural integrity in the demanding PEM fuel cell environment. The sulfonated side chains form aggregated hydrophilic domains that, when fully hydrated, are responsible for the proton transport across the membrane electrolyte.[27] A PFSA membrane must maintain proper water content for effective proton transport; this presents one of the major challenges in PEM fuel cell design and operation. It requires the use of fully humidified fuel–air feeds for optimal fuel cell performance and limits the cell temperature range. Furthermore, the liquid water must be removed promptly to avoid cell flooding. Finally, the mechanical integrity of a PFSA membrane is severely compromised by the significant dimensional changes associated with changing water content.[230,231]

12.3.1.1 Thin Reinforced Membrane for Improved Mechanical Properties

The traditional extrusion casting process was developed for producing thick membranes, normally greater than 125 microns. The extruded membrane is produced with ionomers in the $-SO_3K$ form, and the membrane is later converted to the final $-SO_3H$ form by an acid exchange process. Fuel cell developers have been continuously pushing the limit for thinner and thinner membranes because they reduce the ohmic loss, especially during high-power operations, and reduce the cost. This places tremendous challenges on membrane properties, such as chemical and mechanical stability, gas permeability, and processability. Today, more and more PFSA membranes are produced by solution casting, which is capable of producing thin layers of uniform thickness at a high production rate and low cost.

Membranes based purely on PFSA ionomers typically have low dimensional stability during wet–dry cycles and have a tendency to swell excessively when exposed to organic solvents (such as methanol), which leads to poor mechanical properties and difficulty with handling. This problem is worse for thin membranes, especially when a fuel cell is operating at an elevated temperature under load cycling. In accelerated fuel cell tests, it has been found that the most influential factor for membrane durability is its mechanical strength.[232–234] The failures are typically localized, e.g., local imperfections and stress points formed during MEA fabrication or structural damage caused by cell testing. Therefore, membranes without reinforcement are vulnerable to crack initiation and propagation, which may lead to catastrophic failure.[232–237] Stone and Galis reported that one of the major failures for membranes was ruptures in the thinned area, most likely caused by mechanical stresses induced during rapid changes in membrane water content that occur during fast dynamic operation of fuel cells.[238]

We have identified membrane crossover through edge cracks as one of the major stack failure modes (figure 12.9). The crack in figure 12.9 is at the membrane edge, which is particularly susceptible to damage because of the excessive thermal and mechanical stresses during load cycles and multiple starts and stops. In addition, it is exposed to the chemicals that are leached out of the gaskets or plates and to any coolant material that reaches the MEA side of the system.[36] This could lead to either membrane chemical degradation (pinhole formation) or mechanical failure

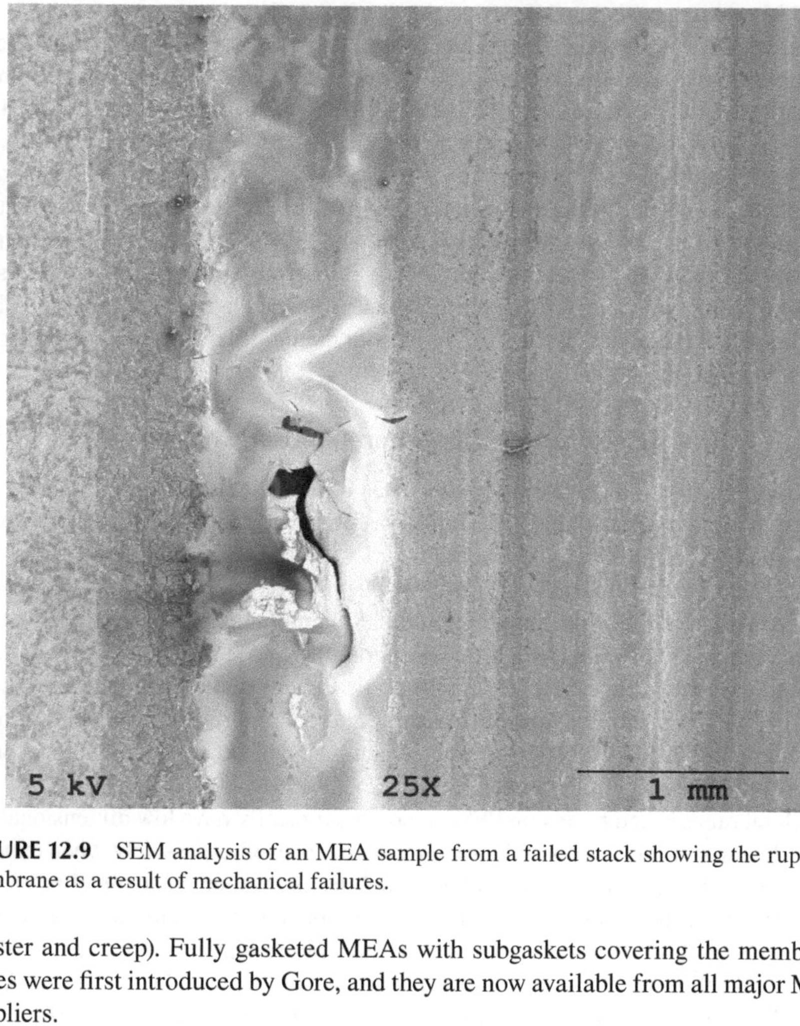

FIGURE 12.9 SEM analysis of an MEA sample from a failed stack showing the ruptured membrane as a result of mechanical failures.

(blister and creep). Fully gasketed MEAs with subgaskets covering the membrane edges were first introduced by Gore, and they are now available from all major MEA suppliers.

Reinforced membranes have been shown to delay crack initiation and propagation.[232–239] PTFE-based porous films, woven fibers, and microfibers are widely used for membrane reinforcement because of their improved chemical stability and excellent structural compatibility with fluorinated ionomers. W. L. Gore introduced Gore-Select® membranes, a class of expanded PTFE-reinforced PFSA membranes as thin as 5 microns.[234,240] These enable a fuel cell to achieve a higher power density with significantly improved durability. Asahi Glass reported a fiber-reinforced PFSA membrane with good mechanical strength and performance.[241] Most recently, Ballard reported a composite membrane with a Solupor® matrix, a highly porous and mechanically robust microporous film that is made of ultra high molecular weight (MW) polyethelene.[238] This film has a thickness of ~20 microns and relatively large pores with an overall porosity of 85%. These properties allow for the fabrication of a composite membrane with high mechanical strength and well-controlled thickness. Performance of this composite membrane is similar to that of traditional membranes but with improved durability. Johnson Matthey presented a reinforced membrane that gave a sixfold increase in durability for a 30-cell stack operated under dynamic operation conditions.[239] DuPont recently introduced Nafion XL™, a reinforced membrane made from a chemically stabilized ionomer (see below). It has achieved a 10-fold increase in lifetime under low RH and potential cycling conditions.[242]

12.3.1.2 Improvement in PFSA Chemical Stability through End-Group Modification

Peroxide radicals from decomposition of H_2O_2 are believed to be responsible for membrane chemical degradation.[29,85,86,243] The generally accepted end-group degradation mechanism, the so-called unzipping mechanism, starts from the end groups of a perfluorinated polymer chain.[85,86] Reactions 19 to 21 illustrate this using the carboxylic end group as an example:

$$R_f\text{-}CF_2COOH + \cdot OH \rightarrow R_f\text{-}CF_2\cdot + CO_2 + H_2O \tag{19}$$

$$R_f\text{-}CF_2\cdot + \cdot OH \rightarrow [R_f\text{-}CF_2OH] \rightarrow R_f\text{-}COF + HF \tag{20}$$

$$R_f\text{-}COF + H_2O \rightarrow R_f\text{-}COOH + HF \tag{21}$$

The rate of PFSA degradation depends strongly on fuel cell operating conditions such as RH, temperature, H_2/O_2 crossover rate, CO concentration, air bleed level, and electrode potential. Fluoride ions (F^-) are generated and are present in the water collected from both anode and cathode outlets. The FRR is a good indicator of the rate of membrane and ionomer degradation and has been used successfully in the past to predict membrane lifetime.[1] The gradual loss of ionomers and the thinning of the membrane eventually lead to a lower membrane performance (increased gas crossover rate) and accelerated mechanical failure (pinholes and shorting). Impurities can also exacerbate the rate of pinhole formation. Figure 12.10 shows an aged membrane with severe thinning. Table 12.1 shows the effect of RH on the FRR and the rate of cell voltage degradation.

Several companies have developed proprietary end-group protection strategies to reduce the number of polymer end groups and their vulnerability. DuPont reported that its modified PFSA membrane exhibited 10 to 25% reduction in FRR compared to a standard Nafion membrane.[29] However, this reduction did not correlate well with the reduction (>90%) in the active end groups, implying that the reactivity of the end groups had been altered or that there are multiple pathways for degradation. More recently, 3M introduced a new ionomer membrane that was claimed to have improved oxidation stability and durability. Schiraldi et al. synthesized a series of model compounds and confirmed that degradation proceeded through the backbone independently of the side chains.[244] The most likely point of attack was shown to be carboxyl end groups such as –COOR and –COR.

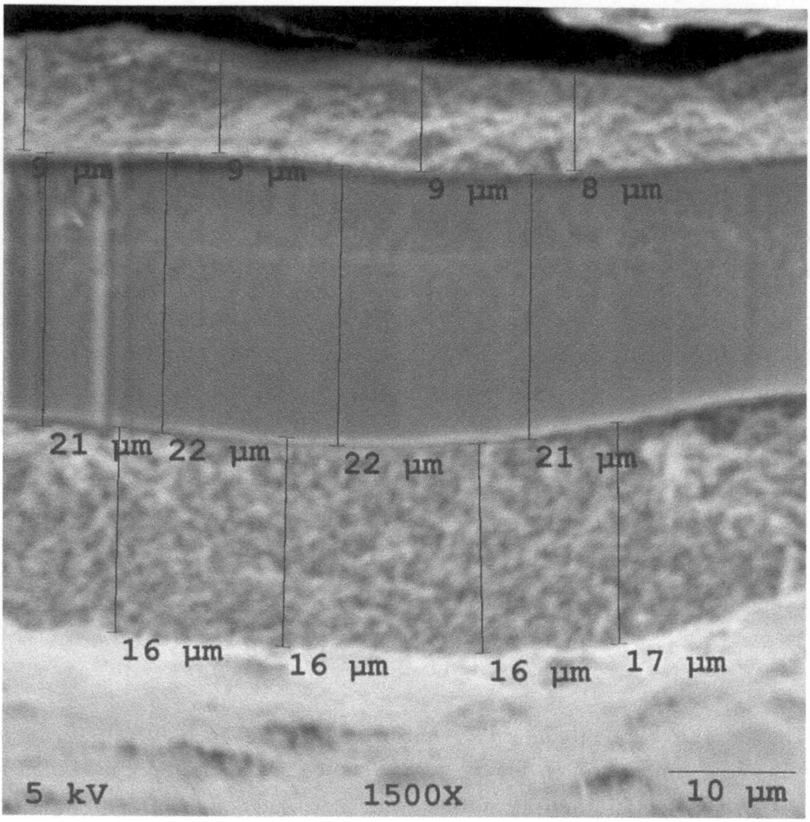

FIGURE 12.10 SEM analysis of an MEA sample from a failed stack showing the thinned membrane (from 45 μm to ~9 μm).

TABLE 12.1

RH Effect on the FRR and the Rate of Voltage Degradation at 65°C, 0.6 A/cm², and 1.2/2.0 Reformate-Air Stoichiometry

RH(%)	Testing Time(h)	Total FRR(10–8/g·h·cm2)	Degradation Rate(μV/h)
40	1,000	4.89	40
60	1,200	3.54	17
80	1,050	3.52	20
100	1,300	2.85	3
120	1,050	2.67	1

12.3.1.3 Modification of PFSA Membrane

Many research groups are actively engaged in modifying existing membrane structures to improve durability and expand the range of operating temperatures while retaining the desirable membrane properties.[30,240,245,246] However, chemical modification is largely limited to the side chains because the perfluorinated backbone offers few opportunities for cross-linking and controlled branching.[247] Even for side chain groups, the opportunities are limited by both the length and availability of perfluorinated precursors. Nonperfluorinated side chains provide more opportunities at the expense of lowered chemical stability. One promising approach is to introduce acidic groups, such as sulfonimide and sulfonyl methide, which are stronger than the sulfonic group.[248–250] DesMarteau synthesized a number of perfluorinated ionomers containing sulfonamide and other acid groups.[248] Its H form (EW ~ 1,100 g/mol) was reported to have an equilibrium water uptake of 116% by weight of its dry ionomer, corresponding to 70 H_2O molecules per acid group.[250]

An active area of research is to prepare composite membranes from PFSA polymers and various organic and inorganic materials.[30,240,245,246] Examples of PTFE-reinforced PFSA composite membranes, such as Gore Select®, were discussed above. Recently, Asahi Glass reported a new perfluorinated polymer composite membrane that can be operated at 120°C and 50% RH for 4,000 h. This membrane reduced the degradation rate by two to three orders of magnitude relative to the degradation rate for conventional MEAs.[251] The primary aims of these endeavors are to elevate the membrane operating temperature and to reduce methanol crossover in DMFC applications. Alberti and his co-workers pioneered the preparation and use of exfoliated layered zirconium phosphates in composite PFSA membranes.[252,253] However, *in situ* formation of inorganic fillers is ideal for membranes manufactured via an extrusion process. For example, Mauritz et al. prepared PFSA composite membranes filled with SiO_2, ZrO_2, SiO_2/TiO_2, and SiO_2/Al_2O_3 using a sol-gel process.[254–260] The proton conductivity of the SiO_2–PFSA composite membranes was found to be ~0.1 S/cm, slightly higher than that of Nafion.[261] Composite membranes can also be made by filling a porous polymer matrix with organic or inorganic fillers, through either direct filtration or *in situ* particle formation.[30] When a sulfonated phenethylsilica sol-gel was used, the resulting filler particles contained both –SiOH and $-SO_3H$ groups. The conductivity of such a composite membrane was found to be three to six times higher than that of Nafion.[262] HPAs are another class of inorganic fillers studied by many research groups because of their electrocatalytic activity, strong acidity, and excellent proton conductivity.[263,264] Most HPAs have a molecular size of about 1 nm in diameter and possess the Keggin structure, ideal for membrane fillers. Malhotra and Datta impregnated a PFSA membrane with HPAs and achieved a proton conductivity of 0.05 S/cm.[265] Tazi and Savadogo prepared composite membranes from recast Nafion and silicotungstic acid.[266] It was found that a fuel cell with this composite membrane had a CD of 810 mA/cm^2 at 0.6 V, compared to 640 mA/cm^2 for a fuel cell with Nafion 117 under similar operating conditions. Staiti et al. prepared doped PFSA–SiO_2 with phosphotungstic and silicotungstic acids. DMFCs made from these membranes were able to operate at 145°C and achieved a significant reduction in the overpotential for methanol oxidation. A membrane doped with phosphotungstic acid was shown to be the best with maximum power densities of 250 and 400 mW/cm^2, using air and oxygen, respectively.[267]

Although much progress has been made on PFSA-based membranes, there remain many challenges. For automotive applications, a membrane is required to have good proton conductivity at an RH as low as 25%. The same membrane should also be fully functional even at external temperatures as low as –40°C.[3] The operating temperature is preferred to be above 120°C for high system efficiency and effective thermal management. For residential applications, the membrane should last for over 40,000 h. For DMFC applications, methanol crossover remains a major problem for PFSA membranes. Finally, chemical synthesis, safety concerns with tetrafluoroethylene, and the cost/availability of perfluoroether co-monomers are manufacturing challenges that still need to be addressed.

12.3.2 Polybenzimidazole Membrane Materials

Plug Power, working with its European partners PEMEAS (now BASF Fuel Cell GmbH) and Vaillant, is actively pursuing a PBI-based HT PEM fuel cell system as a CHP system with high system efficiency and great CO tolerance.[95] This system demonstration project is jointly sponsored by the U.S. Department of Energy and the European Union, one of the first collaborations of this kind between the U.S. and the EU.

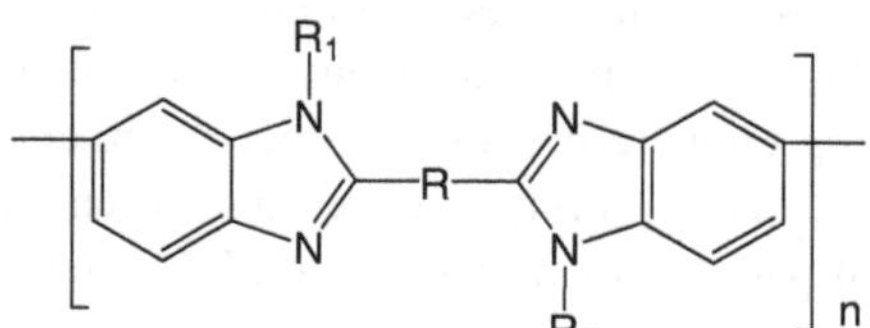

PBI (see chemical structure above) is a hydrocarbon membrane that has been commercially available for decades. Free PBI has a very low proton conductivity (~10^{-7} S/cm) and is not suitable for PEM fuel cell applications.[268] However, the proton conductivity can be greatly improved by doping PBI with acids such as phosphoric, sulfuric, nitric, hydrochloric, and perchloric acids.[269,270] The PA-doped PBI membrane is the most popular one in PEM fuel cell applications because H_3PO_4 is a nonoxidative acid with very low vapor pressure at elevated temperature.[271] Savinell et al. and Wainright et al. first demonstrated the use of PBI-PA for HT fuel cells in 1994.[270–272] Since then, there has been a significant amount of research on the PBI-based membrane because of its low cost and good thermal and chemical stability.[270,272–285] The resulting PBI-PA membrane can be operated at temperatures up to 200°C, with the optimum temperature depending on the acid/PBI ratio.[272,279] With this high operating temperature, PBI-based MEAs exhibit high CO tolerance that allows for a simplified reforming system. The impregnated H_3PO_4 acts as the proton carrier (electrolyte). As such, there is no need for an external water management subsystem, which in turn greatly reduces the system cost and complexity. The PBI membrane also finds applications in portable DMFCs because of its excellent resistance to methanol crossover.[281,282] The primary challenges for PBI-based PEM fuel cells are low power density due to the slow ORR kinetics in a liquid (PA) electrolyte, acid loss, stability of catalyst/catalyst support, including Pt dissolution/agglomeration and carbon corrosion, and mechanical relaxation of the polymer matrix.

Wainright et al.'s early work focused on poly(2,2'-m-phenylene-5,5'-bibenzimidazole) doped with acids. This m-PBI membrane can retain acids at molar ratios of 2 to 8 per repeating unit and retain its proton conductivity (0.04 to 0.08 S/cm) at high temperatures under nonhumidified conditions.[270] The H_2–air fuel cell performance based on this membrane is about 0.45 V at 0.2A/cm^2. Much effort has been made to increase the amount of acid held by PBI membranes because an improved acid doping level leads to an increase in proton conductivity and, presumably, an improvement in fuel cell performance. Wainright et al. found that the proton conductivity was in the range of 10^{-5} to 10^{-4} S/cm at 25°C for PBI membranes with 0.07 to 0.7 H_3PO_4 molecules per repeat unit.[270] For a PBI membrane with four or five H_3PO_4 molecules per repeat unit, the proton conductivity increases to ~10^{-2} S/cm.[270] Li et al. reported an m-PBI-PA complex with 16 moles of H_3PO_4 per repeat unit that exhibited a conductivity of 0.13 S/cm at 160°C.[286] However, a membrane made by simple postpolymerization doping methods loses its mechanical integrity at high acid doping levels.

Xiao et al. have developed a sol-gel process to prepare PBI membranes with high MW and high acid doping levels.[277,278] This sol-gel process uses polyphosphoric acids as the polymer condensation agent, polymer solvent, and membrane casting solvent during the production process, and the process is suitable for large-scale casting production. *In situ* hydrolysis of polyphosphoric acids after casting leads to H_3PO_4 imbibed in the final membrane product.[278] This membrane can retain up to 30 acid molecules per repeat unit and still maintain reasonable mechanical properties because of its high MW. The sol-gel process is used by PEMEAS in the production line of its commercial PBI-PA membranes. Its commercial Celtec-P® MEAs, based on this type of membrane, are claimed to have minimal carbon corrosion and acid loss with the ability to operate for up to 18,000 h.[287]

The proton conductivity of PBI can be increased significantly by grafting PBI with sulfonated groups.[288–290] When fully hydrated, the proton conductivity of these membranes was found to be comparable to those of PBI-PA membranes. However, highly sulfonated PBI membranes are susceptible to embrittlement under dry conditions and they are not suitable for HT applications.[290] On the other hand, PBI grafting is an effective method for replacing the imidazole hydrogen with other functional groups that deactivate the adjacent benzene rings. This makes the fused rings less susceptible to electrophilic attack, thus improving PBI's chemical stability under fuel cell operating conditions. Tang and Sherrington introduced nitrile groups to PBI membranes.[291] Kerres and others attempted to obtain various PBI membranes with a variety of fillers, blends, and sulfonated groups for specific applications.[275,280,283]

12.3.3 Current Status of Hydrocarbon Membranes

In addition to PBI, there are many other hydrocarbon membranes that can also serve as proton-conducting membranes. Most of them have been developed for automotive and DMFC applications.[28,225,292] The driving forces for hydrocarbon membranes are the need for a low-cost membrane electrolyte with a wide operating temperature range (a critical requirement for automotive applications) and, for DMFC applications, low methanol crossover. Other advantages of hydrocarbon membranes over PFSA include easy control of sulfonated group density and distribution for improved proton conductivity, less membrane swelling, lower gas permeability, and absence of HF in the condensed water, which is considered beneficial to the fuel cell hardware and the environment. The disadvantages of hydrocarbon membranes include low chemical stability and peroxide tolerance (and, as a result, the leaching out of membrane main chains and sulfonated groups over time) and embrittlement (with the corresponding loss of mechanical strength, especially under cycling conditions). The design of the polymer backbone and the balance of the hydrophilic and hydrophobic chain groups are keys to improving the performance of hydrocarbon membranes. Some recent activities in hydrocarbon membrane development are highlighted below.

12.3.3.1 Styrene

Styrenic polymers, which are easy to synthesize and modify, were studied extensively in the early literature. One example is BAM® made by Ballard Advanced Materials (see chemical structure below).[293] This membrane is 75 μm thick and has an ion exchange capacity of about 1.1 to 2.6 meg/g. Its chemical stability is not as good as PFSA even with its perfluorinated backbone. Ballard claimed that this membrane could last for several hundred hours under low RH operating conditions. It is no longer in production due to its high cost and the lack of availability of the monomer.

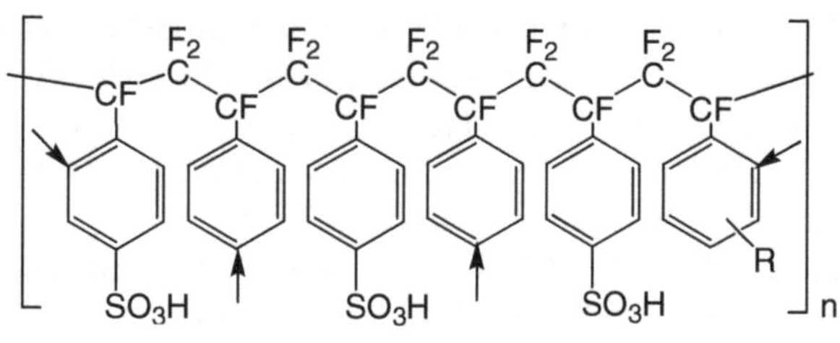

Another example is Dias Analytics' styrenic membrane based on the well-known block copolymer styrene–ethylene/butylene–styrene family.[294] This membrane has good conductivity; 0.07 to 1.0 S/cm when fully hydrated. It showed reasonable performance but had poor oxidative stability due to the susceptibility of its aliphatic backbone to peroxide attack.

12.3.3.2 Poly(Arylene Ether)

Polyarylenes, in particular different types of poly(arylene ether ketone)s, have been the focus of much hydrocarbon membrane research in recent years.[6,28,225] With good chemical and mechanical stability under PEM fuel cell operating conditions, the wholly aromatic polymers are considered to be the most promising candidates for high-performance PEM fuel cell applications. Many different types of these polymers are readily available and with good process capability. Some of these membranes are commercially available, such as poly(arylene sulfone)s and poly(arylene ether sulfone)s under the trade name Udel® by Solvay Advanced Polymers and various types of poly(arylene ether and ether ketone)s under the trade name PEEK™ by Victrex®. In most cases, the sulfonated groups are introduced by subjecting the polyarylenes to direct electrophilic sulfonation. Others are prepared through direct copolymerization of sulfonated monomers, which produces final polymers or co-polymers with improved control of the degree and location of the sulfonated groups. Recent examples from the leading hydrocarbon membrane developers are summarized below.

12.3.3.2.1 BAMG2® Membrane

Made by Ballard Advanced Materials, this membrane contains an aromatic ether (biphenol) segment that is common to poly(ether ketone). This aromatic backbone confers structural flexibility. The sulfone group is stable with respect to oxidation and reduction.

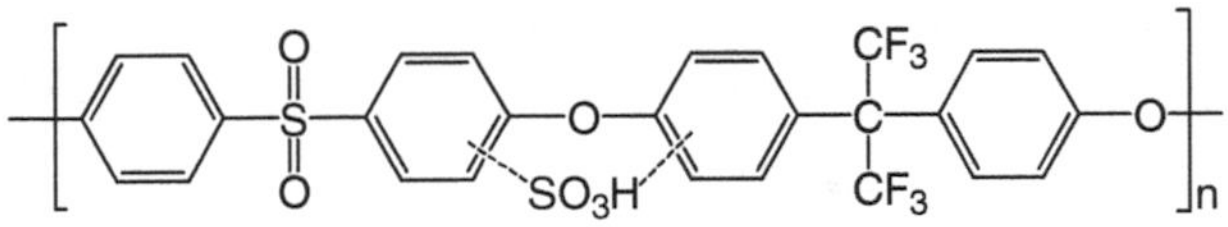

12.3.3.2.2 Poly(aryl Ether Ketone) Random or Block Co-Polymers

These customer-synthesized new polymers are made of chains with either random or block co-polymers on a laboratory scale by Hickner et al.[28] The MW and the ratio of the random and block segments can be well controlled. The sulfonated groups can be introduced directly or modified after polymer synthesis. The preliminary results showed some promise for PEM fuel cell and DMFC applications with low gas/methanol crossover.

12.3.3.2.3 Hoku Membranes

No structural information is available from the manufacturer, but these hydrocarbon membranes are believed to be a part of the poly(arylene ether) family. Hoku Scientific, Inc., reported 2,000-h test data operating with H_2–air.[295]

12.3.3.2.4 PolyFuel Membranes

These membranes are good for DMFC applications because of their low methanol crossover rate.[296–298] The acid–base polymer blend membranes consist of an acidic polymer, a basic polymer, and a third functional polymer for improved membrane conductivity, flexibility, dimensional stability, and reduced methanol crossover.[298] These membranes can be operated at low RH (<50%) and HT (~100°C), which makes them particularly attractive for automotive applications. The conductivity of PolyFuel membranes is slightly higher than that for Nafion. They have 30 to 40% water uptake in boiling water. The membranes have a relatively good tolerance toward chemical degradation, showing ~5% weight loss in an off-cell, 4-h test using Fenton's reagent.[296,297] No publications could be found on the membranes' mechanical properties and durability under the long-term automobile load cycling.

12.3.3.2.5 Honda Membrane

No detailed structural information has been disclosed except that it contains an aromatic main chain and an ion exchange substrate.[299] The aromatic nature of the membrane presumably provides excellent mechanical strength and good thermal stability. It prevents the membrane from softening and deforming at temperatures up to 95°C. This membrane also has excellent dimensional stability and high proton conductivity over a wide temperature range, including at temperatures below 0°C. In addition, it has lower gas permeability than PFSA membranes. MEAs based on this membrane exhibited impressive performance under realistic PEM fuel cell operating conditions.[299]

12.3.3.3 Polyimide Membranes

This class of polymers has great thermal stability and promising short-term performance.[28,300] The six-membered ring of naphthalenic imides is preferred over the five-membered ring of phthanic imides. The latter is susceptible to acid-catalyzed hydrolysis, which leads to chain scission and membrane embrittlement.[229] An example of six-membered ring polyimides is the block sulfonated copolyimides shown below.[301]

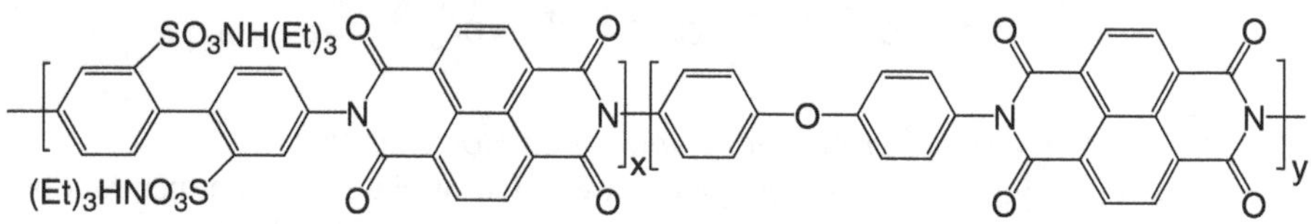

The $-SO_3H$ group can be introduced directly or indirectly. This block copolymer has been shown to be a promising candidate for PEM fuel cell applications, but poor solubility limited the ability to use a casting process. This problem was partially solved through randomized polymerization. However, the resulting membrane displayed a high degree of water swelling and weak mechanical strength. Various hydrophobic segments were altered in the main chain to prevent the membrane from overswelling and, at the same time, create ion-rich domains in the side chains. However, the difficulties in getting an imidazole ring closing reaction to go to completion are expected to cause hydrolysis of the imido-ring in the acidic fuel cell environment, which in turn would lead to membrane failure.

12.3.3.4 Arkema PVDF Membranes

Yi et al. reported a new type of PVDF membrane prepared by blending two very different polymers, a PVDF fluoropolymer such as Kynar® with a sulfonated polyelectrolyte.[302] The new membrane is inexpensive and displayed good performance and durability based on 1,000-h test data.

12.3.3.5 Polyphosphazene Membranes

Polyphosphazene has good chemical and thermal stability.[303] Its polyphosphazene backbone is highly flexible. Various side chains can be introduced to this backbone readily. Cross-linking is needed in order to reduce the dimensional changes in the presence of methanol or water. The membranes have shown reasonable proton conductivity and low methanol crossover. However, an improvement in mechanical strength is needed for practical fuel cell applications.

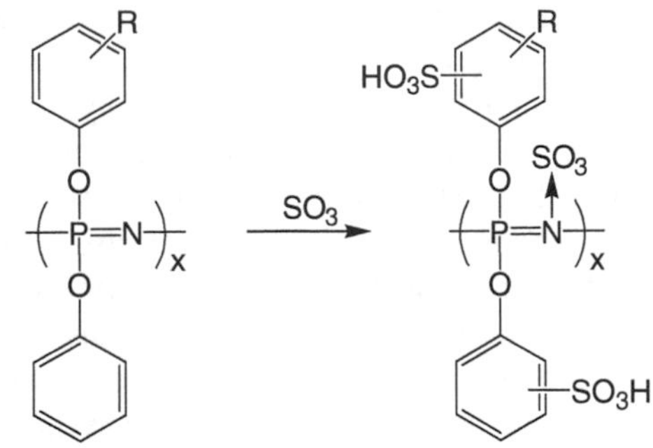

12.4 GAS DIFFUSION LAYER MATERIALS

The GDL received little attention until its importance as a fuel cell component was realized recently.[304] The GDL functional requirements can be summarized as follows: (1) allow uniform transport of reactant gases to the electrode; (2) conduct electrons; (3) remove product water from the electrode; (4) transfer heat to maintain the cell temperature; and (5) provide mechanical support for the MEA. To fulfill these functions, an ideal GDL material should have small gas transfer resistance, good electron conductivity, and good thermal conductivity.[12,305,306] The porosity of a GDL structure is the most important parameter for reactant transfer. Water management is the most challenging problem in GDL and fuel cell design. The ability to remove water is one of the key properties in evaluating GDL performance. If water cannot be removed from the system in a timely fashion, excessive water accumulation will lead to blockage of the reactant pathways and result in local fuel or air starvation. This problem, known as flooding, has been studied through both experiments and modeling.[6,307–309] An example is the study by Nam and Kaviany using network models for the anisotropic solid structure and the liquid water distribution.[310] The results showed that the cell performance can be improved by placing a fine layer (such as a microporous layer (MPL)) between the GDL and the catalyst layer because it creates a saturation jump across the interface.

Commonly used GDL materials are made of porous carbon fibers, including carbon cloth and carbon paper. Carbon cloth is more porous and less tortuous than carbon paper and has a rougher surface. Experimental results showed that carbon cloth GDL has better performance under high-humidity conditions because its low tortuosity (of the pore structure) and rough textural surface facilitate droplet detachment.[311] However, under dry conditions, carbon paper GDL has shown better performance than carbon cloth GDL because it is capable of retaining the membrane hydration level with reduced ohmic loss.

In most fuel cell operations, humidified gases are used to ensure proper membrane hydration. Hence, the ability to remove liquid water becomes the primary concern of GDL selection. PTFE is often used to increase the GDL hydrophobicity.[312–314] Contact angle is commonly used to measure the hydrophobicity (typically in the range between 120 and 140°). However, Gurau et al. suggested that external contact angle measurements were more indicative of the GDL surface roughness than the capillary forces in the GDL pores (which reflects the real measurement of water removal capacity).[315] They presented a new method for estimating the average internal contact angle. Pai et al. found that the GDL hydrophobicity could be improved by a CF_4 plasma treatment.[316]

Recent studies have found that placing a thin, highly hydrophobic MPL between the catalyst layer and the GDL can improve the fuel cell performance. Qi and Kaufman found that an MPL is extremely helpful where the GDL is prone to flooding,[317] in addition to providing better contact with an electrode layer. They attributed most of this effect to water management, noting that the limiting current density was raised and the membrane hydration was improved. They also compared MPLs with different PTFE contents and found that the best performance was achieved with 35% PTFE. Paganin et al. observed that the MPL thickness was more important than the PTFE content, and they suggested that the performance improvement was mainly due to decreased ohmic losses.[318] Separate experimental studies by Kong et al. and Jordan et al. concluded that the optimal MPL pore size was on the order of micrometers, which was believed to be a trade-off between water removal and O_2 diffusion (that required different pore sizes).[319,320] Mathematical models have illustrated the effects of MPL in enhancing the liquid water removal and reducing the water saturation in the electrode.[321–323] The effects of MPL properties such as thickness, porosity, and hydrophobicity were included in the modeling studies. It was found that the presence of the MPL at the cathode side may be more beneficial than at the anode side.[323]

Electrical conductivity is another important factor to be considered in GDL selection. The contact resistance between the GDL and other components dominates the ohmic loss because the bulk resistance of the GDL in the (thin) through-plane direction is negligible. GDL contact resistance has been linked to the extent of GDL compression.[305,306,324] Compression may reduce the contact resistance, but extensive compression not only damages the MEA and GDL structure, but also leads to a higher impermeability and larger mass transfer resistance. Bazylak et al. observed the irreversible damage of GDL due to breakage and deformation of the carbon fiber and PTFE coating.[325] Consequently, the liquid favored the compressed area due to both GDL morphology change and hydrophobicity loss. An optimal GDL compression ratio exists, balance between the need to minimize the contact resistance and the need to reduce the reactant transfer loss. This balance is different for carbon cloth and carbon paper because the magnitude of the contact resistance tends to be smaller for carbon cloth.[325] A numerical model simulation has shown that, in the lateral direction, electron transfer in the GDL directly affected the current density distribution, and hence the water distribution.[326]

In conclusion, the properties of the GDL material have a direct impact on fuel cell operation. These properties need to be optimized to achieve the best performance. However, relatively little progress has been made in this direction so far due to a lack of physical understanding of GDL transport properties.

12.5 BIPOLAR PLATE MATERIALS

Bipolar plates play an important role in fuel cell operation.[33,34,327–330] Generally, the functions of bipolar plates can be summarized as (1) supply and separate reactant gases without introducing impurities; (2) conduct electrons; (3) remove the reaction heat and control the fuel cell operating temperatures; and (4) remove the product water from the system. To fulfill these functions, an ideal bipolar plate material should have minimal gas permeability, high electrical conductivity, high thermal conductivity, and good chemical and electrochemical stability in the fuel cell environment. From a manufacturing perspective, the bipolar plate material should have low density, good mechanical strength, easy processability, and low cost. Nowadays, materials commonly used for bipolar plates are either graphite based or metallic.

Up until the early 1990s, pure graphite was used as the primary bipolar plate material for its good electron conductivity, low density, and high corrosion resistance (e.g., 680 S/cm and 1.78 g/cm^3 for POCO AXF-5Q).[329] However, due to its poor strength and ductility, graphite plates cannot be easily fabricated, and the minimum thickness is limited to about 5 mm. This results in a large stack volume and low efficiency in heat transfer. The brittleness of the graphite limits the methods of fabrication, which leads to a high manufacturing cost.

To provide good mechanical strength and ease of manufacturing, graphite-based conductive polymeric composites have been studied extensively in recent years. Graphite-based composite plates are made from a graphite/carbon powder filler and a polymer resin, either thermoplastic or thermosetting. Polymer resin used in the composite should have good thermal stability, high chemical stability, and low permeability to the reactant gases. The electrical conductivity of graphite composite materials is lower than that of a pure graphite material (with the extent depending on the volume fraction of graphite). However, these composite materials offer low density and low cost. Moreover, the composite plate can be fabricated easily using typical material processing methods such as extrusion or compression molding. In addition, the polymer component can provide appropriate surface characteristics (hydrophobic or hydrophilic) to facilitate water removal from the gas channel. We have demonstrated through neutron radiographic studies that surface alterations have a profound impact on the liquid water removal capability of graphite composite plates (figure 12.11).[331–334]

The main challenge for composite plates is to meet the conflicting needs of thickness, electrical conductivity, mechanical strength, and gas permeability.[335,336] A thin plate with good electrical conductivity requires a high ratio of graphite filler in the composite, which tends to give weaker mechanical strength and higher gas permeability. Oak Ridge National Laboratory developed a carbon–carbon composite bipolar plate using slurry molding of a chopped-fiber preform followed by sealing with chemically vapor infiltrated carbon.[337] The composite is characterized by a low density (0.96 g/cm^3) and a flexural strength about twice that of POCO graphite. Wolf and Willert-Porada developed a composite of liquid crystalline polymers and carbon in which the carbon content was below 40 vol%.[338] Wu and Shaw suggested that a composite with a triple-continuous structure could provide high electrical conductivity and tensile strength simultaneously.[339] In such a composite, the electrical conductivity is provided by the carbon-filled polymer phase, whereas the tensile strength and ductility are provided by the carbon-free polymer phase. Huang et al. proposed a method of stacking and compression molding graphite-filled wet-lay composite sheets in order to achieve higher in-plane conductivity and mechanical strength.[340] Blunk et al. found that alignment of the conductive filler in the composite together with a conductive-tie layer

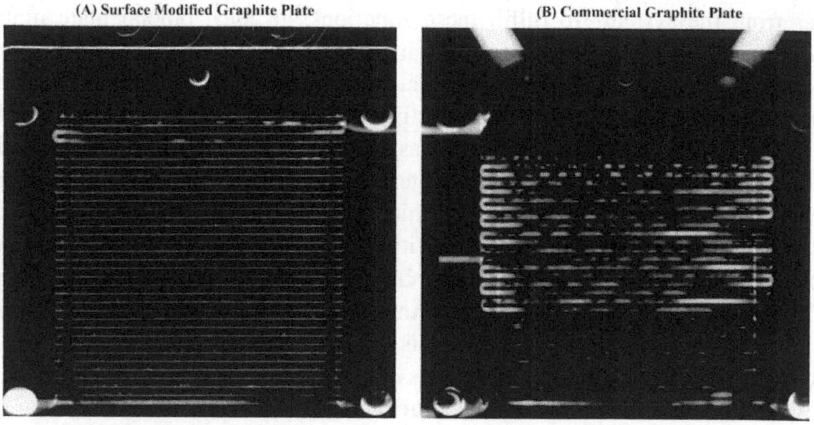

FIGURE 12.11 Comparison of liquid water transport for two 50-cm^2 single-cell PEM fuel cells using commercial graphite composite bipolar plates: (a) surface modified and (b) as received (0.1 A/cm^2, 1.5/2.0 H_2–air stoichiometry, 100% RH).

to reduce the contact resistance at the plate–GDL interface can significantly reduce the filler loading for conductivity requirements, which leads to better mechanical strength.[336] A Pd–Ni-coated polymer composite was shown to be promising because of its excellent electrical and physical performance characteristics.[341]

Metals such as titanium, aluminum, and stainless steel have been considered for bipolar plate materials.[330] With their high mechanical strength and low permeability to gases, metal plates can be much thinner than their graphitic counterparts, and hence they easily meet the conductivity and volume requirements.[3,6] In addition, the metal plate can be fabricated with conventional methods at low cost. For metal plates, a primary concern is surface oxide formation. A stable oxide layer, e.g., Cr_2O_3 at the surface of stainless steel, will form at the metal surface, which increases the contact resistance within the fuel cell. The growth of oxide layers on the surfaces of copper–beryllium alloy and stainless steel (SS316L, SS446) in a fuel cell environment and the effects of these layers have been studied by several groups.[342–345] Although the bulk resistance of metal is much lower than that of graphite, its contact resistance is much higher and it dominates the ohmic loss, i.e., metal plates degrade quickly with the formation of a resistive layer. Another problem with metal plates is the contamination of the membrane and poisoning of the catalyst layer by the soluble cations formed during metal corrosion. Therefore, increasing the corrosion resistance and preventing the resistive oxide formation are major challenges for metal bipolar plates. The most popular approach for solving the surface resistance and corrosion problems is to coat the metal surface with a highly conductive material that is also chemically stable. However, care is needed to avoid localized corrosion at imperfections in the coating. For Ti plates coated with titanium nitride/gold, the fuel cell voltage loss is comparable to that of graphite. However, typical coating materials are noble metals such as Pt and Au. The coating process also creates additional manufacturing cost.

Silva's study showed that although Ni-based alloys have a contact resistance comparable to that of graphite, their corrosion resistance in an acidic medium is unsatisfactory.[345] Nevertheless, nitride-coated stainless steel demonstrates both low contact resistance and good corrosion resistance. Wang et al. used iridium oxide (IrO_2) and Pt to coat Ti bipolar plates, and they found that the cell performance was close to that of graphite bipolar plates.[346] Weil et al. used boronization to increase the corrosion resistance of Ni as a plate material.[347] A composite material comprised of Nylon-6 and SS316L stainless steel alloy fibers was used to fabricate bipolar plates via an injection molding process, but the performance was poor compared to that of graphite plates.[348] Wang and Northwood used a polypyrrole coating to improve the corrosion resistance of SS316L.[349] Two compositions of Fe-based amorphous alloys were developed by Fleury et al.[350] Their study indicated that the contact resistance of the Fe-based amorphous alloys was similar to that of stainless steel, whereas these alloys exhibited better mechanical strength and corrosion resistance than SS316L.

Graphite-based composites and metal/alloy materials both have their own advantages and drawbacks. Current research interests in bipolar plate materials include both graphite composites and coated metals. No doubt progress on these materials will eventually lead to substantial reduction in the volume and cost of the fuel cell stack.

12.6 MATERIALS COMPATIBILITY AND MANUFACTURING VARIABLES

We have demonstrated the importance of system component compatibility and manufacturing variables using examples from our product development experience (figure 12.12).[4,36] Other groups have also shown the effects of non-MEA components on stack life.[351–355] Stack components must be chemically and mechanically stable under fuel cell operating conditions so that they will not leak or leach out species that poison the electrode catalysts,[353] be harmful to membrane stability and its proton conductivity,[354,355] or have adverse effects on the electrode/GDL properties, such as hydrophilic/hydrophobic character.[4] Stanic and Hoberecht linked membrane edge

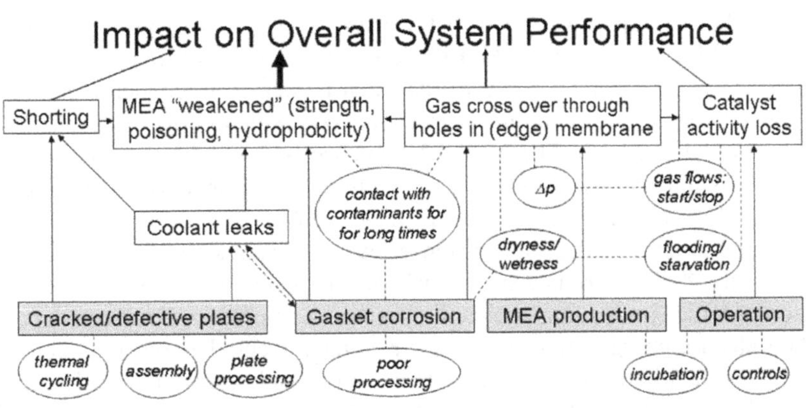

FIGURE 12.12 Impact of system components on fuel cell performance. (From Du, B. et al., *JOM*, 58, 44, 2006. With permission.)

failure and pinhole formation, two primary causes of premature stack failure, to contaminants leached out from gaskets, bipolar plates, hoses, and other components upstream of the stack.[351] Schulze et al. studied the compatibility of the sealing materials and the stack coolant required to maintain stable sealing properties.[352]

12.6.1 Sealing Materials and Coolant Compatibility

We discovered that one widely used commercial fuel cell-grade sealing material did not meet our materials compatibility requirements.[4,36] The decomposition products of the silicon-based sealing material were found at the membrane edges and, to a lesser extent, inside the active area of the membrane, possibly through a diffusion mechanism.[352] This led to a decrease in the conductivity and mechanical integrity of the membrane. Postmortem analyses of MEAs after 500 to 2,000 h of operation also showed calcium deposition in the edge region; this calcium is believed to originate from the fillers of certain siloxane-based gaskets.[36] Furthermore, the silicon component degraded when it came in contact with the coolant and other fluids used in the system. SiO_2 was found to imbibe into the membrane, which led to the loss of the force retention of the sealing material and, subsequently, coolant leakage, plate shorting, and gas crossover. Furthermore, certain fragments of the sealing material and coolant found their way into the GDL, which resulted in an adverse change in its hydrophobic character, thereby increasing the likelihood of flooding.[4,36] Finally, coolant that had leaked into the MEA caused the loss of electrode activity through its decomposition products, such as formic acid and acetic acid.[353] Indeed, we found significant levels of formate and acetate anions in the stack condensate water.[36] These anions originated from both coolant decomposition and plate (binders) leaching. *Ex situ* studies indicated that these anions promoted Ca and Si release from the sealing materials.[36] All of these factors contributed to premature stack failures.

12.6.2 Coolant and Bipolar Plate Compatibility

Coolant leakage can also occur through the bipolar plates if the fillers or binders are susceptible to leaching. This is particularly true for graphite–polymer composite materials. For example, we found that a plate sample with a particular combination of graphitic particles and resin binders emitted 7.2 and 1.7 μg/g plate·day of formate and acetate anions, respectively, when immersed in condensate at 85°C.[36] The coolant decomposition products accelerate corrosion of other components such as the radiator and gasket. These degradation species, in the form of metal ions and organic/inorganic species, lead to MEA contamination, shunt currents in the coolant loop, and electrical shorting of bipolar plates (via local precipitation). One well-documented situation is the shunt current effect on coolant stability and bipolar plate compatibility.[356–362] There is a substantial electric potential across a fuel cell stack that forces ionic species in a coolant loop to move in certain directions depending on their electric charges, thus generating a shunt current. The resulting concentration gradient facilitates the leaching of trace metals and other ionic species from the bipolar plates, sealing materials, and other components in contact with the coolant. Over time, this could lead to the blockage of coolant channels at one end of a stack and, potentially, shorting of bipolar plates under extreme circumstances. Some of the species serve as catalysts for coolant degradation, and this in turn accelerates the leaching process, eventually leading to coolant leakage from bipolar plates and loss of stack performance. To avoid component corrosion induced by shunt currents, it is recommended to use a dielectric coolant system or a coolant loop that is electrically insulated from the rest of the system.[359–362] A thorough materials compatibility study is therefore critical in the selection of the coolant and all components that can come in contact with it. A coolant must be compatible with manifolds, pumps, hoses, radiators, gaskets, and seals in addition to having appropriate physical properties (such as heat capacity, thermal conductivity, thermal expansion, dielectric constant, and viscosity) and meeting the safety and environmental standards (toxicity and flammability, waste recycling, and impact on aquatic species and biodegradability).

12.6.3 Other Component Compatibility Issues

Other components, e.g., hoses and O-rings, can also impact fuel cell performance since the impurities in these parts can leach out and be brought into contact with various fuel cell components through the reactant streams that enter the stack. We identified two contaminants, benzyl alcohol and butyl phthalate, in a stream condensed after passing through a specified length of the hose.[36] The effects of these species on cell voltage were evaluated by introducing them into the cathode inlet air, at a level prorated to reflect the surface areas of hoses and MEAs in a full stack. It was found that the cell voltage degraded with benzyl alcohol in proportion to the level added, and it only recovered partially after the flow was stopped. The voltage did not change significantly when butyl phthalate was introduced.

12.6.4 Component Manufacturing Variables and System Reliability

It is critical to minimize the component manufacturing variations in order to build a reliable PEM fuel cell system. We demonstrated that even within the same commercial MEA sample, the thicknesses of the electrode layers and membrane could vary greatly from one region to another.[4] With the current scale of PEM fuel cell production, commercial-grade fuel cell components often display substantial deviations from their product specifications. Such component manufacturing issues hamper the overall effort toward improving commercial PEM fuel cell system reliability. Without adequate MEA quality control, it is difficult to interpret autopsy results and to link apparent membrane/electrode problems of used MEAs to a particular failure mechanism. Part of the problem is the lack of a nondestructive in-line MEA quality control method to ensure batch-to-batch consistency.

Overall, the component reliability is a challenge to fuel cell manufacturers as well as their component suppliers. The stack is only one of several subsystems in a PEM fuel cell system with hundreds of parts and components. Component compatibility, which includes both chemical and mechanical properties, plays an important role in system reliability and overall performance. To select the best materials/design for a system component, one must first study its properties (physical, chemical, mechanical, and electrochemical) under relevant conditions such as temperature, pressure, and composition. For example, the reactant side of a PEM fuel cell bipolar plate (all sealing materials and plate components) must be able to tolerate high humidity, temperature changes, reactive chemicals, and certain trace hydrocarbons and inorganic species in the gas streams. On the other side of a bipolar plate, the coolant must have stable chemical properties and should not generate any harmful species that could attack the seals, plates, or delivery hoses. At the same time, components in contact with the coolant should not exacerbate coolant degradation.

With the many chemical, mechanical, and physical properties that must be taken into consideration, together with how those properties change with time, any screen for new materials should accept from the outset that some compromise may be necessary to reach a practical and cost-effective solution. Also, it is important to identify the source of any undesirable contaminants: some species may originate from a key raw material, whereas others may come from additives such as flow modifiers or mold release agents used to facilitate the manufacture of complex shapes. With the many options available for gaskets, plates, hoses, and O-rings, a screening method is needed to select materials that meet the durability, performance, and cost requirements for a given application. This screen needs to be rapid yet realistic. Moreover, the method should aim to distinguish between limitations that are inherent to a candidate material and those that are a consequence of the processing method. For example, one should ensure that polymerization of polymeric gaskets and resin-based plates is complete, and that residual monomers are consumed or removed as much as possible. This improvement might be achieved by a suitable postbake. The screen should not be based on chemical tests alone since physical and mechanical properties play an important role. For plates, compression set, compression stress relaxation, the stress–strain curve, and the expansion coefficient are important. For gaskets, additional factors are weight gain, contact pressure for sealing, and load deflection characteristics. In all cases, the reproducibility and reliability of the manufacturing process need to be evaluated. Candidates that show promise based on all the screening criteria should be examined further *in situ* under real fuel cell operating conditions. To avoid false negatives or positives, care should be taken to ensure that any accelerated test conditions truly represent actual fuel cell operating conditions.

12.7 SUMMARY

It is relatively straightforward to select materials for PEM fuel cells that meet two of the three key requirements: cost, durability, and performance. The challenge is to find a combination of materials that will give an acceptable result for the three criteria combined. There are multiple solutions to this problem, partly because each fuel cell application seeks to optimize a unique objective function and partly because there are many possible choices for the constituent materials. Should one select electrodes with a relatively high loading of Pt black, a thick Nafion electrolyte, and design for an MEA to deliver 40,000 h of lifetime? Or, at the other end of the spectrum, should one combine nonnoble metal catalysts with an inexpensive hydrocarbon membrane and accept the need to replace the stack regularly? There is an almost infinite array of possibilities to choose from, which helps explain the large number of active research programs in the area.

The information compiled in this review perhaps serves best to warn the reader of the dangers of suboptimization; i.e., one cannot select the best electrode in isolation from the other components since they function together as an integrated whole. Moreover, it is all too easy to ignore materials interactions that take place at the system level. For example, materials compatibility issues associated with plates, gaskets, and coolant have been found to have a strong impact on both system reliability and stack life. Consequently, systematic screening procedures for materials are needed that give a quick, yet realistic representation of fuel cell behavior. An additional factor, often overlooked, is the impact of variations in the manufacturing of the various components in the stack and system. For example, a small quantity of catalyst deposited inadvertently in the edge region could catalyze reactions between crossover gases and lead to premature failure.

Over the last decade, there has been a significant improvement in the understanding of the interactions that take place in the fuel cell environment — not only the electrochemical reactions, but also the chemical reactions and mechanical forces. New discoveries in materials research, especially those in engineered nanostructured materials, provide exciting opportunities and potential breakthroughs for fuel cell component development. Powerful research tools, such as combinatorial libraries and rapid material screening technologies, make it possible to downselect a few promising candidates from a great number of material combinations for further evaluation under realistic fuel cell operating conditions, which is a rather time-consuming process. It is this knowledge that provides direction for many current research programs and gives hope that several optimized combinations of MEA components will soon be realized.

ACKNOWLEDGMENTS

The authors thank Dr. Ying Wang (MTI Micro Fuel Cells) and Dr. Lisa Xiao (PEMEAS) for reviewing parts of the manuscript and for their valuable comments and suggestions.

REFERENCES

1. Baldwin, R. et al., Hydrogen-oxygen proton-exchange membrane fuel cells and electrolyzers, *J. Power Sources*, 29, 399, 1990.
2. McElroy, J., private communication, 2004.
3. Multi-Year Research, Development and Demonstration Programs for Hydrogen, Fuel Cells & Infrastructure Technologies Program, U.S. Department of Energy, Washington, DC, 2005, http://www.eere.energy.gov/hydrogenandfuelcells/mypp/.
4. Du, B. et al., PEM fuel cells: status and challenges for commercial stationary power applications, *JOM*, 58, 44, 2006.
5. de Bruijn, F., Current status of fuel cell technology for mobile and stationary applications, *Green Chem.*, 7, 132, 2005.
6. Vielstich, W., Lamm, A., and Gasteiger, H.A., Eds., *Handbook of Fuel Cells: Fundamentals, Technology, and Applications*, 1st ed., John Wiley & Sons, West Sussex, England, 2003.
7. EG&G Technical Services, *Fuel Cell Handbook*, 7th ed., U.S. Department of Energy, National Energy Technology Laboratory, Morgantown, WV, 2005.
8. Winter, M. and Brodd, R.J., What are batteries, fuel cells, and supercapacitors? *Chem. Rev.*, 104, 4245, 2004.

9. Bagotzky, V.S., Osetrova, N.V., and Skundin, A.M., Fuel cells: state-of-the-art and major scientific and engineering problems, *Russ. J. Electrochem.*, 39, 919, 2003.
10. Costamagna, P. and Srinivasan, S., Quantum jumps in the PEMFC science and technology from the 1960s to the year 2000. Part I. Fundamental scientific aspects, *J. Power Sources*, 102, 242, 2001.
11. Costamagna, P. and Srinivasan, S., Quantum jumps in the PEMFC science and technology from the 1960s to the year 2000. Part II. Engineering, technology development, and application aspects, *J. Power Sources*, 102, 253, 2001.
12. Mehta, V. and Cooper, J.S., Review and analysis of PEM fuel cell design and manufacturing, *J. Power Sources*, 114, 32, 2003.
13. Gasteiger, H.A. and Mathias, M.F., Fundamental research and development challenges in polymer electrolyte fuel cell technology, in *Proceedings of the 202nd ECS Meeting: Proton Conducting Membrane Fuel Cells III*, Salt Lake City, UT, October 20–25, 2002.
14. Haile, S.M., Fuel cell materials and components, *Acta Mater.*, 51, 5981, 2003.
15. Brandon, N.P., Skinner, S., and Steele, B.C.H., Recent advances in materials for fuel cells, in *Annual Review of Materials Research*, Kreuer, K.-D. et al., Eds., Annual Reviews, Palo Alto, CA, 2003, p. 183.
16. Gasteiger, H.A. et al., Activity benchmarks and requirements for Pt, Pt-alloy, and non-Pt oxygen reduction catalysts for PEMFCs, *Appl. Catal. B*, 56, 9, 2005.
17. Wang, B., Recent development of non-platinum catalysts for oxygen reduction reaction, *J. Power Sources*, 152, 1, 2005.
18. Antolini, E., Recent developments in polymer electrolyte fuel cell electrodes, *J. Appl. Electrochem.*, 34, 563, 2004.
19. Antolini, E., Formation, microstructural characteristics and stability of carbon supported platinum catalysts for low temperature fuel cells, *J. Mater. Sci.*, 38, 2995, 2003.
20. Antolini, E., Formation of carbon-supported PtM alloys for low temperature fuel cells: a review, *Mater. Chem. Phys.*, 78, 563, 2003.
21. Litster, S. and McLean, G., PEM fuel cell electrodes, *J. Power Sources*, 130, 61, 2003.
22. Ralph, T.R. and Hogarth, M.P., Catalysis for low temperature fuel cell. Part I. The cathode challenges, *Platinum Metal Rev.*, 46, 3, 2002.
23. Ralph, T.R. and Hogarth, M.P., Catalysis for low temperature fuel cell. Part II. The anode challenges, *Platinum Metal Rev.*, 46, 117, 2002.
24. Markovic, N.M. and Ross, P.N., New electrocatalysts for fuel cells: from model surfaces to commercial catalysts, *CATTECH*, 4, 110, 2000.
25. Souzy, R. and Ameduri, B., Functional fluoropolymers for fuel cell membranes, *Prog. Polym. Sci.*, 30, 644, 2005.
26. Hickner, M.A. and Pivovar, B.S., The chemical and structural nature of proton exchange membrane fuel cell properties, *Fuel Cells*, 5, 213, 2005.
27. Mauritz, K.A. and Moore, R.B., State of understanding of Nafion, *Chem. Rev.*, 104, 4535, 2004.
28. Hickner, M.A. et al., Alternative polymer systems for proton exchange membranes (PEMs), *Chem. Rev.*, 104, 4587, 2004.
29. Curtin, D.E. et al., Advanced materials for improved PEMFC performance and life, *J. Power Sources*, 131, 41, 2004.
30. Alberti, G. and Casciola, M., Composite membranes for medium-temperature PEM fuel cells, *Ann. Rev. Mater. Sci.*, 33, 129, 2003.
31. Paddison, S.J., Proton conduction mechanisms at low degrees of hydration in sulfonic acid-based polymer electrolyte membranes, *Ann. Rev. Mater. Sci.*, 33, 289, 2003.
32. Schuster, M.F.H. and Meyer, W.H., Anhydrous proton-conducting polymers, *Ann. Rev. Mater. Sci.*, 33, 233, 2003.
33. Li, X. and Sabir, I., Review of bipolar plates in PEM fuel cells: flow-field designs, *Int. J. Hydrogen Energy*, 30, 359, 2005.

34. Hermann, A., Chaudhuri, T., and Spagnol, P., Bipolar plates for PEM fuel cells: a review, *Int. J. Hydrogen Energy*, 30, 1297, 2005.
35. Feitelberg, A.S. et al., Reliability of Plug Power GenSysTM fuel cell systems, *J. Power Sources*, 147, 203, 2005.
36. Guo, Q. et al., Compatibility and durability of fuel cell materials, *ECS Trans.*, 5(1), 187, 2007.
37. Grubb, J.W.T., Fuel Cell, U.S. Patent 2,913,511, November 17, 1959.
38. Grubb, J.W.T. and Niedrach, L.W., Batteries with solid ion-exchange membrane electrolytes, *J. Electrochem. Soc.*, 107, 131, 1960.
39. Grot, W.G., CF_2=$CFCF_2CF_2SO_2F$ and Derivatives and Polymers Thereof, U.S. Patent 3,718,627, February 27, 1973.
40. Kinoshita, K., *Electrochemical Oxygen Technology*, John Wiley & Sons, New York, 1992.
41. Newman, J. and Thomas-Alyea, K.E., *Electrochemical Systems*, 3rd ed., John Wiley & Sons, Hoboken, NJ, 2004.
42. Conway, B.E. and Tilak, B.V., Interfacial processes involving electrocatalytic evolution and oxidation of H_2, and the role of chemisorbed H, *Electrochim. Acta*, 47, 3571, 2002.
43. Du, B. et al., DDP 732 Physical Properties of Current MEAS, Plug Power internal report, April, 2003.
44. Gasteiger, H.A., Panels, J.E., and Yan, S.G., Dependence of PEM fuel cell performance on catalyst loading, *J. Power Sources*, 127, 162, 2004.
45. Du, B., *BNL Low-Pt Anode Catalyst Performance Evaluation*, Plug Power report submitted to Brookhaven National Laboratories, Upton, NY, 2005.
46. Adzic, R. et al., Low Pt loading fuel cell electrocatalysts, 271, *2006 Annual Merit-Review and Peer Evaluation Report*, DOE Hydrogen, Fuel Cells and Infrastructure Technologies Program, Arlington, VA, May 16–19, 2006.
47. Sasaki, K. et al., Ultra-low platinum content fuel cell anode electrocatalyst with a long-term performance stability, *Electrochim. Acta*, 49, 3873, 2004.
48. Zhang, J. et al., Platinum and mixed platinum-metal monolayer fuel cell electrocatalysts: design, activity and long-term performance stability, *ECS Trans.*, 3, 31, 2006.
49. Brankovic, S.R., Wang, J.X., and Adzic, R.R., Metal monolayer deposition by replacement of metal adlayers on electrode surfaces, *Surf. Sci.*, 474, L173, 2001.
50. Brankovic, S.R., Wang, J.X., and Adzic, R.R., New methods of controlled monolayer-to-multilayer deposition of Pt for designing electrocatalysts at an atomic level, *J. Serb. Chem. Soc.*, 66, 887, 2001.
51. Zeis, R. et al., Platinum-plated nanoporous gold: An efficient, low Pt loading electrocatalyst for PEM fuel cells, *J. Power Sources*, 165(1), 65, 2007.
52. Nørskov, J.K. et al., Trends in the exchange current for hydrogen evolution, *J. Electrochem. Soc.*, 152, J23, 2005.
53. Venkataraman, R., Kunz, H.R., and Fenton, J.M., Development of new CO tolerant ternary anode catalysts for proton exchange membrane fuel cells, *J. Electrochem. Soc.*, 150, A278, 2003.
54. Park, K.-W. et al., PtRu-WO_3 nanostructured alloy electrode for use in thin-film fuel cells, *Appl. Phys. Lett.*, 82, 1090, 2003.
55. Hwu, H.H. et al., Potential application of tungsten carbides as electrocatalysts. 1. Decomposition of methanol over carbide-modified W(111), *J. Phys. Chem. B*, 105, 10037, 2001.
56. Chen, K.Y., Sun, Z., and Tseung, A.C.C., Preparation and characterization of high-performance Pt-Ru/WO_3-C anode catalysts for the oxidation of impure hydrogen, *Electrochem. Solid-State Lett.*, 3, 10, 2000.
57. Levy, R.B. and Boudart, M., Platinum-like behavior of tungsten carbide in surface catalysis, *Science*, 181, 547, 1973.

58. Colton, R.J., Huang, J.J., and Rabalais, J.W., Electronic structure of tungsten carbide and its catalytic behavior, *Chem. Phys. Lett.*, 34, 337, 1975.
59. Yang, X.G. and Wang, C.Y., Nanostructured tungsten carbide catalysts for polymer electrolyte fuel cells, *Appl. Phys. Lett.*, 86, 224104, 2005.
60. Barnett, C.J. et al., Electrocatalytic activity of some carburised nickel, tungsten and molybdenum compounds, *Electrochim. Acta*, 42, 2381, 1997.
61. Bodoardo, S. et al., Oxidation of hydrogen on WC at low temperature *Electrochim. Acta*, 42, 2603, 1997.
62. Limoges, B.R. et al., Electrocatalyst materials for fuel cells based on the polyoxometalates $[PMo(12-n)VnO_{40}](3+n)-$ (n = 0–3), *Electrochim. Acta*, 50, 1169, 2005.
63. Haugen, G.M. et al., Increased stability of PFSA proton exchange membranes under fuel cell operation by the decomposition of peroxide catalyzed by heteropoly acids, *ECS Trans.*, 3, 551, 2006.
64. Yandrasits, M.A. et al., Dynamics of PFSA polymer hydration measured in situ by SAXS, *ECS Trans.*, 3, 915, 2006.
65. Oetjen, H.F. et al., Performance data of a proton exchange membrane fuel cell using H_2/CO as fuel gas, *J. Electrochem. Soc.*, 143, 3838, 1996.
66. Springer, T.E. et al., Model for polymer electrolyte fuel cell operation on reformate feed: effects of CO, H_2 dilution, and high fuel utilization, *J. Electrochem. Soc.*, 148, A11, 2001.
67. Baschuk, J.J. and Li, X.G., Modelling CO poisoning and O_2 bleeding in a PEM fuel cell anode, *Int. J. Energy Res.*, 27, 1095, 2003.
68. Gasteiger, H.A., Markovic, N.M., and Ross, P.N., H_2 and CO electrooxidation on well-characterized Pt, Ru, and Pt-Ru. 2. Rotating disk electrode studies of CO/H_2 mixtures at 62°C, *J. Phys. Chem.*, 99, 16757, 1995.
69. Gasteiger, H.A., Markovic, N.M., and Ross, P.N., H_2 and CO electrooxidation on well-characterized Pt, Ru, and Pt-Ru. 1. Rotating disk electrode studies of the pure gases including temperature effects, *J. Phys. Chem.*, 99, 8290, 1995.
70. Ianniello, R. et al., CO adsorption and oxidation on Pt and Pt-Ru alloys: dependence on substrate composition, *Electrochim. Acta*, 39, 1863, 1994.
71. Ruth, K., Vogt, M., and Zuber, R., Development of CO-tolerant catalysts, in *Handbook of Fuel Cells: Fundamentals, Technology, and Applications*, 1st ed., Vielstich, W., Lamm, A., and Gasteiger, H.A., Eds., John Wiley & Sons, West Sussex, England, 2003, p. 489.
72. Camara, G.A. et al., The CO poisoning mechanism of the hydrogen oxidation reaction in proton exchange membrane fuel cells, *J. Electrochem. Soc.*, 149, A748, 2002.
73. Lin, W.F., Iwasita, T., and Vielstich, W., Catalysis of CO electrooxidation at Pt, Ru, and PtRu alloy: an in situ FTIR study, *J. Phys. Chem. B*, 103, 3250, 1999.
74. Wang, K. et al., On the reaction pathway for methanol and carbon monoxide electrooxidation on Pt-Sn alloy versus Pt-Ru alloy surfaces, *Electrochim. Acta*, 41, 2587, 1996.
75. Du, B., Richard, P., and Elter, J.F., CO-air bleed interaction and performance degradation study in proton exchange membrane fuel cells, *ECS Trans.*, 3, 705, 2006.
76. Schmidt, T.J. et al., Electrocatalytic activity of PtRu alloy colloids for CO and CO/H_2 electrooxidation: stripping voltammetry and rotating disk measurements, *Langmuir*, 13, 2591, 1997.
77. Gottesfeld, S.D., Preventing CO Poisoning in Fuel Cells, U.S. Patent 4,910,099, March 220, 1990.
78. Gottesfeld, S. and Pafford, J., A new approach to the problem of carbon monoxide poisoning in fuel cells operating at low temperatures, *J. Electrochem. Soc.*, 135, 2651, 1988.
79. Adcock, P.A. et al., Transition metal oxides as reconfigured fuel cell anode catalysts for improved CO tolerance: polarization data, *J. Electrochem. Soc.*, 152, A459, 2005.

80. Tsiplakides, D. et al., Electrochemical promotion of catalysis: mechanistic investigations and monolithic electropromoted reactors, *Catal. Today*, 100, 133, 2005.
81. Stamenkovic, V. et al., Oxygen reduction reaction on Pt and Pt-bimetallic electrodes covered by CO, *J. Electrochem. Soc.*, 152, A277, 2005.
82. Jusys, Z., Kaiser, J., and Behm, R.J., Simulated "air bleed" oxidation of adsorbed CO on carbon supported Pt. Part I. A differential electrochemical mass spectrometry study, *J. Electroanal. Chem.*, 554/555, 427, 2003.
83. Jusys, Z. and Behm, R.J., Simulated "air bleed" oxidation of adsorbed CO on carbon supported Pt. Part 2. Electrochemical measurements of hydrogen peroxide formation during O_2 reduction in a double-disk electrode dual thin-layer flow cell, *J. Phys. Chem. B*, 108, 7893, 2004.
84. Antoine, O. and Durand, R., RRDE study of oxygen reduction on Pt nanoparticles inside Nafion®: H_2O_2 production in PEMFC cathode conditions, *J. Appl. Electrochem.*, 30, 839, 2000.
85. Pianca, M. et al., End groups in fluoropolymers, *J. Fluorine Chem.*, 95, 71, 1999.
86. Curtin, D.E. et al., Advanced materials for improved PEMFC performance and life, in *8th Grove Fuel Cell Symposium*, London, September 24–26, 2003, PGR12.
87. Du, B., DDP 960 *Pulsed Air Bleed for Improved Fuel Cell Performance*, Plug Power internal report, April, 2005, Latham, NY.
88. Eberle, K. et al., Device and Method for Combined Purification and Compression of Hydrogen Containing CO and the Use Thereof in Fuel Cell Assemblies, U.S. Patent 6,361,896, March 26, 2002.
89. Strasser, P. et al., High throughput experimental and theoretical predictive screening of materials: a comparative study of search strategies for new fuel cell anode catalysts, *J. Phys. Chem. B*, 107, 11013, 2003.
90. Greeley, J. and Mavrikakis, M., Near-surface alloys for hydrogen fuel cell applications, *Catal. Today*, 111, 52, 2006.
91. Sinfelt, J.H., *Bimetallic Catalysts: Discoveries, Concepts and Applications*, John Wiley & Sons, New York, 1983.
92. Rodriguez, J.A. and Goodman, D.W., The nature of the metal-metal bond in bimetallic surfaces, *Science*, 257, 897, 1992.
93. Smotkin, E.S. and Diaz-Morales, R.R., New electrocatalysts by combinatorial methods, *Ann. Rev. Mater. Sci.*, 33, 557, 2003.
94. Stevens, D. et al., 64-Electrode PEM fuel cell studies of CO-tolerant hydrogen oxidation catalysts, *ECS Trans.*, 3, 355, 2006.
95. Staudt, R., Boyer, J., and Elter, J.F., Development, design, and performance of high temperature fuel cell technology, in *Extended Abstracts for 2005 Fuel Cell Seminar*, Palm Springs, CA, November 14–18, 2005, p. 207.
96. Markovic, N.M. and Ross, P.N., Surface science studies of model fuel cell electrocatalysts, *Surf. Sci. Rep.*, 45, 117, 2002.
97. Chatenet, M. et al., Oxygen reduction on silver catalysts in solutions containing various concentrations of sodium hydroxide: comparison with platinum, *J. Appl. Electrochem.*, 32, 1131, 2002.
98. Adzic, R.R. and Wang, J.X., Structures of surface adlayers and oxygen reduction kinetics, *Solid State Ionics*, 150, 105, 2002.
99. Nørskov, J.K. et al., Origin of the overpotential for oxygen reduction at a fuel-cell cathode, *J. Phys. Chem. B*, 108, 17886, 2004.
100. Thompsett, D., Pt alloys as oxygen reduction catalysts, in *Handbook of Fuel Cells: Fundamentals, Technology, and Applications*, 1st ed., Vielstich, W., Lamm, A., and Gasteiger, H.A., Eds., John Wiley & Sons, West Sussex, England, 2003, p. 467.
101. Kitchin, J.R. et al., Modification of the surface electronic and chemical properties of Pt(111) by subsurface 3d transition metals, *J. Chem. Phys.*, 120, 10240, 2004.

102. Zhang, J. et al., Platinum monolayer on nonnoble metal-noble metal core-shell nanoparticle electrocatalysts for O_2 reduction, *J. Phys. Chem. B*, 109, 22701, 2005.
103. Stamenkovic, V.R. et al., Improved oxygen reduction activity on Pt3Ni(111) via increased surface site availability, *Science*, 315, 493, 2007.
104. Xu, Y., Ruban, A.V., and Mavrikakis, M., Adsorption and dissociation of O_2 on Pt-Co and Pt-Fe alloys, *J. Am. Chem. Soc.*, 126, 4717, 2004.
105. Teliska, M. et al., Correlation of water activation, surface properties, and oxygen reduction reactivity of supported Pt–M/C bimetallic electrocatalysts using XAS, *J. Electrochem. Soc.*, 152, A2159, 2005.
106. Wang, Y. and Balbuena, P.B., Design of oxygen reduction bimetallic catalysts: ab-initio-derived thermodynamic guidelines, *J. Phys. Chem. B*, 109, 18902, 2005.
107. Balbuena, P.B. et al., Theoretical analysis of oxygen adsorption on Pt-based clusters alloyed with Co, Ni, or Cr embedded in a Pt matrix, *J. Phys. Chem. B*, 107, 13671, 2003.
108. Tamizhmani, G. and Capuano, G.A., Improved electrocatalytic oxygen reduction performance of platinum ternary alloy-oxide in solid-polymer-electrolyte fuel cells, *J. Electrochem. Soc.*, 141, 968, 1994.
109. Mukerjee, S. et al., Effect of preparation conditions of Pt alloys on their electronic, structural, and electrocatalytic activities for oxygen reduction: XRD, XAS, and electrochemical studies, *J. Phys. Chem.*, 99, 4577, 1995.
110. Ioroi, T. and Yasuda, K., Platinum-iridium alloys as oxygen reduction electrocatalysts for polymer electrolyte fuel cells, *J. Electrochem. Soc.*, 152, A1917, 2005.
111. Bonakdarpour, A. et al., Studies of transition metal dissolution from combinatorially sputtered, nanostructured $Pt_{1-x}M_x$ (M = Fe, Ni; $0 < x < 1$) electrocatalysts for PEM fuel cells, *J. Electrochem. Soc.*, 152, A61, 2005.
112. Bonakdarpour, A. et al., Dissolution of transition metals in combinatorially sputtered, $Pt_{1-x-y}M_xM'_y$ (M, M' = Co, Ni, Mn, Fe) PEMFC electrocatalysts, *J. Electrochem. Soc.*, 153, A1835, 2006.
113. Watanabe, M. et al., Activity and stability of ordered and disordered Co-Pt alloys for phosphoric acid fuel cells, *J. Electrochem. Soc.*, 141, 2659, 1994.
114. Makharia, R. et al., Automotive PEM fuel cell durability via advanced catalysts? in *2006 Fuel Cell Seminar*, Honolulu, HI, November 13–18, 2006.
115. Bouwman, P.J. et al., Platinum-iron phosphate electrocatalysts for oxygen reduction in PEMFCs, *J. Electrochem. Soc.*, 151, A1989, 2004.
116. Bagotzky, V.S. et al., Electrocatalysis of the oxygen reduction process on metal chelates in acid electrolyte, *J. Power Sources*, 2, 233, 1977.
117. van Veen, J.A.R., Colijn, H.A., and van Baar, J.F., On the effect of a heat treatment on the structure of carbon-supported metalloporphyrins and phthalocyanines, *Electrochim. Acta*, 33, 801, 1988.
118. Jasinski, R., A new fuel cell cathode catalyst, *Nature*, 201, 1212, 1964.
119. Biloul, A. et al., Oxygen electrocatalysis under fuel cell conditions: behaviour of cobalt porphyrins and tetraazaannulene analogues, *J. Appl. Electrochem.*, 26, 1139, 1996.
120. Gouérec, P. et al., Oxygen reduction in acid media catalysed by heat treated cobalt tetraazaannulene supported on an active charcoal: correlations between the performances after longevity tests and the active site configuration as seen by XPS and ToF-SIMS, *J. Electroanal. Chem.*, 422, 61, 1997.
121. Bouwkamp-Wijnoltz, A.L., Visscher, W., and van Veen, J.A.R., The selectivity of oxygen reduction by pyrolysed iron porphyrin supported on carbon, *Electrochim. Acta*, 43, 3141, 1998.
122. Faubert, G. et al., Iron catalysts prepared by high-temperature pyrolysis of tetraphenylporphyrins adsorbed on carbon black for oxygen reduction in polymer electrolyte fuel cells, *Electrochim. Acta*, 43, 341, 1998.

123. Gouérec, P., Savy, M., and Riga, J., Oxygen reduction in acidic media catalyzed by pyrolyzed cobalt macrocycles dispersed on an active carbon: the importance of the content of oxygen surface groups on the evolution of the chelate structure during the heat treatment, *Electrochim. Acta*, 43, 743, 1998.
124. Gojković, S.L., Gupta, S., and Savinell, R.F., Heat-treated iron(III) tetramethoxyphenyl porphyrin chloride supported on high-area carbon as an electrocatalyst for oxygen reduction, *Electrochim. Acta*, 45, 889, 1999.
125. Contamin, O. et al., Oxygen electroreduction catalysis: effect of sulfur addition on cobalt tetraazaannulene precursors, *Electrochim. Acta*, 45, 721, 1999.
126. Bouwkamp-Wijnoltz, A.L. et al., On active-site heterogeneity in pyrolyzed carbon-supported iron porphyrin catalysts for the electrochemical reduction of oxygen: an *in situ* Mössbauer study, *J. Phys. Chem. B*, 106, 12993, 2002.
127. Villers, D., Jacques-Bedard, X., and Dodelet, J.-P., Fe-based catalysts for oxygen reduction in PEM fuel cells, *J. Electrochem. Soc.*, 151, A1507, 2004.
128. Tarasevich, M.R., Radyushkina, K.A., and Zhutaeva, G.V., Electrocatalysis of the oxygen reaction by pyropolymers of N4 complexes, *Russ. J. Electrochem.*, 40, 1174, 2004.
129. Ortiz de Montellano, P.R., The mechanism of heme oxygenase, *Curr. Opin. Chem. Biol.*, 4, 221, 2000.
130. Veen, J.A.R.v., Colijn, H.A., and Baar, J.F.v., On the effect of a heat treatment on the structure of carbon-supported metalloporphyrins and phthalocyanines, *Electrochim. Acta*, 33, 801, 1988.
131. Lefèvre, M., Dodelet, J.P., and Bertr, P., Molecular oxygen reduction in PEM fuel cells: evidence for the simultaneous presence of two active sites in Fe-based catalysts, *J. Phys. Chem. B*, 106, 8705, 2002.
132. Bashyam, R. and Zelenay, P., A class of non-precious metal composite catalysts for fuel cells, *Nature*, 443, 63, 2006.
133. Gupta, S. et al., Heat-treated polyacrylonitrile-based catalysts for oxygen electroreduction, *J. Appl. Electrochem.*, 19, 19, 1989.
134. Faubert, G. et al., Activation and characterization of Fe-based catalysts for the reduction of oxygen in polymer electrolyte fuel cells, *Electrochim. Acta*, 43, 1969, 1998.
135. Bouwkamp-Wijnoltz, A.L. et al., Electrochemical reduction of oxygen: an alternative method to prepare active CoN_4 catalysts, *Electrochim. Acta*, 45, 379, 1999.
136. Okada, T. et al., A comparative study of organic cobalt complex catalysts for oxygen reduction in polymer electrolyte fuel cells, *J. Inorg. Organometal. Polym.*, 9, 199, 1999.
137. Bron, M. et al., EXAFS, XPS and electrochemical studies on oxygen reduction catalysts obtained by heat treatment of iron phenanthroline complexes supported on high surface area carbon black, *J. Electroanal. Chem.*, 535, 113, 2002.
138. Wei, G., Wainright, J.S., and Savinell, R.F., Catalytic activity for oxygen reduction reaction of catalysts consisting of carbon, nitrogen and cobalt, *J. New Mater. Electrochem. Syst.*, 3, 121, 2000.
139. Jaouen, F. et al., Oxygen reduction catalysts for polymer electrolyte fuel cells from the pyrolysis of iron acetate adsorbed on various carbon supports, *J. Phys. Chem. B*, 107, 1376, 2003.
140. Fernández, J.L., Walsh, D.A., and Bard, A.J., Thermodynamic guidelines for the design of bimetallic catalysts for oxygen electroreduction and rapid screening by scanning electrochemical microscopy. M-Co (M: Pd, Ag, Au), *J. Am. Chem. Soc.*, 127, 357, 2005.
141. Fernández, J.L. et al., Pd-Ti and Pd-Co-Au electrocatalysts as a replacement for platinum for oxygen reduction in proton exchange membrane fuel cells, *J. Am. Chem. Soc.*, 127, 13100, 2005.
142. Raghuveer, V., Manthiram, A., and Bard, A.J., Pd-Co-Mo electrocatalyst for the oxygen reduction reaction in proton exchange membrane fuel cells, *J. Phys. Chem. B*, 109, 22909, 2005.

143. Ishihara, A. et al., Tantalum oxynitride for a novel cathode of PEFC, *Electrochem. Solid-State Lett.*, 8, A201, 2005.
144. Liu, Y. et al., Zirconium oxide for PEFC cathodes, *Electrochem. Solid-State Lett.*, 8, A400, 2005.
145. Paik, C.H., Jarvi, T.D., and Grady, W.E.O., Extent of PEMFC cathode surface oxidation by oxygen and water measured by CV, *Electrochem. Solid-State Lett.*, 7, A82, 2004.
146. Wang, X., Kumar, R., and Myers, D.J., Effect of voltage on platinum dissolution, *Electrochem. Solid-State Lett.*, 9, A225, 2006.
147. Ferreira, P.J. et al., Instability of Pt/C electrocatalysts in proton exchange membrane fuel cells, *J. Electrochem. Soc.*, 152, A2256, 2005.
148. More, K., Borup, R., and Reeves, K., Identifying contributing degradation phenomena in PEM fuel cell membrane electride assemblies via electron microscopy, *ECS Trans.*, 3, 717, 2006.
149. Borup, R. et al., PEM fuel cell durability with transportation transient operation, *ECS Trans.*, 3, 879, 2006.
150. Yasuda, K. et al., Characteristics of a platinum black catalyst layer with regard to platinum dissolution phenomena in a membrane electrode assembly, *J. Electrochem. Soc.*, 153, A1599, 2006.
151. Xie, J. et al., Microstructural changes of membrane electrode assemblies during PEFC durability testing at high humidity conditions, *J. Electrochem. Soc.*, 152, A1011, 2005.
152. Wu, Y. et al., DDP 703. Hybrid Cell Test Report, Plug Power internal report, February 2003, Latham, NY.
153. Zhang, J. et al., Stabilization of platinum oxygen-reduction electrocatalysts using gold clusters, *Science*, 315, 220, 2007.
154. O'Hayre, R., Barnett, D.M., and Prinz, F.B., The triple phase boundary, *J. Electrochem. Soc.*, 152, A439, 2005.
155. Petrow, H.G. and Allen, R.J., Control of the Interaction of Novel Platinum-on-Carbon Electrocatalysts with Fluorinated Hydrocarbon Resins in the Preparation of Fuel Cell Electrodes, U.S. Patent 4,166,143, August 28, 1979.
156. Ticianelli, E.A. et al., Methods to advance technology of proton exchange membrane fuel cells, *J. Electrochem. Soc.*, 135, 2209, 1988.
157. Kangasniemi, K.H., Condit, D.A., and Jarvi, T.D., Characterization of Vulcan electrochemically oxidized under simulated PEM fuel cell conditions, *J. Electrochem. Soc.*, 151, E125, 2004.
158. Roen, L.M., Paik, C.H., and Jarvi, T.D., Electrocatalytic corrosion of carbon support in PEMFC cathodes, *Electrochem. Solid-State Lett.*, 7, A19, 2004.
159. Stevens, D.A. and Dahn, J.R., Thermal degradation of the support in carbon-supported platinum electrocatalysts for PEM fuel cells, *Carbon*, 43, 179, 2005.
160. Stevens, D.A. et al., Ex situ and in situ stability studies of PEMFC catalysts, *J. Electrochem. Soc.*, 152, A2309, 2005.
161. Reiser, C.A. et al., A reverse-current decay mechanism for fuel cells, *Electrochem. Solid-State Lett.*, 8, A273, 2005.
162. Meyers, J.P. and Darling, R.M., Model of carbon corrosion in PEM fuel cells, *J. Electrochem. Soc.*, 153, A1432, 2006.
163. Du, B. et al., Impact of cold start and hot stop on the performance and durability of a proton exchange membrane (PEM) fuel cell, in *Extended Abstracts of 2006 Fuel Cell Seminar*, Honolulu, HI, November 13–18, 2006, p. 61.
164. Qi, Z. et al., Investigation of PEM fuel cell cathode carbon corrosion under different conditions, in *Extended Abstracts of 2005 Fuel Cell Seminar*, Palm Springs, CA, November 13–18, 2005.
165. Ye, S. et al., Degradation resistant cathodes in polymer electrolyte membrane fuel cells, *ECS Trans.*, 3, 657, 2006, p. 138.

166. Yu, P.T. et al., The effect of air purge on the degradation of PEM fuel cells during startup and shutdown procedures, in *Proceedings of the AIChE 2004 Spring National Meeting*, 37C, New Orleans, April 25–29, 2004.
167. Perry, M.L., Patterson, T., and Reiser, C., Systems strategies to mitigate carbon corrosion in fuel cells, *ECS Trans.*, 3, 783, 2006.
168. Takagi, Y. and Takakuwa, Y., Effect of shutoff sequence of hydrogen and air on performance degradation in PEFC, *ECS Trans.*, 3, 855, 2006.
169. Yu, P.T. et al., The impact of carbon stability on PEM fuel cell startup and shutdown voltage degradation, *ECS Trans.*, 3, 797, 2006.
170. Okada, T., Sugiura, M., and Xie, G., Degradation of carbon supported Pt anode and cathode catalysts in PEM fuel cells, *ECS Trans.*, 3, 667, 2006.
171. Ball, S.C. et al., The effect of dynamic and steady state voltage excursions on the stability of carbon supported Pt and PtCo catalysts, *ECS Trans.*, 3, 595, 2006.
172. Chizawa, H. et al., Study of accelerated test protocol for PEFC focusing on carbon corrosion, *ECS Trans.*, 3, 645, 2006.
173. Taniguchi, A. et al., Analysis of electrocatalyst degradation in PEMFC caused by cell reversal during fuel starvation, *J. Power Sources*, 130, 42, 2004.
174. Baumgartner, W.R.R. et al., Electrocatalytic corrosion of carbon support in PEMFC at fuel starvation, *ECS Trans.*, 3, 811, 2006.
175. Takagi, Y., Sato, Y., and Wang, Z., Effect of anode catalyst support on MEA degradation caused by hydrogen-starved operation of a PEFC, *ECS Trans.*, 3, 827, 2006.
176. Patterson, T.W. and Darling, R.M., Damage to the cathode catalyst of a PEM fuel cell caused by localized fuel starvation, *Electrochem. Solid-State Lett.*, 9, A183, 2006.
177. Kinoshita, K., *Carbon: Electrochemical and Physiochemical Properties*, John Wiley & Sons, New York, 1988.
178. Tada, T., High dispersion catalysts including novel carbon supports, in *Handbook of Fuel Cells: Fundamentals, Technology, and Applications*, 1st ed., Vielstich, W., Lamm, A., and Gasteiger, H.A., Eds., John Wiley & Sons, West Sussex, England, 2003, p. 481.
179. Waje, M.M. et al., Durability investigation of cup-stacked carbon nanotubes supported Pt as PEMFC catalyst, *ECS Trans.*, 3, 677, 2006.
180. Qi, Z. and Pickup, P.G., Novel supported catalysts: platinum and platinum oxide nanoparticles dispersed on polypyrrole/polystyrenesulfonate particles, *Chem. Commun.*, 15, 1998.
181. Qi, Z. and Pickup, P.G., High performance conducting polymer supported oxygen reduction catalysts, *Chem. Commun.*, 2299, 1998.
182. Jia, N. et al., Modification of carbon supported catalysts to improve performance in gas diffusion electrodes, *Electrochim. Acta*, 46, 2863, 2001.
183. Easton, E.B. et al., Chemical modification of proton exchange membrane fuel cell catalysts with a sulfonated silane, *Electrochem. Solid-State Lett.*, 4, A59, 2001.
184. Xu, Z., Qi, Z., and Kaufman, A., Advanced fuel cell catalysts, *Electrochem. Solid-State Lett.*, 6, A171, 2003.
185. Xu, Z., Qi, Z., and Kaufman, A., Superior catalysts for proton exchange membrane fuel cells, *Electrochem. Solid-State Lett.*, 8, A313, 2005.
186. Gullá, A.F. et al., Toward improving the performance of PEM fuel cell by using mix metal electrodes prepared by dual IBAD, *J. Electrochem. Soc.*, 153, A366, 2006.
187. Baker, W.S. et al., Enhanced oxygen reduction activity in acid by tin-oxide supported Au nanoparticle catalysts, *J. Electrochem. Soc.*, 153, A1702, 2006.
188. Trivedi, D.C., Polyanilines, in *Handbook of Organic Conductive Molecules and Polymers*, Nalwa, H.S., Eds., John Wiley & Sons, Chichester, U.K., 1997, p. 505.
189. Andreev, V.N., Belova, N.N., and Timofeev, S.V., Redox conversions of polyaniline in composite polyaniline-Nafion polymer films, *Russ. J. Electrochem.*, 39, 419, 2003.

190. Andreev, V.N., Effect of solution acidity on the electrochemical behavior of Nafion-polyaniline films, *Russ. J. Electrochem.*, 41, 200, 2005.
191. Colomban, P. and Tomkinson, J., Novel forms of hydrogen in solids: the 'ionic' proton and the 'quasi-free' proton, *Solid State Ionics*, 97, 123, 1997.
192. Inzelt, G. et al., Electron and proton conducting polymers: recent developments and prospects, *Electrochim. Acta*, 45, 2403, 2000.
193. Kompan, M.E., Sapurina, I.Y., and Stejskal, J., Overcoming the low-dimension crisis in the active zone of fuel cells, *Technol. Phys. Lett.*, 32, 213, 2006.
194. Mondal, S.K. et al., Electrooxidation of ascorbic acid on polyaniline and its implications to fuel cells, *J. Power Sources*, 145, 16, 2005.
195. Rajesh, B. et al., Electronically conducting hybrid material as high performance catalyst support for electrocatalytic application, *J. Power Sources*, 141, 35, 2005.
196. Rimbu, G.A. et al., The morphology control of polyaniline as conducting polymer in fuel cell technology, *J. Optoelectron. Adv. Mater.*, 8, 670, 2006.
197. Ma, C.A. et al., Electro-oxidation of formaldehyde on polyaniline prepared in 1-ethylimidazolium trifluoroacetate, *Electrochem. Solid-State Lett.*, 8, G122, 2005.
198. Zabrodskiĭ, A.G. et al., Carbon supported polyaniline as anode catalyst: pathway to platinum-free fuel cells, *Technol. Phys. Lett.*, 32, 758, 2006.
199. Wilson, M.S. and Gottesfeld, S., High performance catalyzed membranes of ultra-low Pt loadings for polymer electrolyte fuel cells, *J. Electrochem. Soc.*, 139, L28, 1992.
200. Wilson, M.S. and Gottesfeld, S., Thin-film catalyst layers for polymer electrolyte fuel cell electrodes, *J. Appl. Electrochem.*, 22, 1, 1992.
201. Uchida, M. et al., New preparation method for polymer-electrolyte fuel cells, *J. Electrochem. Soc.*, 142, 463, 1995.
202. Xie, Z. et al., Functionally graded cathode catalyst layers for polymer electrolyte fuel cells, *J. Electrochem. Soc.*, 152, A1171, 2005.
203. Uchida, M. et al., Effects of microstructure of carbon support in the catalyst layer on the performance of polymer-electrolyte fuel cells, *J. Electrochem. Soc.*, 143, 2245, 1996.
204. Xie, J. et al., Porosimetry of MEAs made by "thin film decal" method and its effect on performance of PEFCs, *J. Electrochem. Soc.*, 151, A1841, 2004.
205. Qi, Z. and Kaufman, A., Low Pt loading high performance cathodes for PEM fuel cells, *J. Power Sources*, 113, 37, 2003.
206. Jung, U.H. et al., Improvement of low-humidity performance of PEMFC by addition of hydrophilic SiO_2 particles to catalyst layer, *J. Power Sources*, 159, 529, 2006.
207. Guo, Q., unpublished results, 2006.
208. Weber, A.Z. and Newman, J., Modeling transport in polymer-electrolyte fuel cells, *Chem. Rev.*, 104, 4679, 2004.
209. Wang, C.-Y., Fundamental models for fuel cell engineering, *Chem. Rev.*, 104, 4727, 2004.
210. Chen, H., Johnson, J.K., and Sholl, D.S., Transport diffusion of gases is rapid in flexible carbon nanotubes, *J. Phys. Chem. B*, 110, 1971, 2006.
211. Iijima, S., Helical microtubules of graphitic carbon, *Nature*, 354, 56, 1991.
212. Yoshitake, T. et al., Preparation of fine platinum catalyst supported on single-wall carbon nanohorns for fuel cell application, *Physica B*, 323, 124, 2002.
213. Li, W. et al., Carbon nanotubes as support for cathode catalyst of a direct methanol fuel cell, *Carbon*, 40, 791, 2002.
214. Matsumoto, T. et al., Efficient usage of highly dispersed Pt on carbon nanotubes for electrode catalysts of polymer electrolyte fuel cells, *Catal. Today*, 90, 277, 2004.
215. Xing, Y., Synthesis and electrochemical characterization of uniformly-dispersed high loading Pt nanoparticles on sonochemically-treated carbon nanotubes, *J. Phys. Chem. B*, 108, 19255, 2004.

216. Yang, J., Liu, D.-J., and Wang, X., A novel electrode catalyst for proton exchange membrane fuel cell using aligned carbon nanotubes, in *2005 Fuel Cell Seminar Extended Abstract*, Palm Springs, CA, November 14–18, 2005, p. 240.
217. Liu, D.-J. and Yang, J., Method of Fabricating Electrode Catalyst Layers with Directionally Oriented Carbon Support for Proton Exchange Membrane Fuel Cell, U.S. Patent Application 20060269827, November 30, 2006.
218. Gooding, J.J., Nanostructuring electrodes with carbon nanotubes: a review on electrochemistry and applications for sensing, *Electrochim. Acta*, 50, 3049, 2005.
219. Smalley, R.E., private communication, 2004.
220. Debe, M., Novel catalysts, catalysts support and catalysts coated membrane methods, in *Handbook of Fuel Cells: Fundamentals, Technology, and Applications*, 1st ed., Vielstich, W., Lamm, A., and Gasteiger, H.A., Eds., John Wiley & Sons, West Sussex, England, 2003, p. 576.
221. Debe, M. et al., High voltage stability of nanostructured thin film, *J. Power Sources*, 161, 1002, 2006.
222. Debe, M.K., Steinbach, A.J., and Noda, K., Stop-start and high-current durability testing of nanostructured thin film catalysts for PEM fuel cells, *ECS Trans.*, 3, 835, 2006.
223. Schmoeckel, A.K. et al., Nanostructured thin film ternary catalyst activities for oxygen reduction, in *Extended Abstracts of 2006 Fuel Cell Seminar*, Honolulu, HI, November 13–18, 2006, p. 293.
224. Shah, K., Shin, W.C., and Besser, R.S., Novel microfabrication approaches for directly patterning PEM fuel cell membranes, *J. Power Sources*, 123, 172, 2003.
225. Roziere, J. and Jones, D.J., Non-fluorinated polymer materials for proton exchange membrane fuel cells, *Ann. Rev. Mater. Res.*, 33, 503, 2003.
226. Eisenberg, A. and Yeager, H.L., Eds., *Perfluorinated Ionomer Membranes*, American Chemical Society, Washington, DC, 1982.
227. Heitner-Wirguin, C., Recent advances in perfluorinated ionomer membranes: structure, properties and applications, *J. Membr. Sci.*, 120, 1, 1996.
228. Doyle, M. and Rajendran, G., Perfluorinated membranes, in *Handbook of Fuel Cells: Fundamentals, Technology, and Applications*, 1st ed., Vielstich, W., Lamm, A., and Gasteiger, H.A., Eds., John Wiley & Sons, West Sussex, England, 2003, p. 351.
229. Savadogo, O., Emerging membranes for electrochemical systems: I. Solid polymer electrolyte membranes for fuel cell systems, *J. New Mater. Electrochem. Syst.*, 1, 47, 1998.
230. Yeo, R.S., Dual cohesive energy densities of perfluorosulphonic acid (Nafion) membrane, *Polymer*, 21, 432, 1980.
231. Gebel, G., Aldebert, P., and Pineri, M., Swelling study of perfluorosulphonated ionomer membranes, *Polymer*, 34, 333, 1993.
232. Liu, W. and Crum, M., Effective testing matrix for studying membrane durability in PEM fuel cells. Part 1. Chemical durability, *ECS Trans.*, 3, 531, 2006.
233. Crum, M. and Liu, W., Effective testing matrix for studying membrane durability in PEM fuel cells. Part 2. Mechanical durability and combined mechanical and chemical durability, *ECS Trans.*, 3, 541, 2006.
234. Cleghorn, S., Kolde, J., and Liu, W., Catalyst coated composite membranes, in *Handbook of Fuel Cells: Fundamentals, Technology, and Applications*, 1st ed., Vielstich, W., Lamm, A., and Gasteiger, H.A., Eds., John Wiley & Sons, West Sussex, England, 2003, p. 566.
235. Stucki, S. et al., PEM water electrolysers: evidence for membrane failure in 100 kW demonstration plants, *J. Appl. Electrochem.*, 28, 1041, 1998.
236. Grot, W.G., Laminates of Support Material and Fluorinated Polymer Containing Pendant Side Chains Containing Sulfonyl Groups, U.S. Patent 3,770567, November 6, 1973.
237. Wantanabe, I. et al., Reinforced Ion-Exchange Membrane, U.S. Patent 4,072793, February 7, 1978.

238. Stone, C. and Galis, G., Improved composite membranes and related performance in commercial PEM fuel cells, in *Extended Abstracts of 2006 Fuel Cell Seminar*, Honolulu, HI, November 13-18, 2006, p. 265.
239. Ralph, T.R. and Barnwell, D., Extended PEMFC durability from MEAs based on a new reinforced membrane, *ECS Trans.*, 3, 579, 2006.
240. Nakao, M. and Yoshitake, M., Composite perfluorinated membranes, in *Handbook of Fuel Cells: Fundamentals, Technology, and Applications*, 1st ed., Vielstich, W., Lamm, A., and Gasteiger, H.A., Eds., John Wiley & Sons, West Sussex, England, 2003, p. 412.
241. Miyake, N. et al., Durability of Asahi Kasei Aciplex membrane for PEM fuel cell application, in *Proceedings of the 206th ECS Meeting: Proton Conducting Membrane Fuel Cells IV*, Honolulu, HI, October 3–8, 2004, p. 333.
242. Reichert, D., DuPont marketing presentation, 2007.
243. LaConti, A.B., Hamdan, M., and McDonald, R.C., Mechanisms of membrane degradation, in *Handbook of Fuel Cells: Fundamentals, Technology, and Applications*, 1st ed., Vielstich, W., Lamm, A., and Gasteiger, H.A., Eds., John Wiley & Sons, West Sussex, England, 2003, p. 647.
244. Schiraldi, D.A., Zhou, C., and Zawodzinski, J. T., Chemical durability studies of PFSA polymers and model compounds under mimic fuel cell condition, in *210th Meeting of the Electrochemical Society*, Cancun, Mexico, October 29–November 3, 2006.
245. Jones, D.J. and Rozière, J., Inorganic/organic composite membranes, in *Handbook of Fuel Cells: Fundamentals, Technology, and Applications*, 1st ed., Vielstich, W., Lamm, A., and Gasteiger, H.A., Eds., John Wiley & Sons, West Sussex, England, 2003, p. 447.
246. Lin, J.-C., Kunz, H.R., and Fenton, J.M., Membrane/electrode additives for low-humidification operation, in *Handbook of Fuel Cells: Fundamentals, Technology, and Applications*, 1st ed., Vielstich, W., Lamm, A., and Gasteiger, H.A., Eds., John Wiley & Sons, West Sussex, England, 2003, p. 456.
247. Florjańczyk, Z., Wielgus-Barry, E., and Połtarzewski, Z., Radiation-modified Nafion membranes for methanol fuel cells, *Solid State Ionics*, 145, 119, 2001.
248. DesMarteau, D.D., Novel perfluorinated ionomers and ionenes, *J. Fluorine Chem.*, 72, 203, 1995.
249. Appleby, A.J. et al., Polymeric perfluoro bis-sulfonimides as possible fuel cell electrolytes, *J. Electrochem. Soc.*, 140, 109, 1993.
250. Sumner, J.J. et al., Proton conductivity in Nation117 and in a novel bis[(perfluoroalkyl) sulfonyl]imide ionomer membrane, *J. Electrochem. Soc.*, 145, 107, 1998.
251. Endoh, E., Kawazoe, H., and Honmura, S., Highly durable MEA for PEMFC under high temperature and low humidity conditions, in *Extended Abstracts of 2006 Fuel Cell Seminar*, Honolulu, HI, November 13–18, 2006, p. 284.
252. Alberti, G., Casciola, M., and Costantino, U., Inorganic ion-exchange pellicles obtained by delamination of α-zirconium phosphate crystals, *J. Colloid Interface Sci.*, 107, 256, 1985.
253. Alberti, G. et al., Formation of colloidal dispersions of layered γ-zirconium phosphate in water/acetone mixtures, *J. Colloid Interface Sci.*, 188, 27, 1997.
254. Mauritz, K.A. and Warren, R.M., Microstructural evolution of a silicon oxide phase in a perfluorosulfonic acid ionomer by an in situ sol-gel reaction. 1. Infrared spectroscopic studies, *Macromolecules*, 22, 1730, 1989.
255. Mauritz, K.A. and Stefanithis, I.D., Microstructural evolution of a silicon oxide phase in a perfluorosulfonic acid ionomer by an in situ sol-gel reaction. 2. Dielectric relaxation studies, *Macromolecules*, 23, 1380, 1990.
256. Stefanithis, I.D. and Mauritz, K.A., Microstructural evolution of a silicon oxide phase in a perfluorosulfonic acid ionomer by an in situ sol-gel reaction. 3. Thermal analysis studies, *Macromolecules*, 23, 2397, 1990.

257. Gummaraju, R.V., Moore, R.B., and Mauritz, K.A., Asymmetric Nafion®/silicon oxide hybrid membranes via the in situ sol-gel reaction for tetraethoxysilane, *J. Polym Sci. B*, 34, 2383, 1996.
258. Mauritz, K.A., Organic-inorganic hybrid materials: perfluorinated ionomers as sol-gel polymerization templates for inorganic alkoxides, *Mater. Sci. Eng. C*, 6, 121, 1998.
259. Apichatachutapan, W., Moore, R.B., and Mauritz, K.A., Asymmetric Nafion/(zirconium oxide) hybrid membranes via in situ sol-gel chemistry, *J. Appl. Polym. Sci.*, 62, 417, 1996.
260. Shao, P.L., Mauritz, K.A., and Moore, R.B., Perfluorosulfonate ionomer/mixed inorganic oxide nanocomposites via polymer-in situ sol-gel chemistry, *Chem. Mater.*, 7, 192, 1995.
261. Baradie, B., Dodelet, J.P., and Guay, D., Hybrid Nafion®-inorganic membrane with potential applications for polymer electrolyte fuel cells, *J. Electroanal. Chem.*, 489, 101, 2000.
262. Wang, H. et al., Nafion-bifunctional silica composite proton conductive membranes, *J. Mater. Chem.*, 12, 834, 2002.
263. Savadogo, O., Surface chemistry and electrocatalytic activity of the HER on nickel modified with heteropolyacids, *J. Electrochem. Soc.*, 139, 1082, 1992.
264. Nakamura, O., Ogino, I., and Kodama, T., Temperature and humidity ranges of some hydrates of high-proton-conductive dodecamolybdophosphoric acid and dodecatungstophosphoric acid crystals under an atmosphere of hydrogen or either oxygen or air, *Solid State Ionics*, 3/4, 347, 1981.
265. Malhotra, S. and Datta, R., Membrane-supported nonvolatile acidic electrolytes allow higher temperature operation of proton-exchange membrane fuel cells, *J. Electrochem. Soc.*, 144, L23, 1997.
266. Tazi, B. and Savadogo, O., Parameters of PEM fuel-cells based on new membranes fabricated from Nafion-silicotungstic acid and thiophene, *Electrochim. Acta*, 45, 4329, 2000.
267. Staiti, P. et al., Hybrid Nafion-silica membranes doped with heteropolyacids for application in direct methanol fuel cells, *Solid State Ionics*, 145, 101, 2001.
268. Hoel, D. and Grunwald, E., High protonic conduction of polybenzimidazole films, *J. Phys. Chem.*, 81, 2135, 1977.
269. Savadogo, O. and Xing, B., Hydrogen/oxygen polymer electrolyte membrane fuel cell (PEMFC) based on acid-doped polybenzimidazole (PBI), *J. New Mater. Electrochem. Syst.*, 3, 345, 2000.
270. Wainright, J.S. et al., Acid-doped polybenzimidazoles: a new polymer electrolyte, *J. Electrochem. Soc.*, 142, L121, 1995.
271. Wainright, J.S., Litt, M.H., and Savinell, R.F., High-temperature membranes, in *Handbook of Fuel Cells: Fundamentals, Technology, and Applications*, 1st ed., Vielstich, W., Lamm, A., and Gasteiger, H.A., Eds., John Wiley & Sons, West Sussex, England, 2003, p. 436.
272. Savinell, R. et al., A polymer electrolyte for operation at temperatures up to 200°C, *J. Electrochem. Soc.*, 141, L46, 1994.
273. Ma, Y.L. et al., Conductivity of PBI membranes for high-temperature polymer electrolyte fuel cells, *J. Electrochem. Soc.*, 151, A8, 2004.
274. Zecevic, S.K. et al., Kinetics of O_2 reduction on a Pt electrode covered with a thin film of solid polymer electrolyte, *J. Electrochem. Soc.*, 144, 2973, 1997.
275. Kerres, J.A., Blended and cross-linked ionomer membranes for application in membrane fuel cells, *Fuel Cells*, 5, 230, 2005.
276. Li, Q. et al., PBI-based polymer membranes for high temperature fuel cells: preparation, characterization and fuel cell demonstration, *Fuel Cells*, 4, 147, 2004.
277. Xiao, L. et al., Synthesis and characterization of pyridine-based polybenzimidazoles for high temperature polymer electrolyte membrane fuel cell applications, *Fuel Cells*, 5, 287, 2005.

278. Xiao, L. et al., High-temperature polybenzimidazole fuel cell membranes via a sol-gel process, *Chem. Mater.*, 17, 5328, 2005.
279. Samms, S.R., Wasmus, S., and Savinell, R.F., Thermal stability of proton conducting acid doped polybenzimidazole in simulated fuel cell environments, *J. Electrochem. Soc.*, 143, 1225, 1996.
280. Wycisk, R., Lee, J.K., and Pintauro, P.N., Sulfonated polyphosphazene-polybenzimidazole membranes for DMFCs, *J. Electrochem. Soc.*, 152, A892, 2005.
281. Weng, D. et al., Electro-osmotic drag coefficient of water and methanol in polymer electrolytes at elevated temperatures, *J. Electrochem. Soc.*, 143, 1260, 1996.
282. Wang, J.T., Wasmus, S., and Savinell, R.F., Real-time mass spectrometric study of the methanol crossover in a direct methanol fuel cell, *J. Electrochem. Soc.*, 143, 1233, 1996.
283. Hasiotis, C. et al., Development and characterization of acid-doped polybenzimidazole/sulfonated polysulfone blend polymer electrolytes for fuel cells, *J. Electrochem. Soc.*, 148, A513, 2001.
284. Asensio, J.A., Borros, S., and Gomez-Romero, P., Polymer electrolyte fuel cells based on phosphoric acid-impregnated poly(2,5-benzimidazole) membranes, *J. Electrochem. Soc.*, 151, A304, 2004.
285. Zhai, Y. et al., Degradation study on MEA in H_3PO_4/PBI high-temperature PEMFC life test, *J. Electrochem. Soc.*, 154, B72, 2007.
286. Li, Q., Hjuler, H.A., and Bjerrum, N.J., Phosphoric acid doped polybenzimidazole membranes: physiochemical characterization and fuel cell applications, *J. Appl. Electrochem.*, 31, 773, 2001.
287. Schmidt, T.J. and Baurmeister, J., Durability and reliability in high-temperature reformed hydrogen PEFCs, *ECS Trans.*, 3, 861, 2006.
288. Glipa, X. et al., Synthesis and characterisation of sulfonated polybenzimidazole: a highly conducting proton exchange polymer, *Solid State Ionics*, 97, 323, 1997.
289. Kawahara, M. et al., Synthesis and proton conductivity of sulfopropylated poly(benzimidazole) films, *Solid State Ionics*, 136/137, 1193, 2000.
290. Rozière, J. et al., On the doping of sulfonated polybenzimidazole with strong bases, *Solid State Ionics*, 145, 61, 2001.
291. Tang, H.-G. and Sherrington, D.C., Polymer-supported Pd(II) Wacker-type catalysts. 1. Synthesis and characterization of the catalysts, *Polymer*, 34, 2821, 1993.
292. Kreuer, K.D., Hydrocarbon membranes, in *Handbook of Fuel Cells: Fundamentals, Technology, and Applications*, 1st ed., Vielstich, W., Lamm, A., and Gasteiger, H.A., Eds., John Wiley & Sons, West Sussex, England, 2003, p. 420.
293. Wei, J., Stone, C., and Steck, A.E., Trifluorostyrene and Substituted Trifluorostyrene Copolymeric Compositions and Ion-Exchange Membranes Formed Therefrom, U.S. Patent 5,422,411, June 6, 1995.
294. Ehrenberg, S.G. et al., Fuel Cell Incorporating Novel Ion-Conducting Membrane, U.S. Patent 5,679,482, October 21, 1997.
295. Shindo, D., HOKU hydrocarbon membrane, in *2006 Fuel Cell Seminar*, Honolulu, HI, November 13–18, 2006, p. PT4.
296. Cao, S. et al., UIon Conductive Random Copolymers, U.S. Patent Application 2005/0181256, August 18, 2005.
297. Cao, S. et al., Sulfonated Copolymer, U.S. Patent Application 2004/0039148, February 26, 2004.
298. Nam, K. et al., Acid-Base Proton Conducting Polymer Blend Membrane, U.S. Patent Application 2003/0219640, November 27, 2003.
299. Kanaoka, N., Development of MEA for next generation automotive fuel cells at Honda, in *Extended Abstracts of 2006 Fuel Cell Seminar*, Honolulu, HI, 2006, p. 49.
300. Kreuer, K.D., On the development of proton conducting polymer membranes for hydrogen and methanol fuel cells, *J. Membr. Sci.*, 185, 29, 2001.

301. Genies, C. et al., Soluble sulfonated naphthalenic polyimides as materials for proton exchange membranes, *Polymer*, 42, 359, 2001.
302. Yi, J. et al., Development of a low cost, durable membrane and membrane electrode assembly for fuel cell applications, in *Extended Abstracts of 2006 Fuel Cell Seminar*, Honolulu, HI, November 13–18, 2006, p. 261.
303. Guo, Q. et al., Sulfonated and crosslinked polyphosphazene-based proton-exchange membranes, *J. Membr. Sci.*, 154, 175, 1999.
304. Mathias, M.F. et al., Diffusion media materials and characterization, in *Handbook of Fuel Cells: Fundamentals, Technology, and Applications*, 1st ed., Vielstich, W., Lamm, A., and Gasteiger, H.A., Eds., John Wiley & Sons, West Sussex, England, 2003, p. 517.
305. Lee, W.-K. et al., The effects of compression and gas diffusion layers on the performance of a PEM fuel cell, *J. Power Sources*, 84, 45, 1999.
306. Escribano, S. et al., Characterization of PEMFCs gas diffusion layers properties, *J. Power Sources*, 156, 8, 2006.
307. Nguyen, T.V. and White, R.E., A water and heat management model for proton-exchange-membrane fuel cells, *J. Electrochem. Soc.*, 140, 2178, 1993.
308. Wang, Z.H., Wang, C.Y., and Chen, K.S., Two-phase flow and transport in the air cathode of proton exchange membrane fuel cells, *J. Power Sources*, 94, 40, 2001.
309. Pasaogullari, U. and Wang, C.Y., Liquid water transport in gas diffusion layer of polymer electrolyte fuel cells, *J. Electrochem. Soc.*, 151, A399, 2004.
310. Nam, J.H. and Kaviany, M., Effective diffusivity and water-saturation distribution in single- and two-layer PEMFC diffusion medium, *Int. J. Heat Mass Transfer*, 46, 4595, 2003.
311. Wang, Y., Wang, C.Y., and Chen, K.S., Elucidating differences between carbon paper and carbon cloth in polymer electrolyte fuel cells, *Electrochim. Acta*, 52(12), 3965, 2007.
312. Lim, C. and Wang, C.Y., Effects of hydrophobic polymer content in GDL on power performance of a PEM fuel cell, *Electrochim. Acta*, 49, 4149, 2004.
313. Prasanna, M. et al., Influence of cathode gas diffusion media on the performance of the PEMFCs, *J. Power Sources*, 131, 147, 2004.
314. Williams, M.V. et al., Characterization of gas diffusion layers for PEMFC, *J. Electrochem. Soc.*, 151, A1173, 2004.
315. Gurau, V. et al., Characterization of transport properties in gas diffusion layers for proton exchange membrane fuel cells, *J. Power Sources*, 160, 1156, 2006.
316. Pai, Y.-H. et al., CF_4 plasma treatment for preparing gas diffusion layers in membrane electrode assemblies, *J. Power Sources*, 161, 275, 2006.
317. Qi, Z. and Kaufman, A., Improvement of water management by a microporous sublayer for PEM fuel cells, *J. Power Sources*, 109, 38, 2002.
318. Paganin, V.A., Ticianelli, E.A., and Gonzalez, E.R., Development and electrochemical studies of gas diffusion electrodes for polymer electrolyte fuel cells, *J. Appl. Electrochem.*, 26, 297, 1996.
319. Kong, C.S. et al., Influence of pore-size distribution of diffusion layer on mass-transport problems of proton exchange membrane fuel cells, *J. Power Sources*, 108, 185, 2002.
320. Jordan, L.R. et al., Diffusion layer parameters influencing optimal fuel cell performance, *J. Power Sources*, 86, 250, 2000.
321. Pasaogullari, U. and Wang, C.-Y., Two-phase transport and the role of micro-porous layer in polymer electrolyte fuel cells, *Electrochim. Acta*, 49, 4359, 2004.
322. Pasaogullari, U., Wang, C.-Y., and Chen, K. S., Two-phase transport in polymer electrolyte fuel cells with bilayer cathode gas diffusion media, *J. Electrochem. Soc.*, 152, A1574, 2005.
323. Weber, A.Z. and Newman, J., Effects of microporous layers in polymer electrolyte fuel cells, *J. Electrochem. Soc.*, 152, A677, 2005.

324. Ge, J., Higier, A., and Liu, H., Effect of gas diffusion layer compression on PEM fuel cell performance, *J. Power Sources*, 159, 922, 2006.
325. Bazylak, A. et al., Effect of compression on liquid water transport and microstructure of PEMFC gas diffusion layers, *J. Power Sources*, 163, 784, 2007.
326. Meng, H. and Wang, C.-Y., Electron transport in PEFCs, *J. Electrochem. Soc.*, 151, A358, 2004.
327. Shores, D.A. and Deluga, G.A., Basic materials corrosion issues, in *Handbook of Fuel Cells: Fundamentals, Technology, and Applications*, 1st ed., Vielstich, W., Lamm, A., and Gasteiger, H.A., Eds., John Wiley & Sons, West Sussex, England, 2003, p. 273.
328. Mepsted, G.O. and Moore, J.M., Performance and durability of bipolar plate materials, in *Handbook of Fuel Cells: Fundamentals, Technology, and Applications*, 1st ed., Vielstich, W., Lamm, A., and Gasteiger, H.A., Eds., John Wiley & Sons, West Sussex, England, 2003, p. 286.
329. Roβberg, K. and Trapp, V., Graphite-based bipolar plates, in *Handbook of Fuel Cells: Fundamentals, Technology, and Applications*, 1st ed., Vielstich, W., Lamm, A., and Gasteiger, H.A., Eds., John Wiley & Sons, West Sussex, England, 2003, p. 308.
330. Wind, J. et al., Metal bipolar plates and coatings, in *Handbook of Fuel Cells: Fundamentals, Technology, and Applications*, 1st ed., Vielstich, W., Lamm, A., and Gasteiger, H.A., Eds., John Wiley & Sons, West Sussex, England, 2003, p. 294.
331. Du, B. et al., Neutron radiography as a new tool for *in situ* PEM fuel cell diagnostics, in *2004 Fuel Cell Seminar*, San Antonio, TX, November 1–5, 2004, p. P-94.
332. Du, B. et al., Tuning hydrogen content for improved PEMFC water management: a neutron radiography study, in *2nd International Conference on Green & Sustainable Chemistry*, Washington, DC, June 20–24, 2005, p. TP 319.
333. Du, B. et al., Reformate hydrogen content and PEMFC water management: a neutron radiography study, in *9th Grove Fuel Cell Symposium*, London, October 4–6, 2005, p. GR25.
334. Du, B. et al., Applications of neutron radiography in PEM fuel cell research and development, in *8th World Conference on Neutron Radiography*, Gaithersburg, MD, October 16–19, 2006, p. P005.
335. Blunk, R., Zhong, F., and Owens, J., Automotive composite fuel cell bipolar plates: hydrogen permeation concerns, *J. Power Sources*, 159, 533, 2006.
336. Blunk, R. et al., Polymeric composite bipolar plates for vehicle applications, *J. Power Sources*, 156, 151, 2006.
337. Besmann, T.M. et al., Carbon/carbon composite bipolar plate for proton exchange membrane fuel cells, *J. Electrochem. Soc.*, 147, 4083, 2000.
338. Wolf, H. and Willert-Porada, M., Electrically conductive LCP–carbon composite with low carbon content for bipolar plate application in polymer electrolyte membrane fuel cell, *J. Power Sources*, 153, 41, 2006.
339. Wu, M. and Shaw, L.L., A novel concept of carbon-filled polymer blends for applications in PEM fuel cell bipolar plates, *Int. J. Hydrogen Energy*, 30, 373, 2005.
340. Huang, J., Baird, D.G., and McGrath, J.E., Development of fuel cell bipolar plates from graphite filled wet-lay thermoplastic composite materials, *J. Power Sources*, 150, 110, 2005.
341. Oh, M.H., Yoon, Y.S., and Park, S.G., The electrical and physical properties of alternative material bipolar plate for PEM fuel cell system, *Electrochim. Acta*, 50, 777, 2004.
342. Li, M.C. et al., Electrochemical corrosion characteristics of type 316 stainless steel in simulated anode environment for PEMFC, *Electrochim. Acta*, 48, 1735, 2003.
343. Wang, H., Sweikart, M.A., and Turner, J.A., Stainless steel as bipolar plate material for polymer electrolyte membrane fuel cells, *J. Power Sources*, 115, 243, 2003.
344. Wang, H. and Turner, J.A., Ferritic stainless steels as bipolar plate material for polymer electrolyte membrane fuel cells, *J. Power Sources*, 128, 193, 2004.

345. Silva, R.F. et al., Surface conductivity and stability of metallic bipolar plate materials for polymer electrolyte fuel cells, *Electrochim. Acta*, 51, 3592, 2006.
346. Wang, S.-H., Peng, J., and Lui, W.-B., Surface modification and development of titanium bipolar plates for PEM fuel cells, *J. Power Sources*, 160, 485, 2006.
347. Weil, K.S. et al., Boronization of nickel and nickel clad materials for potential use in polymer electrolyte membrane fuel cells, *Surf. Coatings Technol.*, 201, 4436, 2006.
348. Kuo, J.-K. et al., A novel Nylon-6-S316L fiber compound material for injection molded PEM fuel cell bipolar plates, *J. Power Sources*, 162, 207, 2006.
349. Wang, Y. and Northwood, D.O., An investigation into polypyrrole-coated 316L stainless steel as a bipolar plate material for PEM fuel cells, *J. Power Sources*, 163(1), 500, 2006.
350. Fleury, E. et al., Fe-based amorphous alloys as bipolar plates for PEM fuel cell, *J. Power Sources*, 159, 34, 2006.
351. Stanic, V. and Hoberecht, M., MEA failure mechanisms in PEM fuel cells operated on hydrogen and oxygen, in *Extended Abstracts of 2004 Fuel Cell Seminar*, San Antonio, TX, November 1–5, 2004, p. 85.
352. Schulze, M. et al., Degradation of sealings for PEFC test cells during fuel cell operation, *J. Power Sources*, 127, 222, 2004.
353. Narusawa, K. et al., Deterioration in fuel cell performance resulting from hydrogen fuel containing impurities: poisoning effects by CO, CH_4, HCHO and HCOOH, *JSAE Rev.*, 24, 41, 2003.
354. Makkus, R.C. et al., Use of stainless steel for cost competitive bipolar plates in the SPFC, *J. Power Sources*, 86, 274, 2000.
355. St.-Pierre, J. et al., Relationships between water management, contamination and lifetime degradation in PEFC, *J. New Mater. Electrochem. Syst.*, 3, 90, 2000.
356. Peterson, A., Im, K.S., and Lai, M.C., Numerical simulation of coolant electrolysis in composite plate PEM fuel cell stacks, in *Proc. 3rd Int. Conf. Fuel Cell Sci. Eng. Technol.*, Ypsilanti, MI, May 23–25, 2005, p. 713.
357. Chen, L.D., Schaeffer, J.A., and Seaba, J.P., A Simulink model for calculation of fuel cell stack performance, in *Abstracts of Papers, 228th ACS National Meeting*, Philadelphia, August 22–26, 2004, p. FUEL 143.
358. Katz, M., Analysis of electrolyte shunt currents in fuel cell power plants, *J. Electrochem. Soc.*, 125, 515, 1978.
359. Rapaport, P. and Healy, J.P., Fuel Cell Having Insulated Coolant Manifold, U.S. Patent 6,773,841, August 10, 2004.
360. Roche, R.P. and Nowak, M.P., Integrated Fuel Cell Stack Shunt Current Prevention Arrangement, U.S. Patent 5,079,104, January 7, 1992.
361. Nickols, R.C. and Trocciola, J.C., Corrosion Protection for a Fuel Cell Coolant System, U.S. Patent 3,940,285, February 24, 1976.
362. Katz, M., Smith, S.W., and Reitsma, D., Corrosion Protection for a Fuel Cell Coolant System, U.S. Patent 3,923,546, December 2, 1975.

13 Materials Issues for Use of Hydrogen in Internal Combustion Engines

Russell H. Jones

CONTENTS

13.1 INTRODUCTION

Internal combustion engines (ICEs) offer an efficient, clean, cost-effective option for converting the chemical energy of hydrogen into mechanical energy. The basics of this technology exist today and could greatly accelerate the utilization of hydrogen for transportation. It is conceivable that ICE could be used in the long term as well as a transition to fuel cells. However, little is known about the durability of an ICE burning hydrogen. The primary components that will be exposed to hydrogen and that could be affected by this exposure in an ICE are (1) fuel injectors, (2) valves and valve seats, (3) pistons, (4) rings, and (5) cylinder walls. A primary combustion product will be water vapor, and that could be an issue for aluminum pistons, but is not expected to be an issue for the exhaust system except for corrosion. The purpose of this chapter is to provide a summary of what is known about hydrogen effects on these ICE components, although the amount of data on the actual materials and components in current ICEs is very limited.

13.2 FUEL INJECTORS

The combustion of hydrogen in an internal combustion engine is a technology to help expand the utilization of hydrogen fuel in the near term, before fuel cell technology is fully developed. In order to gain the highest efficiency, the use of direct injection will be needed. Direct injection places greater requirements on the injector than indirect injection. The following discussion deals only with injectors for direct injection. There are several elements to these injectors that could experience degradation in the presence of hydrogen: (1) injector body, (2) actuator, (3) epoxy used to encase the actuator, and (4) electrical contacts. There are little data on the effects of H on epoxies, so only a summary of the H effect on the injector body and actuator material will be presented.

13.2.1 Injector Body

Injector bodies are made primarily from steels such as M2 (UNS T11302), H13 (T20813), and 4140 steel (UNS G41400). The alloy M2 is a high-carbon tool steel with a carbon concentration ranging between 0.8 and 1.05%, while H13 is a tool steel with a carbon concentration of 0.3 to 0.45% and 4140 steel is an alloy steel with a carbon concentration of 0.4%. M2 is a highly alloyed tool steel with about 4% Cr, 5% Mo, 6% W, and 2% V. These elements are all carbide formers, so their combination with high carbon results in a significant volume fraction of carbides in the microstructure. These carbides provide wear resistance, which is needed for the pin and seat of the injector. H13 is a lower-alloy tool steel with approximately 1% Si, 5% Cr, 1% Mo, and 1% V. Alloy 4140 steel contains approximately 1% Cr and 0.2% Mo as the primary alloy additions.

M2 is a high-speed tool steel developed primarily for high-speed cutting tool applications and can be hardened to Rockwell C (HRC) 65 and has excellent retention to softening at temperatures as high as 600°C. This hardness retention results from the stable carbides. The tool steel H13 is generally hardened to HRC 40 to 55 and can retain its hardness to 500°C. The alloy steel 4140 can be hardened to HRC 55 to 60, but this requires rapid quenching from the austenitizing temperature because of the low alloy content, and it does not retain the hardness above about 400°C.

Composition and hardness are factors that directly affect the performance of these steels in hydrogen. Longinow and Phelps[1] have clearly shown a close relationship between the strength and the stress intensity threshold for crack propagation, K_{th}, for 4130, 4145, and 4147 steel. These steels cover the composition range of the 4140 steel. K_{th} decreases from 70 MPa $m^{1/2}$ to 20 MPa $m^{1/2}$ over an ultimate strength range of 800 to 1,100 MPa. There are little data on the effects of hydrogen on tool steels, but Fiddle et al.[2] showed that H11 steel and the alloy steel 4340, which is similar to 4140 steel, exhibited a high sensitivity to hydrogen embrittlement. They used a disc burst test and noted the burst pressure in helium relative to hydrogen, P_{He}/P_{H2}. This ratio is noted as S_{H2}, and these steels had an S_{H2} of 3.5. Materials with an S_{H2} equal to or less than 1 are considered not susceptible to hydrogen embrittlement, a value between 1 and 2 indicates a moderate susceptibility, and values greater than 2 indicate a high susceptibility. For comparison, Type 304 SS has an S_{H2} of 1.8 and the aluminum alloy 7075-T6 a value of 1.

The injector needle and seat will experience impact loading and cyclic loading, so hydrogen embrittlement will cause chips and fractures of these components. Clark[3] and Walter and Chandler[4] showed that the fatigue crack growth rate of steel is increased in the presence of hydrogen gas. Therefore, the needle and seat material for a hydrogen fuel injector must be designed to tolerate hydrogen and impact and cyclic loading.

13.2.2 Actuator Materials

Injectors may use electromagnetic or piezoelectric actuators to provide the active fuel control. Some actuators for direct H injection utilize piezoelectric wafers made of lead zirconium titanate (PZT) embedded into an epoxy or other insulating material. For direct injectors, the actuator is embedded in the hydrogen gas, which has the potential to affect performance by the following processes: (1) change the capacitance of the PZT,[5–7] (2) mechanical failure or cracking of the PZT,[8,9] (3) separation of the PZT wafers, (4) debonding of electrical connections, and (5) degradation of the epoxy or polymer casing materials.

Chen et al.[5] found that the electrical resistance of barium titanate (BTO) decreased by a factor of 10^3 and the capacitance decreased when charged electrolytically with hydrogen. The electrolytic charging was done in a two-electrode cell with a DC voltage of 4.5 V between the cathode and anode in a solution of 0.01 *M* NaOH at 25°C. The cathode current was 0.4 mA/cm^2. The electrochemical potential is not well controlled by this arrangement, but the authors claimed that H_2 evolved from the silver contact electrodes attached to the BTO sample. The authors did not outgas the sample to determine if the effect of the cathodic charge was reversible, so it is not possible to definitely conclude that the property changes are due to H. However, the authors noted that H could be an electron donor in BTO, which would be consistent with the noted property changes. Others[8–10] have concluded that hydrogen becomes incorporated into the lattice of lead zirconium titanate (PZT) as OH^-, and that this causes degradation in its dielectric properties. The details of the mechanism are unclear, but two processes have been suggested: (1) the formation of OH^- releases an electron and increases conductivity, or (2) the H enters an interstitial site and causes the formation of oxygen vacancies.

Hydrogen can also cause fracture of ferroelectric ceramics as demonstrated by Wang et al.[11] and Gao et al.[12] Gao et al.[12] cathodically charged a lead zirconium titanate (PZT-5) with H in a 0.2 mol/l NaOH + 0.25 g/l As_2O_3 solution using varying current densities. They measured an increasing H concentration with increasing current density; however, they do not report how they measured the H concentration. The maximum H concentration observed was 10 wppm, with a charging current density of 300 mA/cm^2. The threshold stress intensity for fracture in hydrogen, K_{IH}, was decreased from about 0.5 MPa m$^{1/2}$ to less than 0.1 MPa m$^{1/2}$ at a H concentration of 10 wppm. There was a larger drop in the K_{IH} at lower H concentrations when the electric field was perpendicular to the crack growth direction than when it was oriented parallel to the crack growth direction. At high H concentrations there was no orientation dependence.

Uptake of H from the gaseous phase will differ kinetically from cathodic charging because of the high surface fugacity of H possible at cathodic potentials. However, the effect of dissolved H will be the same regardless of the source of the H. The results of Chen et al.,[5] Wang et al.,[11] and Gao et al.[12] clearly demonstrated that H has the potential to alter the performance of ferroelectric ceramics, and therefore the performance of H injector actuators. A decrease in the resistance and increase in dielectric loss will clearly lead to failure of an actuator. Hydrogen-induced fracture of a ferroelectric ceramic could lead to electric discharge at the opposing fracture surfaces, and therefore the failure of the actuator. Clearly, a more detailed study of PZT behavior in gaseous H is needed before it can be determined whether there is a stability issue for its use in hydrogen ICE applications, but there is reason to be concerned about its durability.

Hydrogen may cause separation of the piezoelectric wafers and debonding of the electrical connections, but this effect has not been evaluated. Hydrogen could also alter the behavior of the epoxies used as insulation around the piezoelectric components. However, no available data exist on effects of hydrogen on the properties of epoxies. A corollary can be made to the effects of water on epoxies where hydrogen bonding within the epoxy leads to a change in the glass transition temperature.[13]

13.3 HYDROGEN EFFECTS ON INTERNAL ENGINE COMPONENTS

A number of internal components, such as valves, valve seats, cylinder walls, pistons, and rings, will be exposed to hydrogen and water vapor. The potential effects are of two primary types: (1) decarburization of steels and cast iron and (2) hydrogen embrittlement of aluminum pistons. Water vapor could cause excessive corrosion of exhaust systems, but this could be minimized by use of titanium.

13.3.1 Decarburization Effects

Decarburization occurs in steels and cast irons in hydrogen gas by the reaction of H with C in the steel. The decarburization rate is primarily dependent on the diffusion rate of C in the steel, but is also affected by the carbon content of the steel, alloying elements in the steel, such as chromium, impurities in the hydrogen, and of course time and temperature. Carburization of steels, the reverse of decarburization, is usually conducted at temperatures of about 900°C, but decarburization can occur at temperatures as low as 800°C.[14]

Exhaust valves have the highest operating temperature of components in an internal combustion engine, and they typically operate at a maximum of 790°C, while intake valves have a maximum operating temperature of 540°C. Light-duty intake valves are typically made from SAE 1547, which is an iron-based alloy with 1.5% Mn and 0.57% C. For higher-temperature applications, the ferritic stainless steel alloy 422 is used. This alloy has about 8.5% Cr, 3.25% Si, and 0.22% C. Because exhaust valves operate at higher temperatures, materials with a higher alloy content are used. A primary alloy for exhaust valves is 21-2N, which has 21% Cr, 2% Ni, and 2% N. Other alloys used for exhaust applications, depending on the desired operating temperatures, are 21-4N, 23-8N, Inconel 751, Pyromet 31, and Nimonic 80A. Valves used for heavy-duty applications have one of these alloys for the valve head with a hardenable martensitic stem. Valve seats are often made with hard facing alloys such as cobalt-based Stellites or nickel-based Eatonites. These are high-carbon-content alloys having about 2% C. However, much of this carbon is in the form of carbides and is more stable than the carbon in solid solution.

Whether decarburization will be an issue for internal combustion engines burning H_2 is difficult to predict from existing information. Low-alloy carbon steels begin to decarburize at temperatures around the operating temperature of exhaust valves, but exhaust valves and valve seats are made from high-alloy steels, austenitic alloys, and superalloys where the carbon is much more stable than low-alloy carbon steels. The hardenable martensitic valve stems of exhaust valves may experience decarburization over extended periods, and this would lead to accelerated wear because of the softened surface that results from decarburization.

13.3.2 Hydrogen Embrittlement of Pistons

Aluminum pistons in an engine that burns H_2 will be exposed to not only H_2 but also H_2O at temperatures of 80 to 120°C. Aluminum alloys can be totally immune to H_2 embrittlement and H_2-induced crack growth if the natural Al_2O_3 oxide is intact. However, there are processes that can disrupt this film, and it is known that aluminum alloys will absorb H_2 when exposed to H_2O vapor at 70°C. There will also be periods when the engine is cool and condensed water will be present so that aqueous corrosion could occur, but this is not expected to be any different than with an engine with cast aluminum pistons that burns gasoline.

Scully et al.[15] have reviewed the available data on H solubility and permeability in Al and some of its alloys. Their review shows tremendous variability in the available data. However, H is very insoluble in Al at 25°C and 1 atm pressure, with values ranging from 10^{-17} to 10^{-11} atom fraction. They also concluded from data for Al alloys that Li and Mg alloying additions increased the solubility of H in Al because of their chemical affinity for H. A summary of the H diffusivity in Al also revealed a wide range in values, but if it is assumed that the presence of aluminum oxide (Al_2O_3) on the surface is likely under all these tests, the fastest diffusivity is expected to be that closest to bulk diffusivity in Al, because this likely results from material with a defective or thinnest oxide film. There are several studies that resulted in diffusion coefficients at 25°C of about 10^{-7} cm²/sec for Al.

There have been a number of observations of H uptake during corrosion and stress corrosion testing as measured by thermal desorption following exposure. While these observations are less quantifiable than permeation measurements, they do provide direct evidence of H uptake during specific corrosion conditions. Several methods have been used to monitor H uptake during corrosion, including (1) thermal desorption, (2) transmission electron microscopy (TEM) of bubbles, and (3) resistivity change. Charitidou et al.[16] and Haidemenopoulos et al.[16] measured the thermal desorption of H from 2024 Al that had been exposed to the exfoliation corrosion solution according to ASTM G 34-90. Charitidou et al.[16] found that the alloy had absorbed over 1,200 wt ppm after exposure for 40 h following thermal desorption at 600°C, but only about 30 wt ppm was released at 100°C. Haidemenopoulus et al.[17] measured a H release corresponding to 90 wt ppm following 216 h of exposure to the ASTM G34-90 solution when the H extraction was done at 100°C. These two results are very similar considering the longer exposure time in the latter measurement. The H uptake during these tests is significantly greater than that expected in a 3.5% NaCl solution because the G34 solution is extremely aggressive.

The observation of bubbles in Al and Al alloys exposed to water vapor is an indirect method of evaluating H uptake.[18–20] Scamans and Rehal[18] found bubbles that they identified as H bubbles, in pure aluminum and aluminum alloys. The authors do not directly measure H in these bubbles but seem to infer that they are H filled based on the reaction of Al with H_2O to produce H. In an Al-Mg alloy they noted bubbles on grain boundaries and dislocations following only 10 min of exposure to water vapor at 70°C. Alani and Swann[19] also observed bubbles in Al-Zn-Mg alloys exposed to water vapor at 80°C. They proposed that the bubbles were the result of the precipitation of molecular hydrogen and that the cracks observed to emanate from the bubbles resulted from the pressure in the bubbles. However, they also proposed that it was the atomic H dissolved along the grain boundaries that was most embrittling. Scully and Young[21] evaluated the kinetics of crack growth of a low Cu AA 7050 in a 90% relative humidity environment and concluded that crack growth was controlled by H environment-assisted cracking over temperatures of 25 to 90°C.

Aluminum automotive engine pistons are generally made from Alloy 332.0-T-5 and are often cast by the permanent mold technique. For heavy vehicles, alloys 336.0-T551 and 242.0-T571 are used. Permanent mold castings are useful for high-volume production of parts that are larger than feasible for die casting. Stress corrosion cracking is generally not an issue for these alloys. Also, the environment in an engine would not support an aqueous environment that could produce an anodic dissolution type of stress corrosion cracking associated with wrought Al-Mg alloys.

Only recently has it been recognized and accepted that hydrogen induces crack growth and embrittlement of aluminum alloys. It is clear that little happens in dry hydrogen, but that crack growth occurs readily in moist hydrogen. Speidel[22] also demonstrated that the threshold stress intensity for crack growth was relatively low in the presence of moist hydrogen. Values of 5 to 10 MPa $m^{1/2}$ were reported. Threshold stress intensity values this low indicate that small flaws and low stresses are sufficient to produce crack growth and ultimately component failure.

Craig[23] has discussed hydrogen effects in aluminum alloys and notes that the phenomenon is not too different from that in steels. It is possible to find intergranular or transgranular cracking or blistering. Blisters tend to form as a collection of near-surface voids that coalesce to produce a large blister.

Dry hydrogen does not produce hydrogen effects in aluminum because of the slow permeability of hydrogen through the surface aluminum oxide. Anything that disrupts this protective oxide will allow hydrogen uptake. Water vapor provides this breakdown process, although the mechanism by which this occurs has not been presented. In wrought Al-Mg alloys with precipitates of grain boundary beta phase, this breakdown occurs at the beta phase intersecting the surface or crack tip. Once the hydrogen enters the material, it diffuses to locations such as grain boundaries and particles as in other materials. The crack growth rate is therefore a function of the hydrogen uptake and diffusion rate. Jones and Danielson[24] have shown that the diffusivity of hydrogen in aluminum could be as high as 10^{-7} cm^2/sec, although there is a wide disparity in the reported diffusivity values.

13.4 SUMMARY

There is clear evidence that the components of an engine burning hydrogen could experience durability issues because of their exposure to hydrogen or its primary combustion product, water vapor. High-efficiency conversion of hydrogen to mechanical energy will require the use of direct injection of hydrogen. This requires the injectors to be exposed to hydrogen gas, where the tool steel or carbon steel components could experience hydrogen-induced cracking or embrittlement. This is especially a concern for the injector needle and seat, which will also experience impact and cyclic loading. Piezoelectric actuators are one method for providing the fuel injector needle its lift, and there is some evidence that hydrogen could affect the performance of these components. Hydrogen could affect the dielectric properties of the piezoelectric material, the epoxy in which it is encased, or the electrical contacts. Testing is in progress on these components that should provide the data needed on their performance and methods for improving their durability should that be necessary.

Valves and valve seats will be exposed to hydrogen at elevated temperatures and could experience decarburization; however, it is difficult to predict their behavior based on current information. The operating temperatures of exhaust valves and valve seats for gasoline ICEs are at or below that at which decarburization occurs in carbon steels, but they are generally made from alloy steels that have higher decarburization temperatures. Also, the operating temperature of a hydrogen ICE may differ from a gasoline ICE. Gasoline ICEs utilize aluminum pistons, and it is known that aluminum and aluminum alloys experience hydrogen embrittlement when exposed to water vapor at 70°C and above. This operating temperature is certainly within the range of engine operation, so that it is important that this issue be evaluated.

REFERENCES

1. Longinow, A. and Phelps, E.H., Steels for seamless hydrogen pressure vessels, *Corrosion*, 31, 404–412, 1975.
2. Fiddle, J.P., Bernardi, R., Broudeur, R., Roux, C., and Rapin, M., Disk pressure testing of hydrogen environment embrittlement, in *Hydrogen Embrittlement Testing*, STP 543, 221–253, Philadelphia, PA: ASTM International, 1974.
3. Clark, W.G., The effect of hydrogen gas on the fatigue crack growth rate behavior of HY-80 and HY-130 steels, in *Hydrogen in Metals*, I.M. Bernstein and A.W. Thompson, ed., 149–164, Metals Park, OH : ASM, 1974.
4. Walter, R.J. and Chandler, W.T., Cyclic load crack growth in ASME SA-105 grade II steel in high pressure hydrogen at ambient temperature, in *Effect of Hydrogen on Behavior of Materials*, A.W. Thompson and I.M. Bernstein, ed., 273–286, Warrendale, PA, 1976.
5. Chen, W.P., Jiang, X.P., Wang, Y., and Peng, Z., The Metallurgical Society of AIME and H.L.W. Chan, Water-induced degradation of barium titanate ceramics studied by electrochemical hydrogen charging, *J. Am. Ceram. Soc.*, 86, 735–737, 2003.
6. Shimada, T., Wen, C., Taniguchi, N., Otomo, J., and Takahashi, H., The high temperature proton conductor $BaZr_{0.4}Ce_{0.4}In_{0.2}O_{3\text{-Alpha}}$, *J. Power Sources*, 131, 289–292, 2004.
7. Jung, D.J., Morrison, F.D., Dawber, M., Kim, H.H., Kim, K., and Scott, J.F., Effect of microgeometry on switch and transport in lead zironcate titanate capacitors: implications for etching nano-ferritics, *J. Appl. Physics*, 95, 4968–4975, 2004.

8. Seo, S. et al., Hydrogen induced degradation in ferroelectric $Bi_{3.25}La_{0.75}Ti_3O_{12}$ and $PbZr_{0.4}Ti_{0.6}O_3$, *Ferroelectrics*, 271, 283–288, 2002.
9. Krauss, A.R., Studies of hydrogen-induced processes in $Pb(Zr_1\text{-}xTix)O_3$ (PZT) and $SrBi_2Ta_2O_9$ (SBT) ferroelectric film-based capacitors, *Integr. Ferroelectrics*, 271, 1191–1201, 1999.
10. Aggarwal, S. et al., Effect of hydrogen on $Pb(Zr,Ti)O_3$-based ferroelectric capacitors, *Appl. Physics Lett.*, 73, 1973–1975, 1998.
11. Wang, Y., Peng, X., Chu, W.Y., Su, Y.J., Qiao, L.J., and Gao, K.W., Anisotropy of hydrogen fissure and hydrogen-induced delayed fracture of a PZT ferroelectric ceramic, in *Proceedings of the 2nd International Conference on Environment Induced Cracking of Metals*, Banff, Canada, October 2004, in press.
12. Gao, K.W., Wang, Y., Qiao, L.J., and Chu, W.Y., Study on delayed fracture of PZT-5 ferroelectric ceramic, in *Proceedings of the 2nd International Conference on Environment Induced Cracking of Metals*, Banff, Canada, October 2004, in press.
13. Zhou, J. and Lucas, J.P., Hygrothermal effects of epoxy resin. Part II: Variations of glass transition temperature, *Polymer 40,* 5513, 1999.
14. Hotchkiss, A.G. and Webber, H.M., *Protective Atmospheres*, 74, New York: John Wiley & Sons, 1953.
15. Scully, J.R., Young, G.A. Jr., and Smith, S.W., Hydrogen solubility, diffusion and trapping in high purity aluminum and selected Al-base alloy, *Mater. Sci. Forum*, 331–337, 1583, 2000.
16. Charitidou, E., Papapolymerou, G., Haidemenopoulos, G.N., Hasiotis, N., and Bontozoglou, V., Characterization of trapped hydrogen in exfoliation corroded aluminum alloy 2024, *Scripta Mater.*, 41, 1327, 1999.
17. Haidemenopoulos, G.N., Hassiotis, N., Papapolymerou, G., and Bontozoglou, V., Hydrogen absorption into aluminum alloy 2024-T3 during exfoliation and alternate immersion testing, *Corrosion*, 54, 73, 1998.
18. Scamans, G.M. and Rehal, A.S., Electron metallography of the aluminum-water vapor reaction and its relevance to stress corrosion susceptibility, *J. Mater. Sci.*, 14, 2459, 1979.
19. Scamans, G.M., Hydrogen bubbles in embrittled Al-Zn-Mg alloys, *J. Mater. Sci.*, 13, 27, 1978.
20. Alani, R. and Swann, P.R., Water vapour embrittlement and hydrogen bubble formation in Al-Zn-Mg alloys, *Br. Corrosion J.*, 12, 80, 1977.
21. Scully, J.R. and Young, G.A., Jr., The effects of temper, test temperature, and alloyed copper on the hydrogen-controlled crack growth rate of an Al-Zn-Mg-(Cu) alloy, in *Corrosion 2000*, National Association of Corrosion Engineers, Houston, TX, 2000, paper 368.
22. Speidel, M.O., Hydrogen embrittlement of aluminum alloys, in *Hydrogen In Metals,* I.M. Berstein and A.W. Thompson, ed. 174, Metals Park, OH: ASM, 1974.
23. Craig, B., Environmentally induced cracking, in *Metals Handbook*, 9th ed., Vol. 13, *Corrosion*, 169, Metals Park, OH: ASM.
24. Jones, R.H. and Danielson, M.J., Role of hydrogen in stress corrosion cracking of low-strength Al-Mg alloys, in *Hydrogen Effects on Materials Behavior and Corrosion Deformation Interactions*, 861, Warrendale, PA: The Metallurgical Society of AIME, 2003.

Index

D

E

F

I

J

K

L

M

N

O

P

R

Milton Keynes UK
Ingram Content Group UK Ltd.
UKHW021629071024
449327UK00020BA/1252

9 780367 387884